Lipids and Skin Health

Apostolos Pappas
Editor

Lipids and Skin Health

Springer

Editor
Apostolos Pappas
Johnson & Johnson
Skillman, NJ
New Jersey
USA

ISBN 978-3-319-34458-4 ISBN 978-3-319-09943-9 (eBook)
DOI 10.1007/978-3-319-09943-9
Springer New York Heidelberg Dordrecht London

Springer is part of Springer Science+Business Media (www.springer.com)

Contents

Contributors

Michael Anthonavage Presperse, Somerset, NJ, USA

Emanuela Camera Laboratory of Cutaneous Physiopathology and Integrated Center of Metabolomics Research, San Gallicano Dermatological Institute, IRCCS, Rome, Italy

Su-Hyoun Chon The Johnson & Johnson Skin Research Center, CPPW, a division of Johnson & Johnson Consumer Companies, Inc., Skillman, NJ, USA

Steve Cochran Presperse, Somerset, NJ, USA

Béatrice Desvergne Center for Integrative Genomics, University of Lausanne, Lausanne, Switzerland

K. R. Feingold UCSF, San Francisco, USA

Gerhard Fürstenberger Genome Modifications and Carcinogenesis, Deutsches Krebsforschungszentrum, Heidelberg, Germany

Carrie Grueter Vanderbilt University Medical Center, Nashville, USA

Eric Jourdan Laboratoire dermatologique BIODERMA, Lyon, France

P. E. Kolattukudy Burnett School of Biomedical Science, College of Medicine, University of Central Florida, Orlando, USA

Jan Kopecky Department of Adipose Tissue Biology, Institute of Physiology of the Academy of Sciences of the Czech Republic, v.v.i., Prague 4, Czech Republic

Peter Krieg Genome Modifications and Carcinogenesis, Deutsches Krebsforschungszentrum, Heidelberg, Germany

Vasiliki Lagouri Department of Chemistry, University of Athens, Athens, Greece

K Mahmood J&J, New York, USA

Yoshinori Masukawa Global R&D, Kao Corporation, Tokyo, Japan

G. K. Menon ASI Corporation, Fremont, USA

Ashland, Covington, USA

Anna Nicolaou Manchester Pharmacy School, Faculty of Human and Medical Sciences, University of Manchester, Manchester, UK

Lars Norlén Department of Cell and Molecular Biology (CMB), Karolinska Institute, and Dermatology Clinic, Karolinska University Hospital, Stockholm, Sweden

James M. Ntambi Department of Biochemistry and of Nutritional Sciences, University of Wisconsin, Madison, WI, USA

Monica Ottaviani Laboratory of Cutaneous Physiopathology and Integrated Center of Metabolomics Research, San Gallicano Dermatological Institute, IRCCS, Rome, Italy

Apostolos Pappas The Johnson & Johnson Skin Research Center, CPPW, a division of Johnson & Johnson Consumer Companies, Inc., Skillman, NJ, USA

Mauro Picardo Laboratory of Cutaneous Physiopathology and Integrated Center of Metabolomics Research, San Gallicano Dermatological Institute, IRCCS, Rome, Italy

Stephen M Prouty 1075 BRB II/III, Department of Dermatology, Perelman School of Medicine, University of Pennsylvania, Philadelphia, PA, USA

Sabine Rosenberger Genome Modifications and Carcinogenesis, Deutsches Krebsforschungszentrum, Heidelberg, Germany

Nguan Soon Tan School of Biological Sciences, Nanyang Technological University, Singapore, Singapore

Institute of Molecular and Cell Biology, Singapore, Singapore

Georgios N Stamatas Johnson & Johnson Santé Beauté France, Issy-les-Moulineaux, France

Walter Wahli Lee Kong Chian School of Medicine, Nanyang Technological University, Singapore, Singapore

Center for Integrative Genomics, National Research Center Frontier in Genetics, University of Lausanne, Lausanne, Switzerland

Philip W. Wertz University of Iowa, Iowa City, IA, USA

Christos C. Zouboulis Departments of Dermatology, Venereology, Allergology and Immunology, Dessau Medical Center, Dessau, Germany

Part I
Editor's Introduction: Skin Lipids

Chapter 1
Skin Lipids: An Introduction and Their Importance

Apostolos Pappas

It is my intense and outmost pleasure to welcome you to this exciting project on lipids and skin in the form of an academic textbook for the very first time. Although the very first baby steps of lipids and skin together took place about 50 years ago in the laboratories of skin lipid pioneer scientists such as professors Downing, Strauss, and Nicolaides, nowadays it seems that together they could rapidly turn into one of the most emerging and intriguing fields of dermatology, lipid science, and metabolism.

Skin is the largest and the most visible—to us and to the others—organ of the human body but in addition to that, the most desired to be understood by the average consumer. Is it perhaps my biased opinion or perception? The jury is out to discover the truth and the validity of this statement. On the other hand, the majority of academic focus has been on cardiovascular health, metabolic diseases, and cancer which pose the most serious threats to our society and pose socio-economic issues far more severe than any disease that stems from the skin (perhaps melanoma being the most aggressive, which is, however, also classified as cancer). Certainly, academic laboratories can secure more funding from NIH, NSF, and private funding institutions to work towards curing devastating diseases responsible for the loss of many human lives, such as the multiple forms of cancer, diabetes, multiple sclerosis, and autoimmune diseases, rather than wrinkles, pimples, and age spots for the sake of an example. Perhaps one can agree that the majority of the healthy population that is not concerned daily with issues such as famine or clean water is focusing more on skin conditions than on any other disease. However, we are still far from solving completely all issues with devastating diseases in order to allocate even more resources in quality of life, prevention, and wellness, which certainly gain more ground as the years go by.

Skin, however, constitutes our wall to our environment. This wall is perhaps as important as the immune system since it is constantly repelling and confronting any chemical, physical, and microbiological invasion. Our skin though, as every other

A. Pappas (✉)
The Johnson & Johnson Skin Research Center, CPPW, a division of Johnson & Johnson Consumer Companies, Inc., Skillman, NJ 08558, USA
e-mail: APappas@its.jnj.com

A. Pappas (ed.), *Lipids and Skin Health*, DOI 10.1007/978-3-319-09943-9_1

wall, is somehow composed of certain "bricks" which are stuck together by a special "mortar"; a very popular model used to describe the outer layer of the epidermis (the stratum corneum). In the case of skin, though, that specific mortar is nothing other than a mixture of lipids; and sure enough the better and stronger that mortar is, the better and stronger the wall becomes. So these are the epidermal lipids made by the epidermal cells, the keratinocytes and somehow as you would read in this book they are extruded out of the cell to possess the extracellular milieu and keep the differentiated corneocytes together and form that rigid wall that is called epidermis.

That alone can outline the importance of lipid synthesis in skin since the skin will never offer the good barrier properties unless it has the proper lipid mixture. However, lipids are also found on the surface of the skin as a result of the sebaceous gland activity. Once considered the appendage of skin, the sebaceous gland is now considered one of the major endocrine sites of the skin, potent enough to synthesize an amazing diversity of hormones, growth factors, and transcriptional factors. The sebaceous biology is remarkable, complex, and unique, can hardly be compared to the biology of any other cell type. Nowadays we know that sebum is not there only to trouble us with acne but also to condition the skin and offer an extra lipid layer of protection; hopefully in the future more studies would fully decode its role.

Of course there is the third and forgotten layer of skin: the subcutaneous layer or hypodermis that is mainly composed of fat cells, the adipocytes, which are grouped together in lobules separated by connective tissue. Although it is believed that this layer serves roles as padding or as an energy reserve, providing some minor thermoregulation as well, many aspects remain unexplored, especially on how this layer communicates with the dermis and its cell types. This could especially be useful to understand cellulite. So far the major components of this layer, the differentiated adipocytes, have only been explored for facial volume loss; as autologous fat grafting has become popular to restore facial volume loss in addition to other facial fillers.

The field of lipids and skin took only baby steps in our fast-paced scientific society since any advancement was impaired by the analytical limitations and challenges that these extremely hydrophobic molecules offer. Not having excellent spectroscopic properties and being more hydrophobic than other tissue lipids, they posed many analytical challenges. Even nowadays outdated technologies such as thin layer chromatography are still used to analyze them and offer valuable solutions whereas modern technologies still need stringent validation. However, better analytical techniques are being developed and they would help to increase our understanding on their role and clarify their complexity. For example, it was only few years ago that the classes of ceramides were expanded from 6 to 9; with new and modern analytical techniques they became 12 and more than 360 species of skin ceramides were identified. Therefore, the field of epidermal surface lipids is still open for many new discoveries and is constantly enhanced by advances in analytical techniques.

It is not by coincidence that in an era where genomics is the past and proteomics and metabolomics the present, the near future is for lipidomics, even though it still poses challenges to most biologists and analytical chemists. Undoubtedly, future scientists will find themselves in need to incorporate all the acquired learning from the various "-omics" fields to advance further the new era of lipidomics. Once scientists decode the role of lipids by more efficient, accurate, and reliable analysis, the academic community will eventually shed light not only on intriguing dermatological diseases such as acne, atopic dermatitis, ichthyosis, and psoriasis, but also aging and many more conditions that scientists so far do not currently associate with the epidermal surface lipids.

On the front of cell biology we have seen a tremendous effort and progress in skin lipids since more and more laboratories generate immortalized cell lines previously unavailable (as sebocytes) and hopefully in the near future we would see even more of them and perhaps even better 3D models that will be able to incorporate a variety of cells besides epidermal and dermal cells, such as sebocytes, immune system cells, and even preadipocytes.

Skin lipids contribute to normal skin functions as the barrier function and the maintenance of healthy skin and hair. Consequently, they contribute not only to many diseases, but also to aging as well as the conditioning and defense status of this organ. This book constitutes an effort to sum up all the primary and relevant references that one needs to review to understand the complex and diverse roles of lipids in skin. May this book inspire scientists, dermatologists, nutritionists, and people from all medical disciplines to invest more time in connecting the two areas of research, in lipids and in skin, for more books and research to come and bridge any possible gaps of knowledge.

Part II
Skin Basics

Chapter 2
Skin Basics; Structure and Function

Gopinathan K. Menon

Introduction

Skin in Health While healthy skin is often taken for granted and not thought about in our daily life, skin diseases have a profound effect on physical comfort, and quality of personal and social life, impacting on daily functioning and socioeconomic status. Jowett and Ryan (1985) found that in a survey, 64 % of the patients with skin problems reported experiencing embarrassment, anxiety, a lack of confidence, and depression. Further, almost 40 % found challenges with employment experiences of limited opportunities, and functional and interpersonal difficulties in the workplace. Although all systemic health issues do not lead to skin diseases, skin in general is a mirror for overall health, and a window on hereditary diseases of connective tissue (Holbrook and Byers 1989)—this is also evident from the practice of traditional medicine around the world, where the healers often used skin changes for diagnosing various maladies. Even with modern medicines, skin reactions characterize drug eruptions, several allergies, and as in the famous case of a political leader, assassination attempt by poisoning. Skin is also the primary surface for medical treatment and nursing care, ranging from transdermal drug delivery, attachment of devices for monitoring physiological functions, etc.

Functional Aspects of Skin in Health

Skin is the body's protective barrier against a whole battery of environmental aggressors—both of natural and anthropogenic origins. It primarily protects against desiccation, and thus makes life on land possible (Attenborough 1980). Every class

G. K. Menon (✉)
ASI Corporation, 1005, Route 202/206 Brdge water, NJ 08807, USA
e-mail: Gmenon@ashland.com

Ashland, 1005, Route 202/206 Brdge water, NJ 08807, USA

A. Pappas (ed.), *Lipids and Skin Health*, DOI 10.1007/978-3-319-09943-9_2

of terrestrial organism has evolved an appropriate barrier against fluid loss from the body and almost all employ modifications of the cutaneous surface for this purpose (Hadley 1989; Lillywhite 2006). The formation of the barrier layer and its maintenance and renewal is the function of epidermal keratinocytes, although other cell types interacting with keratinocytes also play a significant role in regulation of this function. Besides, skin supports a large commensal microflora that is important in keeping in check the colonization of skin by pathogenic microbes. A basic description of the structural organization of the mammalian skin, emphasizing the different kinds of barriers and their location is given below:

Structural Organization of the Mammalian Skin

As can be expected, the structural details of skin have been well studied and published with excellent accounts of the histology, (Montagna 1967), histochemistry and ultrastructure (Matoltsy 1986; Holbrook and Odland 1975; Monteiro-Riviere 2010), and barrier functions of skin (Scheuplein and Blank 1971; Elias and Feingold 2006) having been available over several decades, that even a short listing of this voluminous literature is a daunting task, which I will not even attempt. Histologically, the skin shows distinct compartments such as the epidermis, the dermis, and the hypodermis. A basement membrane serving as an anchor to the epidermis separates this compartment from the dermis, both structurally and functionally.

In this chapter, the epidermis is discussed first, as much of the barrier functions reside within this layer.

Epidermis

Epidermis is predominantly made up of Keratinocytes (Fig. 2.1), but also houses the melanocytes as well as dendritic cells such as Langerhans cells. Nerve fibers have been demonstrated to innervate all nucleated layers of epidermis (Hilliges et al. 1995). The nucleated cells of epidermis has three layers, the stratum basale (where stem cells as well as postmitotic, transiently amplifying cells are located) the stratum spinosum (or prickly layer), and the stratum granulosum. The keratinocytes elaborate structural proteins such as the epidermal keratin, the natural moisturizing factors (NMF), and the barrier lipids, proliferate to heal the wounds or replace the corneocytes that are lost by exfoliation, transport water, glycerol, and urea through the aquaporins, receive melanin from the melanocytes, and house the antigen presenting Langerhans (sentinel) cells. The barrier lipids are elaborated as epidermal lamellar bodies which also contain antimicrobial peptides, thus serving as the basis of permeability and antimicrobial barriers. The epidermis also secretes a variety of chemokines, growth factors, etc., for cellular communication within the epidermis as well as with dermal cells (fibroblasts, mast cells). This layer also stimulates production of the dermal matrix, or when appropriate, its degradation. Having no direct blood supply, transport of nutrients within this layer is conducted

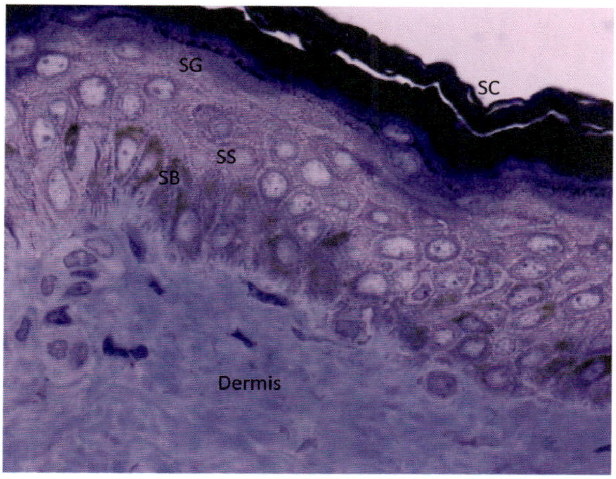

Fig. 2.1 Histology of pigmented skin, plastic embedded, semi-thick sections stained with Toluidine blue. (*SB* Stratum basale; *SS* Stratum spinosum; *SG* Stratum sranulosum; *SC* Stratum corneum)

by diffusion through intercellular fluids, once they have passed the selective barrier of the basement membrane separating epidermis from the vascular dermis. Sensory nerve fibers do extend into the epidermal compartment, and secrete trophic neuropeptides that influence keratinocyte physiology, as well as play some roles in dysfunctions associated with sensitive skin. Keratinocytes are possibly one of the most studied cell type—and have been used extensively in basic research as well as tissue engineering of reconstituted skin used for skin transplants as in case of burns. Besides, epidermal stem cells have been a hot topic of research for more than 2 decades (Watt 2001; Fuchs and Segre 2000; Blanpain and Fuchs 2009)

Stratum Basale (SB) The cells are cuboidal to columnar in shape, with a large nucleocytoplasmic ratio. The cells adhere to the basement membrane via hemidesmosomes, and with their neighboring cell with desmosomes. Tonofilaments are prominently seen in the cytoplasm, radiating from the hemidesmosomes as well as desmosomes. Other cellular organelles such as ER, mitochondria, etc., are also seen at the ultrastructural level (Fig. 2.2). In pigmented skin, many melanosomes can be seen "capping" the nucleus of the cells of SB (Fig. 2.3). Basal cells are characterized by presence of Keratins 5 and 14.

Stratum Spinosum The spinous or "prickly layer" is named so due to the large number of desmosomes that connect adjacent cells, and the prominent tonofilaments that radiate from the desmosomes. Several layers of cells are seen in this suprabasal layer. In histological preparations, large intercellular spaces (preparation artifacts) that are seen between the adjacent desmosomes gives this layer a distinct appearance. At the electron microscopic level of observation, large numbers of desmosomes mark this layer, and epidermal lamellar bodies begin to appear in the cytoplasm of cells. These organelles ranging from 0.2 to 0.5 μ size contain cholesterol, glycolipids, and fatty acids, in addition to a battery of enzymes and antimicrobial peptides. Thus the synthesis of proteins (Keratins 1 and 10) as well as lipids mark

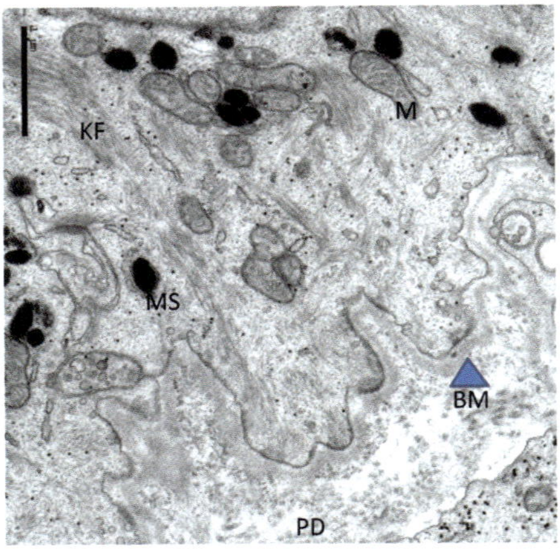

Fig. 2.2 Dermal-epidermal junction, showing papillary dermis (*PA*), Basement membrane (*BM, Arrowhead*), and basal cells (*BC*) shown partially. KF (*Keratin filaments*) and Mitochondria (*M*) as well as Melanosomes (*MS*) are seen

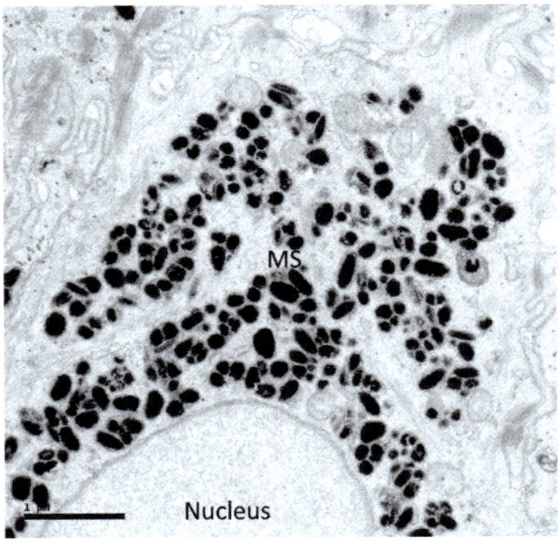

Fig. 2.3 TEM showing the microparsol of melanosomes over nucleus

the progressive differentiation of the keratinocytes that move up being "pushed" by the newly proliferating cells below.

Stratum Granulosum (SG) This layer is named for the presence of Keratohyalin granules that appear within the cells along with the flattening or elongation of the keratinocytes. Keratohyalin granules can be clearly identified in histological preparations (Fig. 2.1). With Electron microscopy, these granules appear irregularly shaped, and embedded within the keratin filaments. Immunoreactivity shows that they contain Profilaggrin, (whose breakdown product, Filaggrin is needed for aggregation of keratin filaments), Loricrin as well as Involucrin, two proteins that are important for the formation of the cornified envelope. Synthetic activities—both for structural proteins and barrier lipids—reach their peak at this layer. In the uppermost layer of SG, about 20 % of the cell volume is occupied by the epidermal lamellar bodies (LB), which are membrane bound structures containing stacked

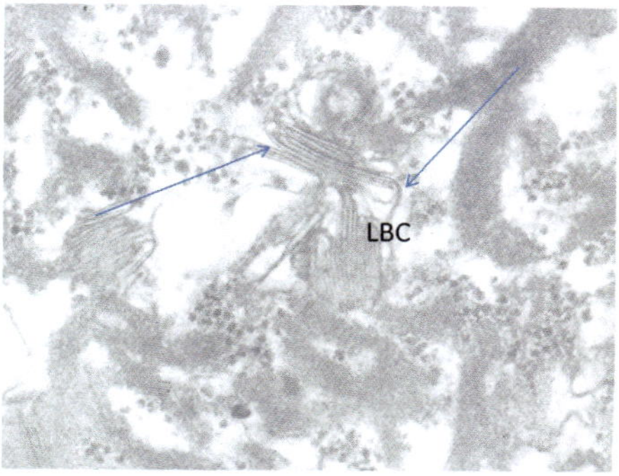

Fig. 2.4 Lamellar body contents (*LBC*) appear as pleated sheets (*arrows*) when the limiting membrane is disrupted. (Modified from Elias and Menon 1991)

disc like contents. The discs can be compared to flattened liposomes, but they are actually more like pleated sheets—connected with each other (Fig. 2.4) folded like an accordion (Elias and Menon 1991). The LBs arise from the Trans-Golgi network, and remain interconnected with each other—forming a Lamellar body secretory system, positioned for secretion at a basal rate, or for a coordinated massive secretory response (Elias et al. 1998), when needed as following barrier disruption or at terminal differentiation of individual cells. As for proteins, during the process of terminal differentiation, the enzyme transglutaminase mediates the cross-linking of involucrin and loricrin, forming a thickened cornified envelope (CE) just inside of the plasma membrane. Profilaggrin is cleaved by proteases to produce filaggrin, which aggregates the keratin filaments. About the same time cellular organelles begin to get degraded through activity of proteases such as Caspases involved in the programmed cell death of keratinocytes. Thus the fully differentiated keratinocytes becomes corneocytes, the cells that make up the stratum corneum. Loss of function mutations in Filaggrins are now recognized as the basis of several skin dysfunctions ranging from sensitive skin/atopic dermatitis to several forms of Ichthyotic conditions termed disorders of cornification.(Irvine et al. 2011). In addition, tight junctions between the adjacent cells in outermost layer of SG add another barrier under the SC (Brandner 2009; Brandner and Proksch 2006)—and it is believed that they help in the polarized secretion of lamellar bodies. However, gene knock-out rodents for tight junction (TJ) proteins also fail to develop a functional permeability barrier, pointing to a highly significant role(s) of the TJs in the overall epidermal barrier formation (Proksch et al. 2008).

Stratum Corneum (SC) At this stage, the corneocyte is an extremely flattened (30–40 μ wide) and thin (about half a Micron in thickness) ghost of a cell filled with keratin bundles (Fig. 2.5). Further breakdown of Filaggrin leads to the production of a mixture of aminoacids collectively termed: natural moisturization factors (NMF) that allows each cell to hold moisture needed to plasticize the keratin within (Rawlings and Matts 2005). The secreted lamellar body contents, which fill the interstices of the corneocytes within the strata, undergo further catabolic modifications mediated by co-secreted enzymes, to give rise to an equimolar mixture of Cholesterol: Ceramides and fatty acids, (Fig. 2.6) which forms a lamellar structure

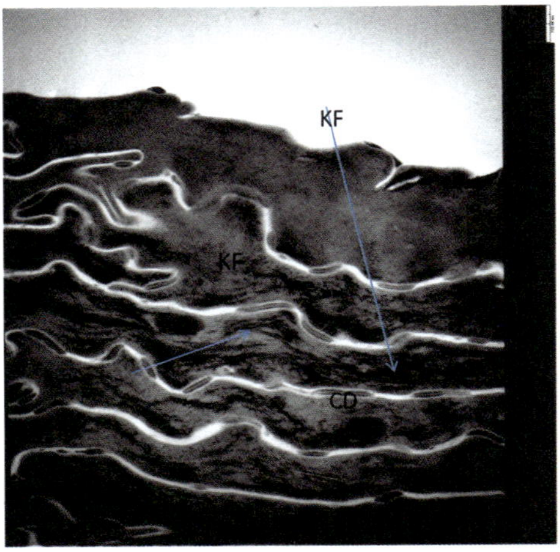

Fig. 2.5 Human stratum corneum: individual cells appear as "bags" filled with Keratin filaments (*KF*, *arrows*) and connected with other corneocytes through corneodesmosomes (*CD*) that appear as spot-weldings. Postfixation with O_SO_4, (hence the intercellular lipids are not stained)

Fig. 2.6 Stratum corneum lipids: "The Big Three"

Species	Approximate wt.%	Molar ratio
Ceramides	50	1
Cholesterol	25	1
Free Fatty Acids	25	1

(Wertz 2000) that occlude the intercellular space within SC—forming the permeability barrier (Fig. 2.7). Thus, the SC is transformed to a "brick and mortar" like model—a composite material—displaying properties that are more than the sum of its components. During the terminal differentiation, desmosomes are also modified to "corneodesmosomes": that make the corneocytes very cohesive, until the cells reach the outer most layers when programmed proteolysis of corneodesmosomes release individual cells or small units of cells that exfoliate from the surface. Such shedding of corneocytes signal the underlying cells to differentiate and join the SC-maintaining homeostasis/Autopoesis of the stratum corneum. (Hoath 2001) The number of cell layers in SC may be 10–18 cell layers varying with anatomic location. As the visible and terminally differentiated barrier layer (10 μm thick), it is often said that the SC stands between life and certain death. The characteristic, loose basket-weave appearance of this layer in histological preparations is indeed an artifact created during the dehydration and paraffin embedding protocol. Much research has been conducted on this layer, with an amazing array of techniques in the past 3 decades or so (see Elias and Feingold 2006) and what has emerged is a view that SC is not an inflexible brick wall, but an interactive boundary layer, sensing and responding to environmental changes like a "smart material"

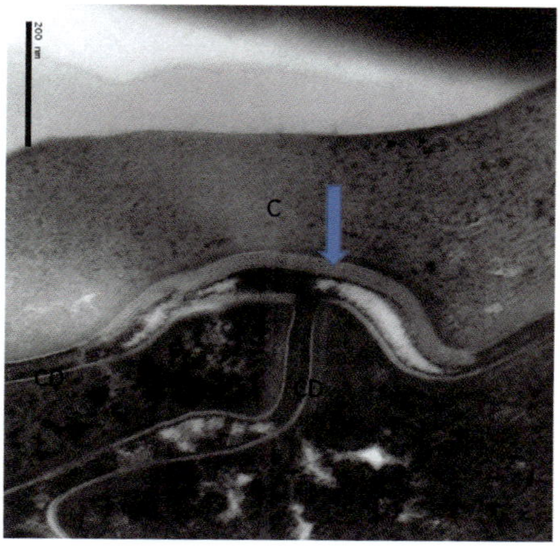

Fig. 2.7 Highly magnified view of Human SC, showing the relationship of extracellular lipid lamellae (*Arrow*) and corneodesmosomes (*CD*). Postfixation with RuO_4 to reveal the lipid structures

(Menon and Elias 2001). As the Stratum corneum is the perceived (oily, dry, pigmentary changes) and perceiving (touch, temperature) surface of skin it is literally and figuratively, the touch screen of the body.

Diverse and Interdependent Barriers in Epidermis

Although permeability barrier is the primary basis of successful terrestrial adaptation, several other environmental stressors pose challenges that need to be met by additional barriers (Menon and Kligman 2009). Skin, being the versatile organ system it is, has evolved to exploit the same structural and biochemical machinery to forge these additional barriers. An antimicrobial barrier composed of antimicrobial peptides (AMPs) such as beta defensins and cathelicidins (reviewed in Gallo and Hopper 2012) has been intensely studied in recent past. The AMPs are packaged in the same lamellar bodies that deliver the barrier lipids to the SC, closely linking the permeability and antimicrobial/innate immunity barriers. Adaptation to counter UV radiation and consequent DNA damage is reflected in the constitutive darker pigmentation in populations originally inhabiting latitudes closer to equator, where ambient UV radiation is higher. The fair-skinned races (skin types I and II) have their origin in higher latitudes where selective advantages of maximizing the low levels of UVR for Vitamin D synthesis must have been the selective force (Webb 2006; Yuen and Jablonski 2010). As the number of melanocytes are same for the dark and light skinned individuals, the activity of melanocytes and the type of melanosomes determine the level of melanin in the epidermis (Costin and Hearing 2007). The "microparsol" of melanosomes capping the nuclei of the basal keratinocytes must provide basic protection to the genomic material of these cells. Facultative pigmentation (tanning) also indicate the essential adaptive value of melanin. Additionally, efficiency of the permeability barrier repair response following experimental perturbations has been found to correlate with the degree of pigmentation of the subjects (Reed et al. 1995).

UV radiation induced oxidative damage also needs to be countered by the skin, and several biochemical (antioxidants such as Co Enzyme Q10, Vitamin E, etc.) and physiological (p53 related) mechanisms have evolved. It is worthwhile to note that the same biochemical pathway for the synthesis of cholesterol (crucial for permeability barrier) underlie the production of CoQ10 (Bentinger et al. 2010)—another example of the versatile nature of the interdependent skin barrier CoQ10 levels in the epidermis is 10 fold higher in the epidermis than in the dermis (Shindo et al. 1994). Although health implications of the use of statins, as far as CoQ10 levels have received attention (Mabuchi et al. 2005), as the epidermal cholesterol synthesis is autonomous, and as statins may not reach the epidermis (due to being metabolized in the liver), an effect on epidermal CoQ10 levels may be minimal at worst.

Immune Barriers Sentinel cells such as the Langerhans cells (LC), positioned in the stratum granulosum, connect the epidermal barrier to the immune system. LCs internalize allergens that traverse the SC, leave the epidermis, and migrate to the lymph nodes where they sensitize the T-cells. Langerhans cells have also been reported to send their dendrites through the tight junction barriers, (Kubo et al. 2009) and in this search mode, function in "sampling" the intra epidermal environment for allergens that cross the SC barrier.

Additionally, Birbek granules of Langerhans cells contain "langerin" found to be a barrier for the transmission of HIV virus (de Witte et al. 2007).

Increasing presence of xenobiotics in our environment, due to human activity (Biocides, pharmaceuticals, industrial waste such as plastics, etc.) is a major toxicological concern. Physiological defense of skin include CYP 450 enzymes, located within the epidermal and dermal cells (Baron et al. 2001) These phase 1 enzymes catalyze introduction of functional groups into hydrophobic organic molecules. Subsequently, phase 2 enzymes help in elimination of the xenobiotics by conjugating these chemicals with hydrophilic molecules such as GSH and glucuronic acid.

While this enzyme based defense help detoxify many of the chemicals, they can also activate some of the carcinogenic compounds. Thus every cell type in the epidermis partake in the myriad defensive barrier functions of skin.

The Dermal-Epidermal Junction (DEJ) or the Basement Membrane Zone (BMZ)

The BMZ is a 0.5–1.0 μm thick band situated between the epidermis and dermis, histologically identified with periodic acid-Schiff staining. Transmission electron microscopy unraveled the complex structural components in the BMZ, (Fig. 2.2) whose major function is to anchor the epidermis to the dermis. The different regions of BMZ are (1) hemidesmosomes (that connect stratum basale with the basement membrane), (2) the *lamina lucida* appears electron lucent and has fine anchoring fibers connecting the hemidesmosomes. It is composed of laminins, which are heterodimers of various combinations of alpha, beta, and gamma submits, secreted by the keratinocytes. Another component is Fibronectin, which is associated with collagen fibers, and has important biological roles as well (Mosher and Furcht 1981), (3) the *lamina densa* (electron dense appearance), the next layer is 35–45 nm thick, and is composed mainly of type IV collagen, perlecan (heparan sulfate proteoglycan), and possibly laminin; (4) the *sub-lamina densa*, located below the lamina densa, the fibrillar structures connecting lamina densa to dermal plaque like structures, termed anchoring fibrils, are composed mainly of type VII collagen, secreted both by keratinocytes and fibroblasts (Marinkovich et al. 1993).

In addition to facilitating adherence of epidermis to dermis, BMZ also functions in structural support, regulation of permeability of substances from dermis to epidermis, and in embryonic differentiation. Flattening of the DEJ is a feature of aged skin, but in younger individuals, photoaged skin shows much more prominent flattening of the DEJ than sun-protected sites, and activity of Matrix Metalloproteinases (MMP) has been implicated in this alteration. Mutations that affect the components of BMZ cause heritable, blistering skin diseases.

The Dermis

Dermis, accounts for about 90 % of the weight of skin, and forms the foundation of this organ system. Two distinct zones are seen in the dermis. A Superficial papillary dermis subjacent to the BMZ appears as a loose network of connective tissue with thin collagen fibrils. Below this layer is a compact, deeper reticular layer, displaying a dense connective tissue matrix with thick and regularly oriented bundles of fibrils. The primary cell type in the dermis are fibroblasts, which produce the extracellular structural proteins, collagen, elastin as well as the glycosaminoglycans (GAGs such as Hyaluronic acid) the major water holding components of the dermis. Together, these components are known as the extracellular matrix (ECM). The early views on Dermis that is functioning merely as a structural foundation, have now been fully discarded, and the biological significance of the ECM, as well as the connective tissue fibers in skin health and disease is now fully appreciated (Gustafsson and Fassler 2000) and the physical, chemical, and biological roles of its components continue to be unraveled.

Collagen

Collagen is the most abundant protein in the dermis. Skin contains collagen Types I, III, and V, in addition to Type IV in the basement membrane. Type III is more abundant in the papillary dermis, while the reticular dermis has an abundance of Type I. Also associated with dermis is a family of secreted proteoglycans that include decorin, biglycan, fibromodulin, lumican, epiphycan, and keratocan—and mutations in these proteins can lead to defective fibrillogenesis and functions. Collagen cross-linking is achieved via the enzyme Lysyl oxidase, and reduced activity of the enzyme can lead to collagen defects, as in Cutis laxa (reviewed by Myllyharju and Kivirikko 2004).

Patients with the Ehlers–Danlos syndrome lack collagen type III and show fragility of the skin, fragility of the blood vessels, hyperelasticity of the skin, and hypermobility of the joints. Fibroblasts of Ehlers–Danlos patients synthesize only type I Collagen, and hence the defects are associated with the lack of Collagen III (Pope et al. 1975)

Elastin

Elastin fibers are much less abundant compared to Collagen, but they are critical for the resiliency of the skin, where the stretched skin returns to its original position with a snap. Functional significance of elastin is exemplified in congenital cutis laxa, a rare syndrome with significant skin laxity (in addition to pulmonary and cardiovascular complications), where fibroblasts production of tropoelastin is highly reduced (Zhang et al. 1999).

Histologically, the elastin fibers of human skin have been categorized into three types: Oxytalan, elaunin, and elastic. Oxytalan (most superficial, very thin oriented perpendicular to the DEJ) originates from a plexus that stains like elaunin (thin) fibres, which are in turn connected with the thicker elastic fibers of the reticular dermis. Ultrastructurally, oxytalan fibers are formed by bundles of tubular microfibrils. In the deeper dermis, the fibers display an amorphous material in the core of the bundles. Such an amorphous material is sparse in elaunin fibers, but abundant and rather compact in the elastic fibers (Cotta-Pereira et al. 1976). Microfibrils are constituents of elastic and oxytalan fibers that impart mechanical stability as well as elasticity to tissues. Their core is made of the glycoprotein fibrillin, microfibril associated proteins (MFAPs) and microfibril associated glycoproteins (MAGPs) which link microfibrils to elastin and other ECM components, and to cells. Much attention has been paid to the role of Elastosis and "elastotic tissue" (highly altered in appearance and function) in the photoaging of skin (see Uitto 2008).

Glycosaminoglycans (GAG)

GAGs are the major organic ECM components and the most abundant polysaccharides in the body. They include Hyaluronic acid (HA), heparan sulfate (HS), chondroitin sulfate, and dermatan sulfate. HA and HS are probably the most well-studied components in skin extracellular matrix (Lamberg & Stoolmiller 1974). HA is the key water holding molecule in the dermis, and is now well recognized as having several signaling functions that influence fibroblasts, as well as epidermal differentiation. HS is also abundant in the epidermis, basement membrane

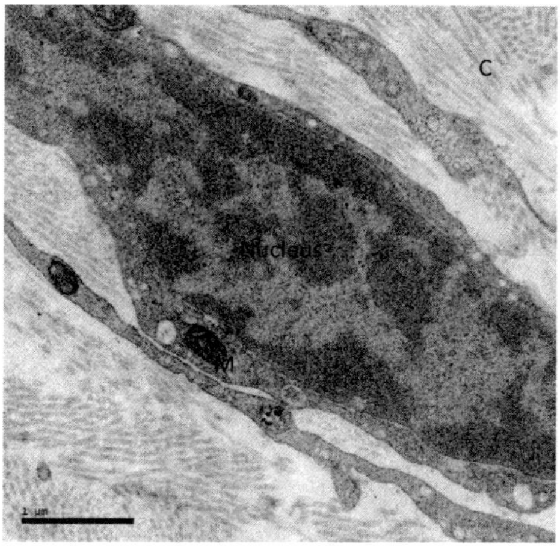

Fig. 2.8 Fibroblasts in human dermis show their spatial association with collagen bundles. Collagen (*C*) appears in longitudinal as well as *cross-sectional* profile

zone, and dermis. GAGS have the ability to bind and interact with a multitude of molecules, including collagen, various growth factors, enzymes, and enzyme inhibitors. An age-related decrease in dermatan sulfate and heparin sulfate (but not HA) was reported for human subjects (Van Lis et al. 1973).

Fibroblasts

Fibroblasts are responsible for production and remodeling of all crucial dermal components, and are integral to epithelial–mesenchymal interactions crucial in development, repair, and regeneration of the skin. Fibroblasts are not a homogeneous population of cells, (Fig. 2.8) and could be functionally and metabolically different in different anatomic locations (Sorrel and Caplan 2004) This is only to be expected, as epidermal differentiation varies in different anatomical locations such as lips, palm, and the sole, directed by local epithelial–mesenchymal interactions. Differences in fibroblast physiology from fetal and adult skin provide a clue to the scarless wound healing in the fetus (Lorenz et al. 1994). Typically, fibroblasts produce and degrade hyaluronic acid and a variety of other matrix molecules, as well as collagen and elastin, and participate in the dermal homeostasis and wound healing.

Other Components

The dermis, with its abundant blood and lymph vessels, neuronal components, mast cells, dendritic cells (that act as sentinels of the immune system) houses the sweat glands, and pilosebaceous follicles as well as adipocytes. Adipocytes synthesize and store fat, and in addition secrete a slew of cytokines and hormones that have vast implication to human health, metabolic syndrome, obesity, and longevity.

Skin Appendages

No description of skin is complete without its appendages—hair, nails, sebaceous and sweat glands. However, as extended descriptions of these are beyond the scope of this chapter, these are covered only very superficially. Hair follicle anchors the hair in the dermis, and is composed of five layers of concentrically arranged epithelial cells. The three inner layers (medulla, cortex, and cuticle) contribute to the hair shaft, the two outer layers contributing to the hair sheath. The medulla is the core of the hair shaft, while the bulk of the structure is made up of the cortex. The cuticle covers the hair with an overlapping pattern. The next two layers form the internal root sheath, and the external—outer root sheath which does not participate in hair formation. Due to the huge significance that human society has placed on the appearance and health of hair, an enormous amount of information is available on its physical, biophysical, and biological aspects including development, pigmentation, renewal, and loss (see Forslind et al. 2004).

Closely associated with the hair follicle are the sebaceous glands (derived from epithelial cells) made up of sebocytes (see Wheately 1986). Sebum functions in delivering Vitamin E, the only lipophilic antioxidant to the skin surface Readers are directed to the chapter on Sebaceous glands in this book for the details of its structure, physiology and roles played in skin physiology.

Sweat glands are composed of coiled tubes, located within the dermis, and are present over the majority of the body surface—secreting sweat, which has roles in thermoregulation, as well as an excretory role. There are two kinds of sweat glands—eccrine and apocrine. The former is found all over the skin, but in abundance on the palms, soles, axillae, and forehead, and are innervated by cholinergic nerves. Under psychological and/or thermal stress, they secrete a watery sweat containing chloride, lactic acid, urea, and several glycoproteins, some of which contribute to immunity (Sato 1977). A recent finding that proinflammatory cytokines in sweat activate keratinocytes as a danger signal in eczematous lesions (Dai et al. 2013) also show the importance of an intact skin barrier, where the keratinocytes are not normally exposed to such secreted factors. The larger apocrine glands often open into hair follicles, and are seen in axillae, ano-genital region, and areolae. They become active at puberty, and produce a protein rich secretion, which although odorless, are acted upon by bacteria that create a characteristic odor. These glands are under adrenergic control.

Hypodermis

The hypodermis or subcutis although is not considered by many as a part of the skin, it is one of the three layers of skin which helps the attachment of the skin to the muscles and bone, it also supply skin with blood vessels and nerves. The hypodermis is formed of loose connective tissue and contains mainly adipocytes and also fibroblasts and macrophages. The adipocytes or fat cells play significant roles in health, as it contains about 50% of the body fat, serving as an energy store and a padding for the skin, important in sculpting of body contours. An interesting discussion on the significance of this layer has been presented by Paus et al. (2007) and within this book in part "Subcutaneous Fat," in chapter "Facial Subcutaneous Adipocytes" wherein the aesthetic issues (cellulite, lipodystrophy in aging) and health implications (metabolic syndrome) of subcutaneous fat cells are discussed quite lucidly, to which I draw the attention of readers.

Summary of Skin Functions

To summarize, skin primarily performs a protective function, including physical protection (protect from physical impact, dehydration, seasonal effects of heat and cold, oxidant damage, UV radiation) as well as excretion (sweat, storage excretion through melanin/hair), thermoregulation (TEWL, sweat), acts as an endocrine organ (peripheral metabolism of hormones, Vitamin D synthesis), and a sense organ (touch, heat, cold) (Tobin 2010). Additionally, it has a major role in physical attraction and reproduction (through secondary sexual characters, pheromones) and intra-species communication (expression of emotions/nonverbal cues), and is the basis of a multi-billion dollar personal care industry globally. The forgoing information is woefully inadequate account of the skin, but may be enough to show that the mini-essay "My organ is more important than your organ" (Bernard 1990) is not a preposterous, provocative, and vain statement, but is, if anything an understatement.

References

Attenborough D. Life on earth. Boston: Little Brown and Company; 1980.

Baron JM, Holler D, Schiffer R, Frankenberg S, Neis M, Merk HF, Jugert FK. Expression of multiple cytochromeP450 enzymes and multidrug resistance-associated transport proteins in human skin keratinocytes. J Invest Dermatol. 2001;116:541–8.

Bentinger M, Tekle M, Dallner G. Coenzyme Q-biosynthesis and functions. Biochem Biophys Res Comm. 2010;396:74–9.

Bernard JD. My organ is more important than your organ. Arch Dermatol. 1990;126:827.

Blanpain C, Fuchs E. Epidermal homeostasis: a balancing act of stem cells in the skin. Nature Rev Mol Cell Biol. 2009;10:207–17.

Brandner J. Tight junctions and tight junction proteins in mammalian epidermis. Eur J Pharm Biopharm. 2009;72:289–94.
Brandner JM, Proksch E. Epidermal barrier function: role of tight junctions. In: Elias PM, Feingold KR, editors. Skin barrier. New York: Taylor Francis; 2006.
Costin G-E Hearing VJ. Human skin pigmentation: melanocytes modulate skin color in response to stress. FASEB J. 2007;21:976–94.
Cotta-Pereira G, Rodrigo G, Bittencourt-Sampaio S. Oxytalan, elaunin and elastic fibers in the human skin. J Invest Dermatol. 1976;66:143–8.
Dai X, Okazaki H, Hanakawa Y, Murakami M, Tohyama M, et al. Eccrine sweat contains IL-1α, IL-1β and IL-31 and activates epidermal keratinocytes as a danger signal. PLoS ONE. 2013;8(7):e67666. doi:10.1371/journal.pone.0067666.
de Witte L, Nabatov A, Pion M, Fluitsma D, de Jong MAWP, deGruijl T, Piguet V, van Kooyk Y, Geitenbeek TB. Langerin is a natural barrier to HIV-1 transmission by Langerhans cells. Nat Med. 2007;13:367–71.
Elias PM, Feingold KR. Permeability barrier homoeostasis. In: Elias PM, Feingold KR, editors. Skin barrier. New York: Taylor Francis; 2006.
Elias PM, Menon GK. Structural and lipid biochemical correlates of the epidermal permeability barrier. In: Elias PM, editors. Skin lipids. London: Academic; 1991. pp. 1–27. (Adv Lipid Res. Vol 24).
Elias PM, Cullander C, Mauro T, et al. The secretory granular cell: the outermost granular cell is a specialized secretory cell. J Invest Dermatol Symp Proc. 1998;3:87–100.
Forslind B, Lindberg M, Norlen L, Editors. Skin, hair and nails. New York: Marcel Dekker; 2004. p. 483.
Fuchs E, Segre JA. Stem cells: a new lease on life. Cell. 2000;100:143–55.
Gallo RL, Hopper LV. Epithelial antimicrobial defence of the skin and intestine. Nat Rev Immunol. 2012;12:503–16.
Gustafsson E, Fassler R. Insights into extracellular matrix functions from mutant mouse models (mini review). Exp Cell Res. 2000;261:52–68.
Hadley NF. Lipid water barriers in biological systems. Prog Lipid Res. 1989;28:1–33.
Hilliges M, Wang L, Johansson O. Ultrastructural evidence for nerve fibers within all vital layers of the human epidermis. J Invest Dermatol. 1995;104:134–7.
Hoath S. The skin as a neurodevelopmental interface. Neo Rev. 2001;2:269–81.
Holbrook KA, Byers PH. Skin is a window on heritable diseases of connective tissue. Am J Med Gen. 1989;34:105–21.
Holbrook KA, Odland GF. The fine structure of developing human epidermis:light, scanning, and transmission electron microscopy of the periderm. J Invest Dermatol. 1975;65:16–38.
Irvine AD, McLean WHJ, Leung DYM. Filaggrin mutations associated with skin and allergic diseases. N Eng J Med. 2011;365:1315–27.
Jowett S, Ryan T. Skin disease and handicap: an analysis of the impact of skin conditions. Soc Sci Med. 1985;20:425–8.
Kubo A, Nagao K, Yokouchi M, Sasaki H, Maagi M. External antigen uptake by Langerhans cells with reorganization of epidermal tight junction barriers. J Exp Med. 2009;206:2937–46.
Lamberg SI, Stoolmiller AC. Glycosaminoglycans, a biochemical and clinical review. J Invest Dermatol. 1974;63:433–49.
Lillywhite HB. Water relations of tetrapod integument. J. Exp Biol. 2006;209:202–26.
Lorenz PH, Lin RY, Longaker MT, Whitby DJ, Adzik SN. The fetal fibroblast: the effector cell of scarless fetal skin repair. Plast Reconstr Surg. 1994;96:1251–59.
Mabuchi H, Higashikata T, Kawashiri M, Katsuda S, Mizuno M, Nohara A, Inazu A, Koizumi J, Kobayashi J. Reduction of serum Ubiquinol-10 and Ubiquinone-10 levels by atorvastatin in hypercholesterolemic patients. J Atheroscler Thromb. 2005;12:111–9.
Marinkovich PM, Keene DR, Rimberg CS, Burgeson RE. The cellular origin of the dermal-epidermal basement membrane. Dev Dynamics. 1993;197:255–67.
Matoltsy GA. Structure and function of the mammalian epidermis. In: Bereiter-Hahn J, et al. editors. Biology of the integument. Berlin: Springer; 1986. pp. 255–71.

Menon GK, Elias PM. The epidermal barrier and strategies for surmounting it: an overview. In: Hengge UR, Volc-Platzer B, editors. The skin and gene therapy. Berlin: Springer; 2001. pp. 3–26. (605).
Menon GK, Kligman AM. Barrier functions of human skin: a holistic view. Skin Pharmacol Physiol. 2009;22:178–89.
Montagna W. Comparative anatomy and physiology of the skin. Arch Dermatol. 1967;96:357–63.
Monteiro-Riviere NA. Structure and function of skin. In: Monteiro-Riviere N, editors. Skin toxicology. New York: Informa Healthcare; 2010. pp. 1–18.
Mosher DF, Furcht LT. Fibronectin: review of its structure and possible functions. J Inves Dermatol. 1981;77:175–80.
Myllyharju J, Kivirikko KI. Collagens, modifying enzymes and their mutations in humans, flies and worms. Trends Genet. 2004;20:33–43.
Paus R, Lapière CM, Maquoi E. Viewpoint 7. Exp Dermatol. 2007;16:67–70.
Pope FM, Martin GR, Lichenstein JR, Penttinen R, Gerson B, Rowe DW, McKusick VA. Patients with Ehlers-Danlos syndrome type IV lack type III collagen. PNAS. 1975;72:1314–16.
Proksch E, Brandner JM, Jensen J-M. The skin: an indispensable barrier. Exp Dermatol. 2008;17:1063–72.
Rawlings AV, Matts PJ. Stratum corneum moisturization at the molecular level: an update in relation to the dry skin. J Invest Dermatol. 2005;124:1099–110.
Reed T, Ghadially R, Elias PM. Skin type, but neither race or gender, influence epidermal permeability function. Arch Dermatol. 1995;131:1134–8.
Sato K. The physiology, pharmacology and biochemistry of the eccrine sweat gland. Rev Physiol Biochem Pharmacol. 1977;79:51–131.
Scheuplein RJ Blank IH. Permeability of the skin. Physiol Rev. 1971;51:702–47.
Shindo Y, Witt E, Han D, Epstein W, Packer L. Enzymic and non-enzymic antioxidants in epidermis and dermis of human skin. J Invest Dermatol 1994;102:122–4.
Sorrel MJ, Caplan AI. Fibroblast heterogeneity: more than skin deep. J Cell Sci. 2004;117:667–75.
Tobin DJ. Sensing the environment. In: Monteiro-Reviere N, editors. Skin toxicology. New York: Informa Healthcare; 2010. pp. 172–91.
Uitto J. The role of elastin and collagen in cutaneous aging: intrinsic aging versus photoexposure. J Drugs Dermatol. 2008;7(2 Suppl):12.
Van Lis JMJ, Kruiswijk T, Mager WH, Kalsbeek GL. Glycosaminoglycans in human skin. Brit J Dermatol. 1973;88:355–61.
Watt FM. Stem cell fate and patterning in mammalian epidermis. Curr Opin Genet Dev. 2001;11:410–7.
Webb AR. Who, what, where and when- influences on cutaneous vitamin D synthesis. Prog Biophys Mol Biol. 2006;92:17–25.
Wertz PW. Lipids and barrier function of the skin. Acta Derm Veneriol. 2000;208:1–5.
Wheately VR. The physiology and pathophysiology of the skin. Vol. 9. The sebaceous glands. London: Academic; 1986. p. 2971.
Yuen AWC, Jablonski NG. Vitamin D: in the evolution of human skin colour. Med Hypothesis. 2010;74:39–44.
Zhang M-C, He L, Giro MG, Yong SL, Tiller GE, Davidson JM. Cutis laxa arising from frameshift mutations in exon 30 of the elastin gene (ELN). J Biol Chem. 1999;274:981–6.

Chapter 3
Skin Barrier, Structure, and Properties

Lars Norlén

Core Messages

a. The skin offers the ultimate barrier for the living organism.
b. Lipids in the horny layer constitute the most important component of the skin's barrier.
c. The lipids are in a gel-like state.
d. The lipids are organized as stacked bilayers of fully extended ceramides with cholesterol associated with the ceramide sphingoid moiety.
e. The lipids' organization makes the skin's barrier simultaneously impermeable and robust.

Introduction

Terrestrial life was only made possible through the adaptive evolution of a waterproof barrier in the integument of organisms. In man, like in all other land-living higher vertebrates, this barrier is constituted by a uniquely organized lipid material situated between the cells of the horny layer of skin.

Not having a native reference for the molecular organization of the horny layer's lipid material has until recently limited our fundamental understanding of the skin both in healthy and disease states, and hampered technological progress in areas such as development of treatments for skin disease, transdermal drug delivery, dealing with toxicity from topical exposure to chemicals, and the development of non-invasive diagnostic sensors.

However, a recent breakthrough has come from the use of very high magnification cryo-electron microscopy of vitreous skin sections (CEMOVIS) combined

L. Norlén (✉)
Department of Cell and Molecular Biology (CMB),
Karolinska Institute, and Dermatology Clinic, Karolinska University Hospital,
von Euler's v 1, 171 77, Stockholm, Sweden
e-mail: lars.norlen@ki.se

A. Pappas (ed.), *Lipids and Skin Health*, DOI 10.1007/978-3-319-09943-9_3

with electron microscopy (EM) simulation, which has allowed for direct molecular structure determination of the horny layer's lipid material in situ and in its *near native state* (Iwai et al. 2012). Remarkably, the lipid material is organized in an arrangement not previously described in a biological system—stacked bilayers of fully extended ceramides with cholesterol molecules associated with the ceramide sphingoid moiety (Iwai et al. 2012).

A brief account of the structure–function relationships of the skin barrier is presented in the following sections.

History

The fact that the skin's barrier function essentially is located to the epidermis, was already clear in the 1800s (Homolle 1853; Duriau 1856) (Fig. 3.1). The structure and function of the skin barrier has intrigued researchers ever since.

In the 1940s, Winsor and Burch (1944) showed, by sandpapering the skin surface, that the skin's barrier actually resides in the epidermal horny layer. Until then it had been thought that the horny layer's only function was to protect the body from physical and chemical injury by virtue of its toughness. It now became clear that it also was responsible for the maintenance of body homeostasis. In the 1950s and 1960s, Berenson and Burch (1951) and Onken and Moyer (1963) showed that horny layer impermeability essentially was a function of its lipid content, and Brody (1966) managed to locate the horny layer lipids to the extracellular space.

The first EM visualizations of the lipid material's stacked lamellar morphology came with Breathnach et al. (1973) and Elias and Friend (1975) in the early 1970s. Then, trying to determine the molecular composition and organization of this lipid material became the subject of much activity.

Determining the lipid composition was difficult, as contamination sources were numerous and the amount of contaminating lipid normally was large in comparison with the amount of actual horny layer lipids studied. The problem was approached by studying young pigs with inactive sebaceous glands (Wertz and Downing 1983), and by studying human epidermal cysts that could be collected without contamination (Wertz et al. 1987).

The introduction of ruthenium tetroxide staining in the mid 1980s revealed the lipid material's characteristic broad: narrow: broad electron lucent band staining pattern (Madison et al. 1987).

Soon afterward, using small-angle X-ray diffraction on isolated horny layer, White et al. (1988), Garson et al. (1991) and Bouwstra et al. (1991) reported the presence of one shorter (ca 4.5 nm) and one longer (ca 6.5 nm) diffraction peak related to lipids. Also, McIntosh (2003) observed an asymmetric distribution of cholesterol molecules within model systems composed of extracted horny layer lipids.

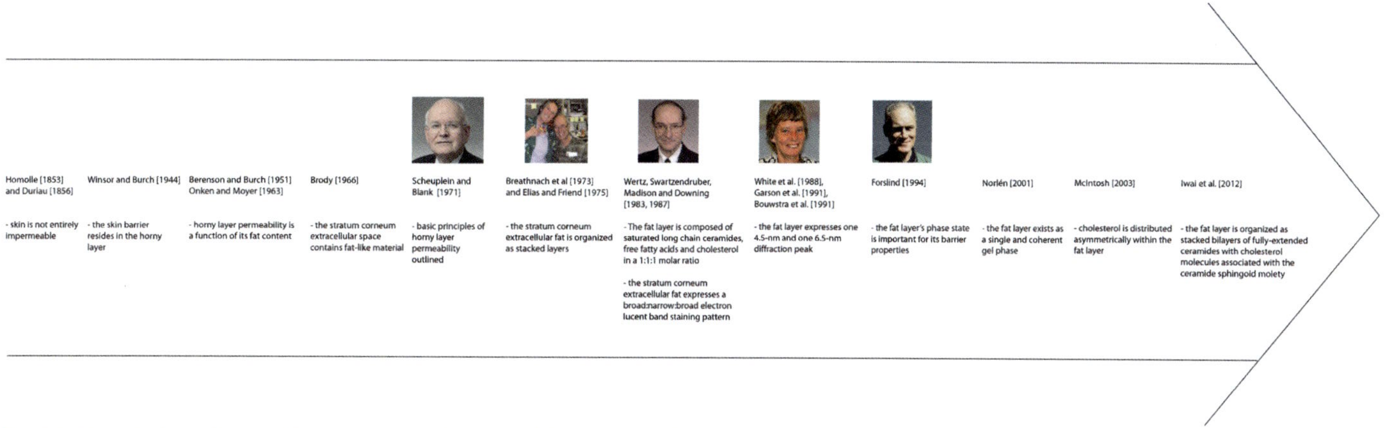

Fig. 3.1 Historical timeline of skin barrier research. Adapted from Norlén (2013), with permission

Fig. 3.2 Molecular arrangement of the skin barrier. The horny layer's lipid material is organized as stacked bilayers of fully extended ceramides with cholesterol molecules associated with the ceramide sphingoid moiety (Iwai et al. 2012). *Green spheres* represent hydrogen and carbon atoms in ceramides, cholesterol, and free fatty acids. *Red spheres* represent oxygen atoms. Adapted from Norlén (2013), with permission

In the early 1990s, the importance of the lipid material's phase state for its barrier properties began to be stressed (Forslind 1994). In 2001, Norlén proposed that the lipid material exists as a single and coherent gel phase (Norlén 2001b), which was later supported by the experimental work of Iwai et al. (2012).

In 2012, Iwai et al. showed experimentally that the horny layer's lipid material is organized as stacked bilayers of fully extended ceramides with cholesterol molecules associated with the ceramide sphingoid moiety (Figs. 3.2 and 3.3).

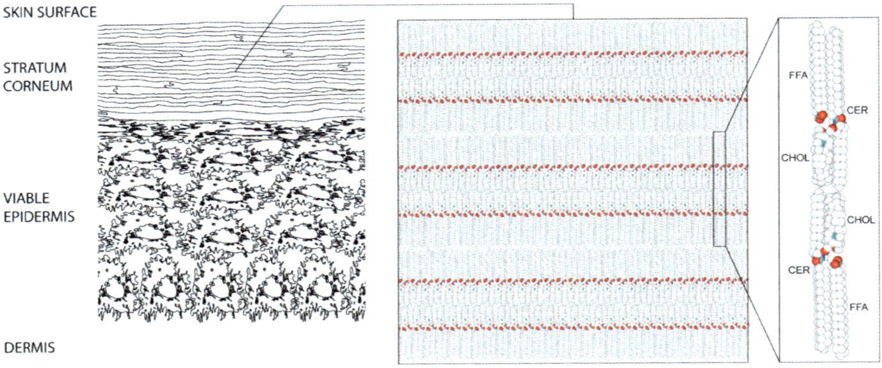

Fig. 3.3 Schematic drawing of skin. *Left part*: schematic cellular-scale drawing of epidermis. *Mid part*: molecular-scale drawing of the lamellar lipid material occupying the space between the cells of the horny layer. *Right part*: atomic model of the lipid material's repeating unit, composed of two mirrored subunits, each composed of one fully extended ceramide (*CER*), one cholesterol (*CHOL*), and one free fatty acid (*FFA*) molecule. Adapted from Norlén (2012), with permission

Skin Lipid Composition and Phase State

Essential for determining the lipid material's molecular organization and physical properties was to obtain an idea of its molecular composition. This was difficult given the contamination sources and the horny layer's morphological complexity, where extracted horny layer lipids are derived not only from the extracellular lipid material but also from the cell plasma membrane (also termed corneocyte "lipid envelope"), and possibly also from intracellular sources. However, by minimizing contamination sources and by comparing the experimental and interindividual variability between different extracted lipid fractions, a consensus was eventually reached with respect to the rough molecular composition of the horny layer's lipid material (Wertz and Norlén 2003).

The lipid material consists of a heterogeneous mixture of saturated, long-chain ceramides, free fatty acids, and cholesterol in a roughly 1:1:1 molar ratio (Wertz and Norlén 2003). More than 300 different species have been identified in the ceramide fraction alone (Masukawa et al. 2009).

The most characteristic features of the lipid composition (Wertz and Norlén 2003) are (i) extensive compositional heterogeneity with broad, but invariable, chain length distributions (20–32 C; peaking at 24 C) in the ceramide fatty acid and free fatty acid fractions, (ii) almost complete dominance of saturated very long hydrocarbon chains (C20:0–C32:0), and (iii) large relative amounts of cholesterol (about 30 mol%).

These are the same as the factors that typically stabilize lipid gel phases. It has therefore been proposed that the horny layer's lipid material exists as a single and coherent gel phase (Norlén 2001b). The viscous gel-like behavior of the lipid material was recently demonstrated experimentally by its remarkable malleability in situ (Fig. 3.4; Iwai et al. 2012).

Molecular Structure Determination Directly In Situ

Recently, the lipid material's molecular organization was determined with the aid of high-resolution CEMOVIS (Iwai et al. 2012). This technique yields high-resolution images of the horny layer's lipid material in situ and in its *near native state*. When combined with molecular modeling and EM simulation, CEMOVIS has proved to be remarkably effective in identifying the molecular organization of the horny layer lipids.

A short description of the new experimental procedure is provided in the following section, which could also be generalized to determine the molecular organization of other biomolecular structures in skin, such as the corneocyte keratin network

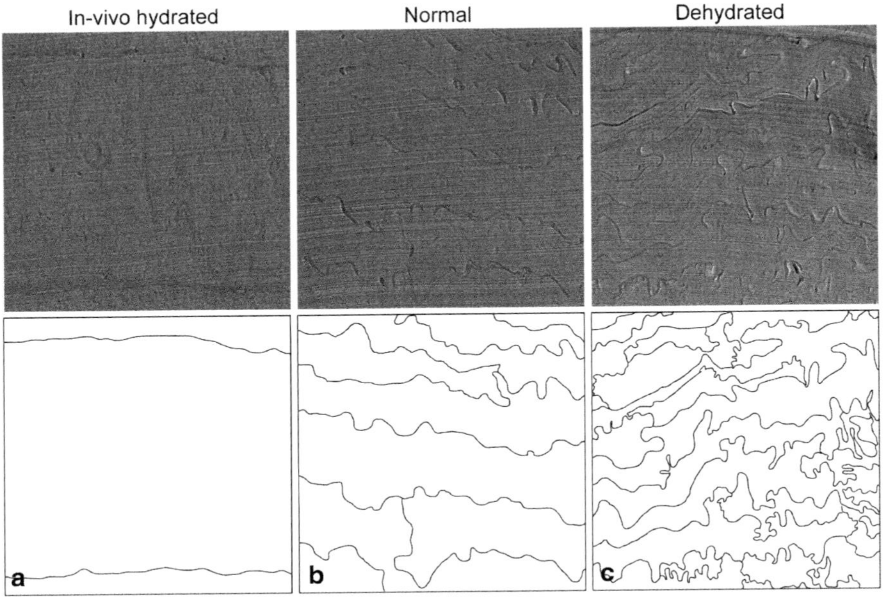

Fig. 3.4 The horny layer's lipid material is, despite its crystalline-like character, malleable. The horny layer's extracellular lipid material is folded locally. The folding decreases on hydration and increases on dehydration. Low magnification CEMOVIS micrographs of the horny layer after hydration in vivo (**a**), at normal in vivo conditions (**b**), and after hydration in vivo followed by dehydration ex vivo (**c**). The lower panel illustrates the folded pattern of the extracellular space. Image side lengths: 5 μm. Adapted from Iwai et al. (2012) (online suppl. material), with permission

and the corneocyte lipid envelope. For a more detailed description, see Norlén et al. (2014).

The Procedure

The structure determination procedure involves four stages (Norlén et al. 2014): CEMOVIS to yield high resolution (ca 1 nm) images of the lipid material (Fig. 3.5); construction of candidate molecular models for the lipid material (Fig. 3.6); simulation of electron micrographs resulting from the proposed molecular models (Fig. 3.6); and confrontation of the simulated micrographs with those observed experimentally to identify a molecular organization that is consistent with the observed CEMOVIS data (Fig. 3.7).

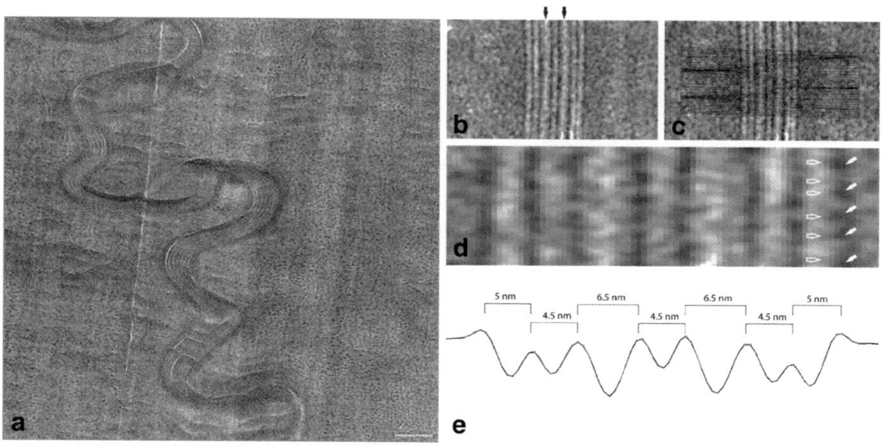

Fig. 3.5 The CEMOVIS intensity pattern of the horny layer's extracellular lipid material consists of folded *stacked layers*. **a** Medium magnification CEMOVIS micrograph of the interface between two cells in the *midpart* of the horny layer. Note that in CEMOVIS, the tissue is unstained, and that the pixel intensity is directly related to the local electron density of the sample. The *stacked lamellar pattern* represents the extracellular lipid material. *Dark* ~10 nm dots represent keratin intermediate filaments filling out the intracellular space. **b** High magnification CEMOVIS micrograph of the extracellular space in the *midpart* of the horny layer. The averaged intensity profile of the lipid material was obtained by fuzzy distance-based image analysis. The *red stars* in **b** represent the manually chosen start and end points for fuzzy distance based path growing. **c** The *red line* represents the traced out path. *Stacked lines* mark extracted intensity profiles. **d** Enlarged area of the central part of **b**. **e** Reversed averaged pixel intensity profile obtained from the extracted area in **c**. Peaks in **e** correspond to *dark bands* and valleys to *lucent bands* in **d**. *Black arrows* in **b** denote electron lucent narrow bands at the centre of the 6.5 nm bands. Section thickness ~50 nm (**a**–**d**). Scale bar **a**: 100 nm. Pixel size in (**a**–**d**): 6.02 Å. Adapted from Iwai et al. (2012), with permission

CEMOVIS

In CEMOVIS, a biological specimen is vitrified by means of ultra rapid (~20 ms) cooling (below −140 °C) under high pressure (~2000 bar) (Al-Amoudi et al. 2004). The vitrified sample is then cut into ultrathin (20–30 nm nominal thickness) sections in a cryo-microtome, and finally directly observed in a cryo-electron microscope without further preparation. In this way, the nanostructure of biological tissues can be studied in their native (preserved down to the molecular level) hydrated state, without chemical fixation or staining. Another crucial advantage for later structure determination is that the resulting micrograph pixel intensity is directly related to the local electron density of the native specimen.

Biomolecules generally possess small differences in electron density, as they are essentially composed of atoms with similar atomic weight (carbon, nitrogen, and oxygen). However, for orderly arranged molecular assemblies, such as lipid tails

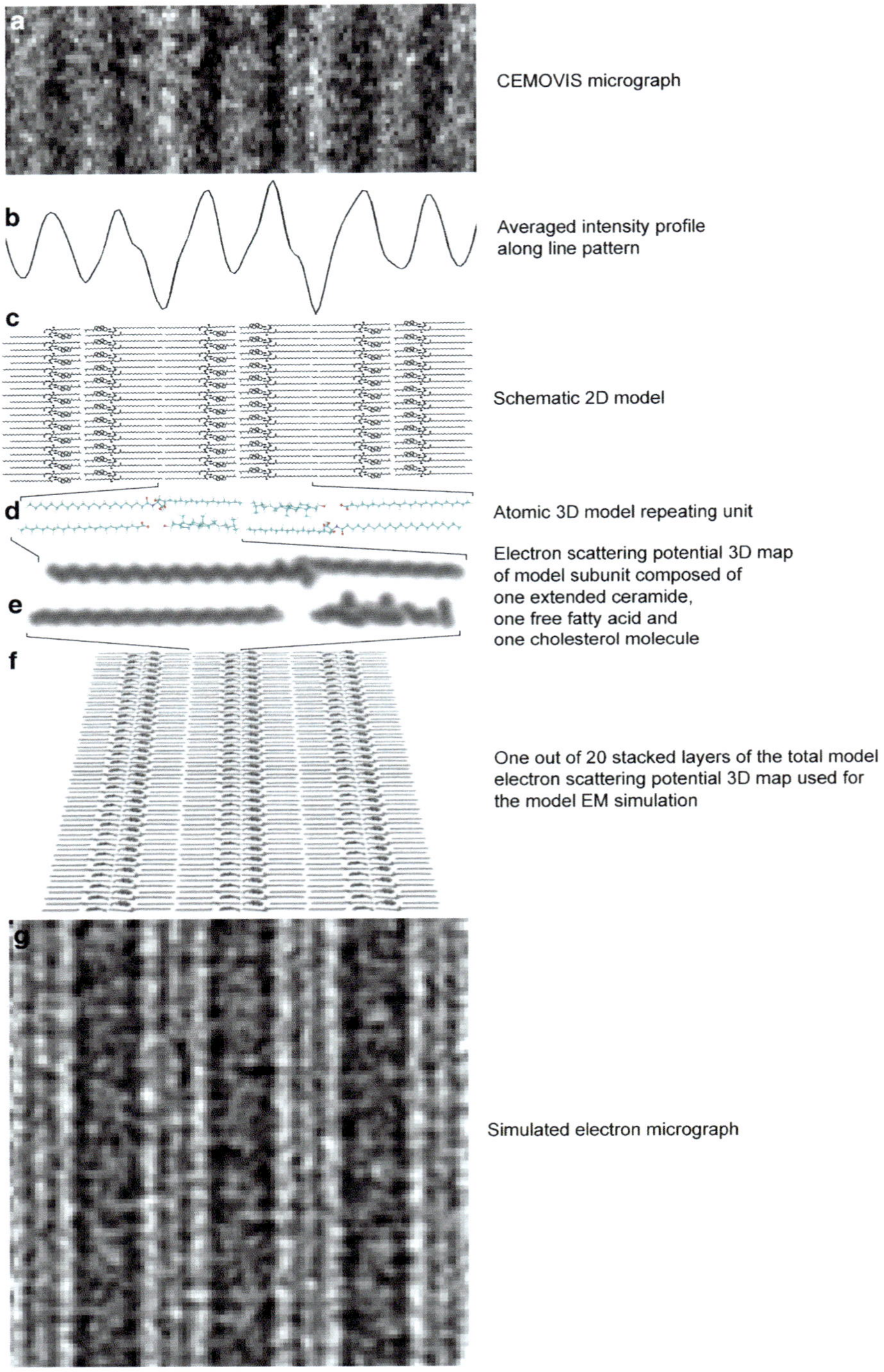

Fig. 3.6 Electron microscopy (EM) simulation of the horny layer's extracellular lipid material. **a** High magnification CEMOVIS micrograph of the extracellular space in the *midpart* of the horny

and headgroups in membranes, even small differences in shape and atomic composition may be amplified because of interference effects that appear in the image phase contrast. During EM image acquisition, phase contrast is made visible using defocus. At high magnification, complex interference patterns can be resolved in the CEMOVIS micrographs. Exploiting this, CEMOVIS micrographs can be recorded repeatedly at very high magnification at the same position of the skin sample while increasing stepwise the microscope's defocus, thus ensuring that differences in the recorded micrographs are due exclusively to the different defocuses used (Iwai et al. 2012).

Modeling and Simulation of the Horny Layer Lipid organization

Objectively interpreting high-resolution data, whether in EM or in X-ray diffraction, ideally involves the use of a simulator, to compare experimental real data with simulated data of candidate molecular models.

As a starting point for building different candidate models of the horny layer's lipid material, the predominant lipid component of each of the three major lipid classes, namely, C24 ceramide NP (the predominant ceramide in the horny layer), cholesterol, and a C24 free fatty acid (the predominant free fatty acid) were modeled in-silico in their low energy equilibrium conformations. This allowed the construction of 16 distinct candidate molecular organizations. The individual molecules were packed in a small unit cell and then repeatedly reproduced in *x*-, *y*-, and *z*-axes to yield a periodic structure. This periodic structure then served as a model organization for each of the 16 candidate models (Figs. 3.8, 3.9, and 3.10).

For each of the 16 candidate molecular models, simulated electron micrograph images were generated using a newly developed EM simulator (Rullgård et al. 2011) (Figs. 3.8, 3.9 and 3.10). First, an electron scattering potential map was generated from each candidate model. Second, the interaction between the potential map and the electron beam, the optical transformation effect of the microscope's lens system, and the image formation on the microscope's detector were simulated. Parameters defining optical properties of the microscope, e.g., acceleration voltage, aberration constants, defocus, and the detector's point spread function, were set to mimic the

layer. **b** Corresponding averaged intensity profile obtained by fuzzy distance based path growing (cf. Fig. 3.1b, c). **c** Schematic 2D illustration of ceramides (tetracosanylphytosphingosine (C24:0)) in fully extended conformation with cholesterol associated with the ceramide sphingoid part and free fatty acids (lignoceric acid (C24:0)) associated with the ceramide fatty acid part. **d** Atomic 3D model of the repeating unit composed of two mirrored subunits, each composed of one fully extended ceramide, one cholesterol and one free fatty acid molecule. **e** Calculated electron scattering potential of one model subunit. **f** Calculated electron scattering potential 3D maps of the topmost layer out of 20 superimposed layers used to generate the simulated electron micrograph **g**. Defocus (**a, g**): −2.5 μm. Pixel size in (**a, g**): 3.31 Å. Adapted from Iwai et al. (2012), with permission

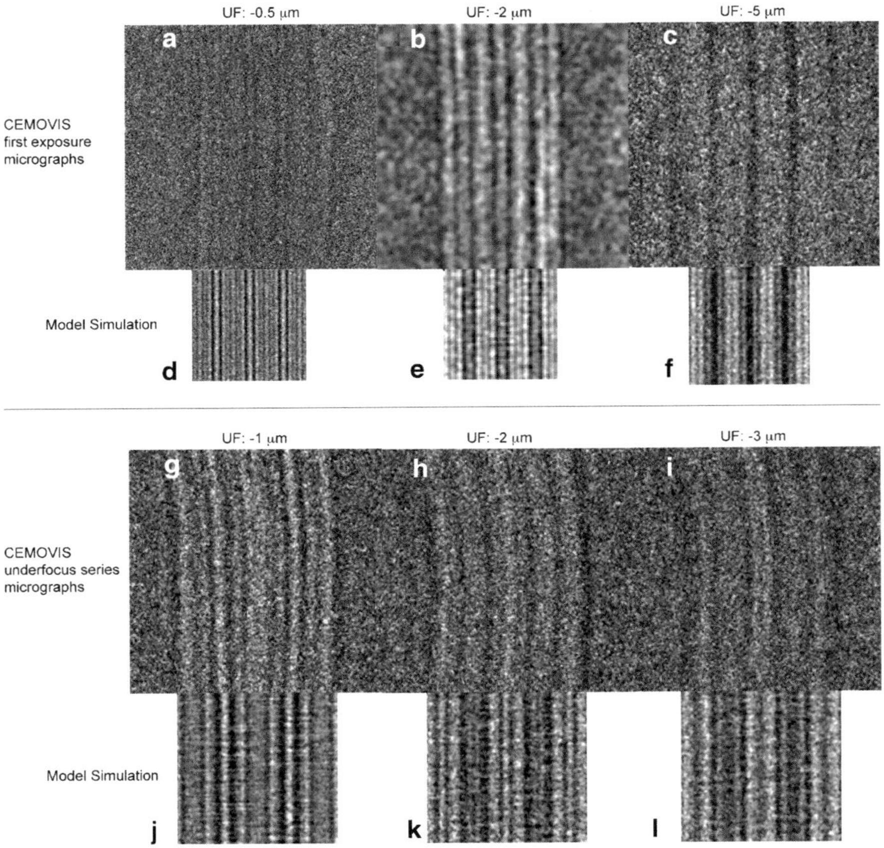

Fig. 3.7 Electron microscopy (EM) simulation of alternating fully extended ceramides with selective localization of cholesterol to the ceramide sphingoid part accurately accounts for the observed CEMOVIS intensity patterns as well as for the interference pattern changes observed in sequential CEMOVIS micrograph defocus-series obtained at very high magnification (≤1.88 Å pixel size). (**a**–**c**) High magnification CEMOVIS micrographs (first exposition images) of the extracellular space in the *midpart* of the horny layer obtained at −0.5 μm (**a**), −2 μm (**b**), and −5 μm (**c**) defocus. At very low defocuses (−0.5 μm) (**a**) CEMOVIS intensity patterns can only be observed at very high magnification (≤1.88 Å pixel size). At very high defocuses (−5 μm) (**c**) image resolution is low but still allows for resolution of the ~11 nm repeating unit. The slightly larger lamellar repeat distance in (**b**) (~12 nm) compared to **a** and **c** (~11 nm) may be due to more pronounced compression of the vitreous skin section during cryo-sectioning along the lamellar plane in (**b**) compared to in (**a**) and in (**c**). (**d**–**e**) represents corresponding atomic 3D model (cf. Fig. 3.3 *right part*; Fig. 3.6d) EM simulation images recorded at −0.5 μm (**d**), −2 μm (**e**) and −5 μm (**f**) defocus. (**g**–**i**) sequential CEMOVIS micrograph defocus-series obtained at very high magnification (1.88 Å pixel size). Note the fine changes in interference patterns caused by gradually increasing

conditions in the real CEMOVIS experiments. Finally, the simulated micrographs were compared with the real CEMOVIS micrographs. This approach proved to be highly discriminating between different lipid models (Iwai et al. 2012).

Skin Lipid Structure

The lipid organization that emerged from the high-resolution CEMOVIS data analysis of the horny layer's lipid material is a bilayer structure of fully extended (splayed chain) ceramides with the sphingoid moieties interfacing. Both cholesterol and the free fatty acid were distributed selectively: cholesterol at the ceramide sphingoid end and the free fatty acid at the ceramide fatty acid end (Iwai et al. 2012; Figs. 3.2 and 3.3).

A notable feature of the horny layer's lipid material is that it comprises a bilayer rather than an arrangement of stacked monolayers. The preference for the bilayer organization could result from the biological processes involved in its formation. The lipid material is formed from stacked bilayers of glycosylceramides in the hairpin conformation (i.e., with both lipid tails pointing in the same direction) in a hydrated environment. They subsequently undergo dehydration and deglycosylation into ceramides (Holleran et al. 1993; Norlén 2001a; Al-Amoudi et al. 2005). During dehydration and deglycosylation, the initial hairpin glycosylceramide bilayer organization may carry over into the extended (splayed chain) ceramide bilayer organization, by a flip of one of the two ceramide alkyl chains together with cholesterol (Iwai et al. 2012). The extension of the two ceramide chain moieties in opposite directions aids the separation of both the ill-matched cholesterol and free fatty acids, and the ill-matched ceramide C24 fatty acid and C18 sphingoid moieties, into different bands within the close-packed dehydrated fully extended ceramide-based bilayer structure.

the microscope's defocus during repeated image acquisition at a fixed position. Due to electron beam damage after repeated electron exposure, the image contrast is lower in micrographs (**h**–**i**) compared to micrograph (**g**), which was acquired first. In micrograph (**i**) some shrinkage can be observed. This is probably due to mass loss during repeated electron exposure. Also, the curvature of the lamellar pattern is slightly increased in micrographs (**h**–**i**) compared to in micrograph (**g**), which likewise may be caused by nonhomogeneous mass loss during repeated electron exposure. (**j**–**l**) represents corresponding atomic 3D model (cf. Fig. 3.3 *right part*; Fig. 3.6D) EM simulation images recorded at −1 μm (**j**), −2 μm (**k**), and −3 μm (**l**) defocus. It is shown that the atomic 3D model in Fig. 3.5 accurately accounts not only for the major features of the CEMOVIS micrographs (**a**–**f**) but also for the interference intensity pattern changes observed upon varying the microscope's defocus during image acquisition at very high magnification (**g**–**l**). Pixel size in (**c**, **f**): 3.31 Å, in (**b**, **e**): 6.02 Å, and in (**a**, **d**, and **g**–**l**): 1.88 Å. Adapted from Iwai et al. (2012), with permission

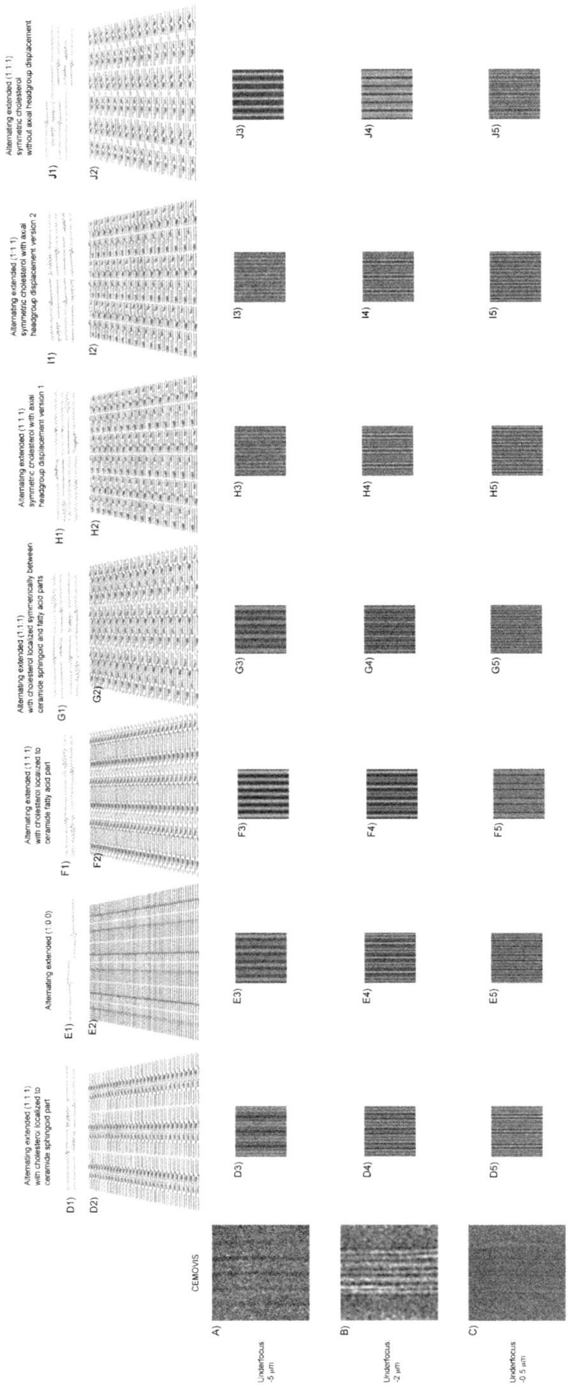

Fig. 3.8 Electron microscopy (EM) simulation results from seven fully extended ceramide bilayer models with varying cholesterol distribution. (**a**–**c**) CEMOVIS micrographs of the horny layer's

Skin Lipid Formation

In order to understand the structure–function relationships of the skin barrier in vivo, horny layer lipid formation may yield important clues, as the horny layer's lipid material may not represent an equilibrium system but rather a "frozen-in" or "immobilized" open biological system. Horny layer lipid formation is also central from a dermatological standpoint, since malformation of the lipid material may be an etiological factor in barrier-deficient skin disease, such as eczema, psoriasis, and "dry skin" (Norlén 2012).

In 2001, it was proposed (Norlén 2001a), and later supported experimentally (Al-Amoudi et al. 2005), that skin lipid formation proceeds via (1) membrane synthesis in the trans-Golgi of a membrane system with cubic-like symmetry, followed by (2) morphologically continuous (non-fusion dependent) secretion of the cubic-like membrane system into the extracellular space, (3) phase transition from cubic-like to lamellar membrane morphology, (4) dehydration, (5) condensation, and (6) lipid chain rearrangement from a folded (hairpin) to an extended (splayed chain) stacked bilayer conformation (Iwai et al. 2012). CEMOVIS specifically supports the proposed continuity of the lipid secretion system as well as the proposed structural association of nonlamellar and lamellar lipid morphologies (Norlén et al. 2003; Al-Amoudi et al. 2005). However, structure determination of the intermediate stages of horny layer lipid formation may require access to native molecular resolution tomographic 3D data in situ (molecular tissue TOVIS [cf. Norlén et al. 2009]), a developing technology that may not yet have reached its full potential (Norlén 2012).

Skin Lipid Function

A stacked, fully extended (splayed chain) ceramide bilayer arrangement (Figs. 3.2 and 3.3) with high cholesterol content (ca 30 mol%) and a highly heterogeneous, but characteristic, lipid composition of saturated, long-chain ceramides, and free

extracellular lipid material acquired at −5 μm (**a**), −2 μm (**b**), and −0.5 μm (**c**) defocus. (**d3–j5**) Corresponding simulated electron micrographs obtained from seven fully extended ceramide models. (**d1–j1**) Repeating units for each simulated model. (**d2–j2**) calculated electron scattering potential 3D maps of the topmost layer out of 20 superimposed layers used to generate each individual simulated micrograph (**d3–j5**). In model (**d**), cholesterol is selectively localized to the ceramide sphingoid part. In model (**e**), cholesterol has been removed to evaluate whether the simulation method could discriminate the presence (**d**) or absence (**e**) of cholesterol. In model (**f**), cholesterol is selectively localized to the ceramide fatty acid part. In models (**g–j**), cholesterol is homogenously distributed between the ceramide sphingoid and fatty acid parts. Contrary to models (**g, j**), models (**h, i**) express axial headgroup displacement of cholesterol and free fatty acids. Models (**h, i**) differ in that model (**h**) expresses a pairwise lateral distribution of ceramides while model (**i**) expresses a homogeneous lateral distribution of ceramides. Note that except for the position of the lipid headgroups, the localization of cholesterol within the fully extended ceramide structure largely determines the electron scattering properties of the models. Adapted from Iwai et al. (2012) (online suppl. material), with permission

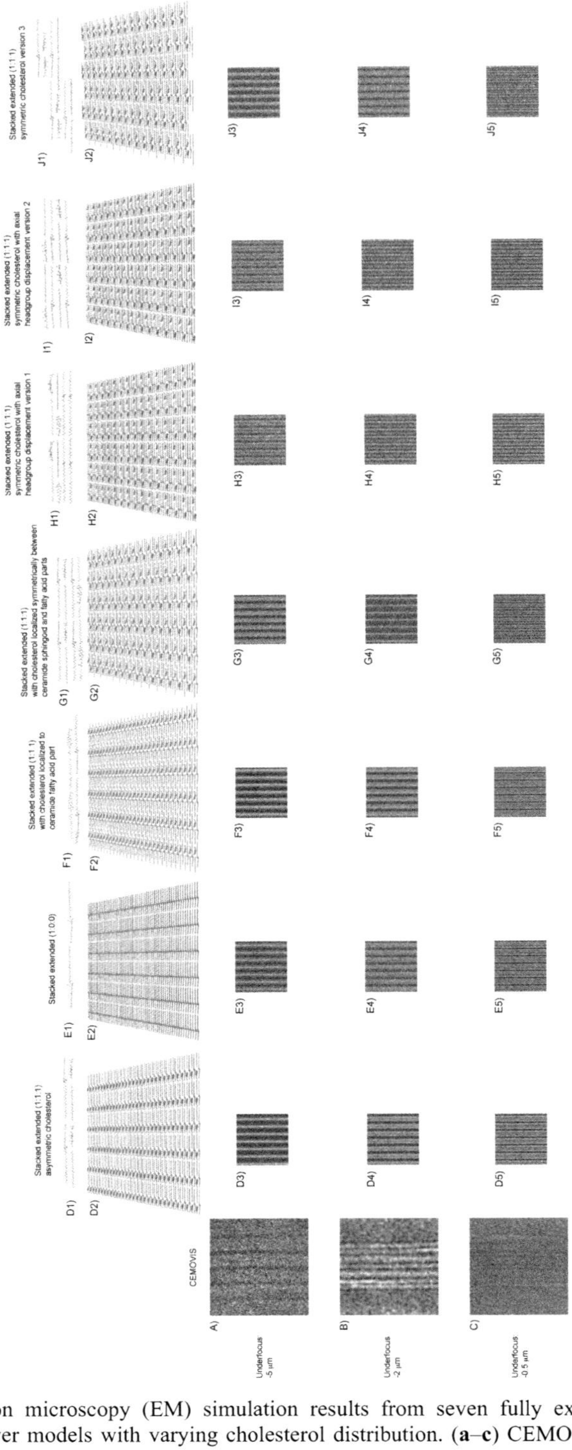

Fig. 3.9 Electron microscopy (EM) simulation results from seven fully extended ceramide stacked monolayer models with varying cholesterol distribution. (**a**–**c**) CEMOVIS micrographs

fatty acids, may represent an ideal barrier material for skin. This is because it may be largely impermeable to water as well as to both hydrophilic and lipophilic substances, because of its condensed chain packing and its alternating lipophilic (alkyl chain) and hydrophilic (headgroup) regions. Likewise, it may be resistant to both hydration and dehydration because of its lack of exchangeable water between lipid leaflets. It may also be resistant toward temperature and pressure changes because of its heterogeneous lipid composition and high cholesterol content, which stabilize gel-like chain packing and thereby prevents both lateral domain formation and induction of "pores" or nonlamellar morphologies, upon environmental stress. Further, this bilayer arrangement may account for horny layer cell cohesion without advocating specialized intercellular adhesion structures, such as desmosomes. The absence of desmosomes in the horny layer may hence allow for sliding of horny layer cells to accommodate skin bending. Finally, as the interaction between the individual layers of the lipid material involves only hydrocarbons, the layers may be relatively free to slide with respect to one another, making the lipid material pliable. The fully extended ceramide bilayer arrangement, with high cholesterol content and heterogeneous saturated long-chain lipid composition, may thus meet the barrier needs of skin by being simultaneously impermeable and robust (Iwai et al. 2012).

Summary

Terrestrial life was only made possible through the adaptive evolution of a waterproof barrier in the integument of organisms. In man, like in other land-living higher vertebrates, this barrier is constituted by a uniquely organized lipid material situated between the cells of the horny layer of skin. It was recently shown that this lipid material is organized as stacked bilayers of fully extended ceramides.

---

of the horny layer's extracellular lipid material acquired at −5 µm (**a**), −2 µm (**b**), and −0.5 µm (**c**) defocus. (**d3–j5**) Corresponding simulated electron micrographs obtained from seven stacked fully extended ceramide models. (**d1–j1**) Two repeating units for each simulated model. (**d2–j2**) Calculated electron scattering potential 3D maps of the topmost layer out of 20 superimposed layers used to generate each individual simulated micrograph (**d3–j5**). In model (**d**), cholesterol is selectively localized to the ceramide sphingoid part. In model (**e**), cholesterol has been removed to evaluate whether the simulation method could discriminate the presence (**d**) or absence (**e**) of cholesterol. In model (**f**), cholesterol is selectively localized to the ceramide fatty acid part. In models (**g–j**), cholesterol is distributed homogenously between the ceramide sphingoid and fatty acid parts. Contrary to models (**g, j**), models (**h, i**) express axial headgroup displacement of cholesterol and free fatty acids. Models (**h, i**) differs in that model (**h**) expresses a pairwise lateral distribution of ceramides while model (**i**) expresses a homogeneous lateral distribution of ceramides. Note that except for the position of the lipid headgroups, the localization of cholesterol within the fully extended ceramide structure largely determines the electron scattering properties of the models. Adapted from Iwai et al. (2012) (online suppl. material), with permission

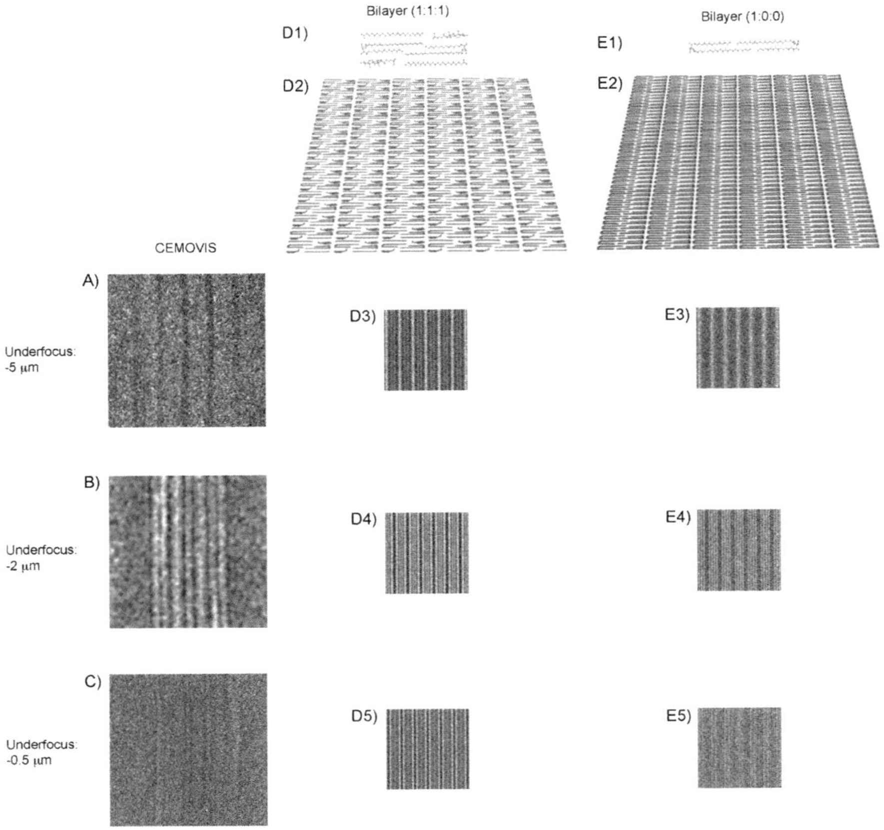

Fig. 3.10 Electron microscopy (EM simulation results from two-folded ceramide bilayer models with and without the presence of cholesterol. (**a–c**) CEMOVIS micrographs of the horny layer's extracellular lipid material acquired at −5 μm (**a**), −2 μm (**b**), and −0.5 μm (**c**) defocus. (**d3–e5**) Corresponding simulated electron micrographs obtained from two-folded ceramide models. (**d1–e1**) Repeating units for each simulated model. (**d2–e2**) Calculated electron scattering potential 3D maps of the topmost layer out of 20 superimposed layers used to generate each individual simulated micrograph (**d3–e5**). In model (**d**), cholesterol is present. In model (**e**), cholesterol has been removed to ascertain if the simulation method could distinguish the presence (**d**) or absence (**e**) of cholesterol. Note that the presence of cholesterol within the folded ceramide structure largely determines the electron scattering properties of the models. Adapted from Iwai et al. (2012) (online suppl. material), with permission

The physical state of the skin's lipid material may be that of a single and coherent gel phase. Further, the lipid material may be formed via a phase transition from cubic-like to stacked lamellar morphology.

The skin's lipid material is responsible for both the skin's low permeability and its robustness toward environmental stress.

References

Al-Amoudi A, Chang J-J, Leforestier A, McDowall A, Michel Salamin L, Norlén L, Richter K, Sartori Blanc N, Studer D, Dubochet J. Cryo-electron microscopy of vitreous sections. EMBO J. 2004 Sep 15;23(18):3583–8.

Al-Amoudi A, Dubochet J, Norlén L. Nanostructure of the epidermal extracellular space as observed by cryo-electron microscopy of vitreous sections of human skin. J Invest Dermatol. 2005;124:764–77.

Berenson GS, Burch GE. Studies of diffusion of water through dead human skin: the effect of different environmental states and of chemical alterations of the epidermis. Amer J Trop Med. 1951;31:842–53.

Bouwstra JA, Gooris GS, Van der Spek JA, Bras W. Structural investigations of human stratum corneum by small-angle X-ray scattering. J Invest Dermatol. 1991;97:1005–12.

Breathnach AS, Goodman T, Stolinski C, Gross M. Freeze fracture replication of cells of stratum corneum of human epidermis. J Anat. 1973;114:65–81.

Brody I. Intercellular space in normal human stratum corneum. Nature. 1966;209:472–6.

Dubochet J, Adrian M, Chang J-J, Homo J-C, Lepault J, McDowall AW, Schultz P. Cryo electron microscopy of vitrified specimens. Q Rev Biophys. 1988 May;21(2):129–228.

Duriau F. Recherches expérimentales sur l'absorption et l'exhalation par le tégument externe. Arch Gen Med T. 1856;7:161–73.

Elias PM, Friend DS. The permeability barrier in mammalian epidermis. J Cell Biol. 1975;65:180–91.

Forslind B. A domain mosaic model of the skin barrier. Acta Derm Venereol (Stockh.). 1994;74:1–6.

Garson JC, Doucet J, Leveque J-L, Tsoucaris G. Oriented structure in human stratum corneum revealed by X-ray diffraction. J Invest Dermatol. 1991;96:43–9.

Holleran WM, Takagi Y, Menon G, Legler G, Feingold KR, Elias PM. Processing of epidermal glycosylceramides is required for optimal mammalian cutaneous permeability barrier function. J Clin Invest. 1993;91:1656–64.

Homolle A. Expériences physiologiques sur l'absorption par la tégument externe chez l'homme dans le bain. Union Med. 1853;7:462.

Iwai I, Han H, den Hollander L, Svensson S, Öfverstedt L-G, Anwar J, Brewer J, Bloksgaard Mølgaard M, Laloeuf A, Nosek D, Masich S, Bagatolli L, Skoglund U, Norlén L. The human skin barrier is organized as stacked bilayers of fully-extended ceramides with cholesterol molecules associated with the ceramide sphingoid moiety. J Invest Dermatol. 2012;132:2215–25.

Madison KC, Swartzendruber DC, Wertz PW, Downing DT. Presence of intact intercellular lamellae in the upper layers of the stratum corneum. J Invest Dermatol. 1987;88:714–8.

Masukawa Y, Narita H, Sato H, Naoe A, Kondo N, Sugai Y, Oba T, Homma R, Ishikawa J, Tagaki Y, Kitahara T. Comprehensive quantification of ceramide species in human stratum corneum. J Lipid Res. 2009;50:1708–19.

McIntosh TJ. Organization of skin stratum corneum extracellular lamellae: diffraction evidence for asymmetric distribution of cholesterol. Biophys J. 2003;85:1675–81.

Norlén L. Skin barrier formation: the membrane folding model. J Invest Dermatol. 2001a May 6;117(4):823–9.

Norlén L. Skin barrier structure and function: the single gel-phase model. J Invest Dermatol. 2001b Oct;117(4):830–6.

Norlén L. Skin lipids. In Roberts GCK, editor. Encyclopedia of biophysics. Berlin: Springer; 2012.

Norlén L. Current understanding of skin barrier morphology. Skin Pharm Phys. 2013;26:213–6.

Norlén L, Al-Amoudi A. Stratum corneum keratin structure, function, and formation: the cubic rod-packing and membrane templating model. J Invest Dermatol. 2004 Oct;123(4):715–32.

Norlén L, Öktem O, Skoglund U. Molecular cryo-electron tomography of vitreous tissue sections: current challenges. J Microsc. 2009;235:293–307.

Norlén L, Anwar J, Öktem O. Accessing the molecular organization of the stratum corneum using high resolution electron microscopy and computer simulation. In: Computational Biophysics of the Skin (Chap. 10). Ed. B. Querleux, Pan Stanford Publishing, Singapore; 2014. pp. 289–331.
Onken HD, Moyer CA. The water barrier in human epidermis. Arch Dermatol. 1963;87:584–90.
Rullgård H, Öfverstedt L-G, Masich S, Daneholt B, Öktem O. Simulation of transmission electron microscope images of biological specimens. J Microsc. 2011 Sep;243(3):234–56.
Scheuplein RJ, Blank IH. Permeability of skin. Physiol Rev. 1971 Oct;51(4):702–47.
Wertz P, Norlén L. "Confidence intervals" for the "true" lipid compositions of the human skin barrier? In: Forslind B, Lindberg M, editors. Skin, hair, and nails: structure and function. New York: Marcel Dekker; 2003. pp. 85–106.
Wertz PW, Downing DT. Ceramides of pig epidermis: structure determination. J Lipid Res. 1983;24:759–65.
Wertz PW, Swartzendruber DC, Madison KC, Downing DT. Composition and morphology of epidermal cyst lipids. J Invest Dermatol. 1987;89:419–24.
White SH, Mirejovsky D, King GI. Structure of lamellar lipid domains and corneocyte envelopes of murine stratum corneum. An x-ray diffraction study. Biochem. 1988;27:3725–32.
Winsor T, Burch GE. Differential roles of layers of human epigastric skin on diffusion of water. Arch Intern Med. 1944;74:428–44.

Chapter 4
Polyunsaturated Fatty Acid Oxygenated Metabolites in Skin

Anna Nicolaou

Core Messages

- Cutaneous polyunsaturated fatty acids play a vital role in skin health and, through their oxygenated metabolites, are involved in a plethora of homeostatic and pathophysiological reactions.
- Cyclooxygenase, lipoxygenase and cytochrome P450 monooxygenase isoforms mediate the production of prostaglandins and other eicosanoids, octadecanoids and docosanoids many which have been identified in the skin and contribute to keratinocyte proliferation, melanocyte dendricity, photocarcinogenesis, allergy and inflammation.
- Non-enzymatic oxidations produce an array of regio- and/or stereo-isomers of eicosanoids and highly reactive keto-aldehydes with less potent bioactivities.
- Nutritional supplementation with polyunsaturated fatty acids has the potential to alter the cutaneous lipid profile and this has been explored as means of developing chemoprotective and therapeutic interventions.

Introduction

Fatty acids are of particular importance to skin biology. Not only they play a vital role in the integrity of the epidermal barrier, but are also contributing to cellular development and communications, in particular through the production of oxygenated metabolites. These potent mediators are formed through enzymatic or free radical-mediated reactions and include the eicosanoids, octadecanoids, docosanoids and a number of related lipid species. They are produced by resident and infiltrating skin

A. Nicolaou (✉)
Manchester Pharmacy School, Faculty of Human and Medical Sciences,
University of Manchester, Stopford Building, Oxford Road, Manchester M13 9PT, UK
Email: anna.nicolaou@manchester.ac.uk

A. Pappas (ed.), *Lipids and Skin Health*, DOI 10.1007/978-3-319-09943-9_4

cells including epidermal keratinocytes, melanocytes and Langerhans cells, dermal fibroblasts, leukocytes and macrophages, and are important for skin physiology and homeostasis, with particular importance for cutaneous inflammation and immunity (McCusker and Grant-Kels 2010; Kendall and Nicolaou 2013; Ziboh et al. 2002).

The polyunsaturated fatty acid (PUFA) precursors of these bioactive lipids are found in low concentration in the epidermis and dermis, and their levels are strongly dependent on dietary sources. Early studies by Burr and Burr have demonstrated the importance of fatty acid metabolism for skin health (Burr and Burr 1929), while more recent findings have shown that human epidermis is lacking Δ5 and Δ6 desaturase activites (Chapkin and Ziboh 1984; Ziboh and Chapkin 1988). Therefore, the epidermis relies upon the bloodstream for its supply of the essential fatty acids linoleic acid (LA; 18:2n−6) and α-linolenic acid (ALA; 18:3n−3), as well as their elongated and desaturated PUFA derivatives including dihomo-gamma-linolenic acid (DGLA, 20:3n−6), arachidonic acid (AA; 20:4n−6), eicosapentaenoic acid (EPA; 20:5n−3) and docosahexaenoic acid (DHA; 22:6n−3). While the concentration of LA is approximately 12 % of total epidermal fatty acids (reflecting its importance for efficient barrier function), the epidermal concentration of AA is approximately 3 % of total fatty acids, with EPA and DHA each contributing less that 1 % of total fatty acids (Schurer et al. 1994; Zuo et al. 20082008; Hansen and Jensen 1985; Ziboh and Chapkin 1988).

Cutaneous PUFA are found esterified in membrane phospholipids and, to a lesser extent, in cholesterol esters and ceramides. A number of stressors including reactive oxygen species (ROS), ultraviolet radiation (UVR), mechanical trauma, tumour promoters and bacterial toxins can activate skin lipases and phospholipases (PL) that in turn hydrolyse membrane PUFA making them available for further metabolism. Of particular importance is PLA_2, a family of enzymes hydrolysing phospholipid-esterified AA and other PUFA (Dennis et al. 2011). Cutaneous isoforms relevant to the production of eicosanoids include the cytosolic PLA_2 ($cPLA_2$) enzyme whose expression and activity can be stimulated by oxidative stress such as UVR-produced ROS (Chen et al. 1996; Hanson and DeLeo 1990; Uozumi et al. 1997; Gresham et al. 1996), and the secretory PLA_2 ($sPLA_2$) that has been identified in keratinocytes and sites of cutaneous inflammation such as psoriatic tissue (Andersen et al. 1994; Sjursen et al. 2000; Chan and Mauro 2011; Maury et al. 2000) and atopic dermatitis (Schafer and Kragballe 1991). Epidermal keratinocytes have a defence mechanism against ROS and this may be linked to $cPLA_2$ activation (Bito and Nishigori 2012). Other relevant lipolytic enzymes include the phosphatidylinositol (PI)-specific phospholipase C (PLC) that releases diacylglycerol (DAG) which can be further metabolised by lipases to generate AA and, consequently, contribute to the formation of eicosanoids; these enzymes are also involved in cutaneous inflammatory conditions as shown by their role in UVR-induced inflammation and psoriasis (Bergers et al. 1990; Oka et al. 2011; Fisher et al. 1990; Carsberg et al. 1995).

This chapter will review the enzymatic and non-enzymatic pathways leading to oxygenated metabolites of cutaneous PUFA, and discuss the bioactivities and relevant physiological and pathophysiological situations where these potent autacoids are of particular interest.

Cyclooxygenase-Derived Lipid Mediators

One of the most potent classes of cutaneous eicosanoids are the prostanoids, group of bioactive lipids formed by the enzyme prostaglandin endoperoxide H synthase, commonly known as cyclooxygenase (COX) (Smith et al. 2000, 2012). Skin cells express both COX isoforms, i.e. the constitutive COX-1 and the inducible COX-2. The COX-mediated reaction converts AA to the unstable endoperoxide prostaglandin H_2 (PGH_2) that is then isomerised to individual series-2 prostaglandins (PG), thromboxanes (TX) or prostacyclin (PGI_2) via tissue or cell specific prostanoid synthases (Fig. 4.1). COX isoforms can also metabolise the 20 carbon fatty acids DGLA and EPA, to form series-1 and -3 prostanoids, respectively (reviewed in (Massey and Nicolaou 2011)). Studies on human and animal skin have shown the prevalence of PGE_2, PGE_1, PGE_3, PGD_2, $PGF_{2\alpha}$, PGI_2 and TXB_2 (Ziboh 1992; Sugimoto et al. 2006; Rhodes et al. 2009).

PGE_2 is the dominant cutaneous prostanoid and is produced by almost all cutaneous cell types (Black et al. 1978; Pentland and Needleman 1986; Ziboh 1992; Cho et al. 2005; Sugimoto et al. 2006; Rhodes et al. 2009; Gledhill et al. 2010). It is formed via the cytosolic (cPGES), and microsomal (mPGES-1 and mPGES-2) prostaglandin E synthases (Kudo and Murakami 2005; Murakami and Kudo 2006). Evidence for linked expression of the inducible mPGES-1 and COX-2 isozymes indicates the presence of an efficient system for increased PGE_2 production upon stimulation (Ueno et al. 2005). PGD_2 is produced via the hematopoetic-type (H-PGDS) or the lipocalin-type (L-PGDS) prostaglandin D synthases (Urade and Eguchi 2002), while its further non-enzymatic hydrolysis gives rise to anti-inflammatory cyclopentanone PGs such as PGJ_2 and 15d-PGJ_2 (Surh et al. 2011; Scher and Pillinger 2005). PGD_2 is produced by Langerhans cells, dermal mast cells and melanocytes, and it is believed to be involved in immune and allergic responses (Shimura et al. 2010; Ujihara et al. 1988; Ikai and Imamura 1988). Interestingly, Langerhans cells and mast cells express H-PGDS while melanocytes express L-PGDS reflecting their neural crest origin (Takeda et al. 2006; Masoodi et al. 2010; Shimura et al. 2010). $PGF_{2\alpha}$ can be produced via the PGH_2 9,11-endoperoxide reductase (prostaglandin F synthase; PGFS) or through further metabolism of PGE_2 via the enzyme PGE 9-ketoreductase. The epimeric form 9α,11β-PGF_2 (11-epi-$PGF_{2\alpha}$) is formed from PGD_2 via the PGD 11-ketoreductase (Watanabe 2002, 2012). $PGF_{2\alpha}$ has been found in skin extracts and keratinocytes in vitro (Sugimoto et al. 2006; Ziboh et al. 1977; Rhodes et al. 2009) and has been associated to inflammatory and immune responses, pigmentation and hair growth (Colombe et al. 2007; Scott et al. 2004, 2007). Prostacyclin synthase (PGIS) forms PGI_2, a very unstable prostanoid that is readily hydrolysed to the stable but inactive metabolite 6-keto-$PGF_{1\alpha}$ (Cathcart et al. 2010). Finally, thromboxane A synthase (TXAS) converts PGH_2 to TXA_2, another unstable mediator that is quickly transformed to the stable but inert TXB_2 (Cathcart et al. 2010). Both PGI_2 and TXB_2 have been found in whole skin extracts and cultured skin cells at low concentrations and it is possible that they may derive from infiltrat-

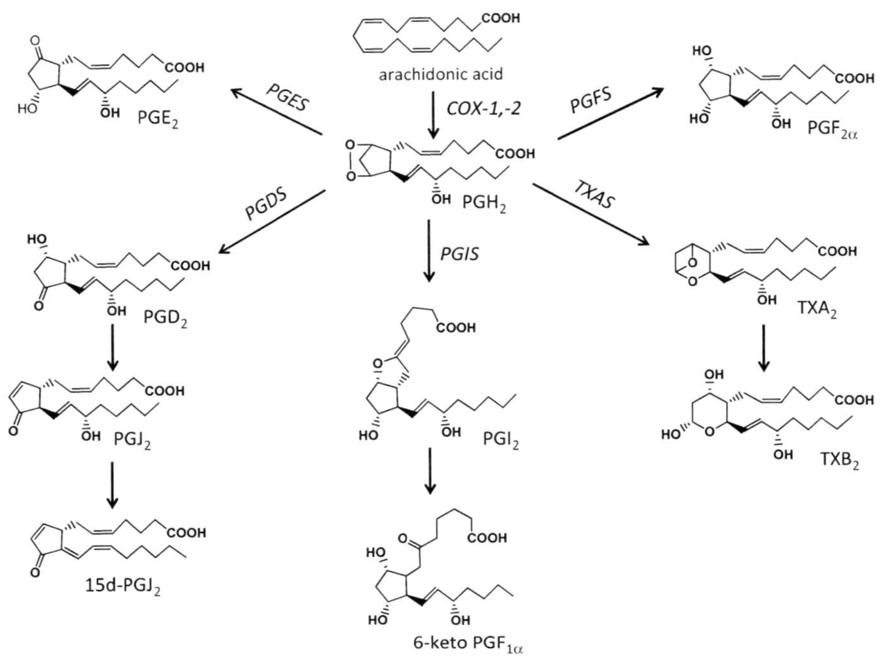

Fig. 4.1 Schematic representation of cyclooxygenase-mediated pathways leading to the biosynthesis of prostanoids from arachidonic acid. *COX* cyclooxygenase, *PGES* prostaglandin E synthase, *PGDS* prostaglandin D synthase, *PGFS* prostaglandin F synthase, *PGIS* prostacyclin synthase, *TXAS* thromboxane A synthase

ing leukocytes or vascular endothelial cells and not resident skin cells (Pentland and Needleman 1986; Sugimoto et al. 2006).

Prostanoids mediate their activities via a series of G-protein coupled receptors designated EP, DP, FP, IP and TP. They are sub-divided to eight groups as defined by pharmacological studies and related signal transduction pathways, namely: EP1, EP2, EP3, EP4, DP1, FP, IP and TP. However, recent studies have showed the presence of a second PGD receptor, DP2 (also termed CRTh2), as well as a number of receptor heterodimers (Woodward et al. 2012; Breyer et al. 2001). Almost all prostanoid receptors have been identified in skin cells (Scott et al. 2005; Weber et al. 2002; Muller et al. 2000; Black et al. 2008; Andoh et al. 2007; Nakajima et al. 2010; Bryniarski et al. 2008; Konger et al. 2005; Honda et al. 2009; Satoh et al. 2006; Kanda et al. 2010).

Finally, these active autacoids are not stored in cells but produced upon request and exported via a number of PG transporters and the mulitspecific organic anion transporters (Schuster 2002). Prostaglandin levels are actively controlled through oxidation via the NAD^+-dependent 15-ketoprostaglandin dehydrogenases (15-PGDH) followed by reduction via the Δ^{13}-15-ketoprostaglandin reductases (Tai et al. 2002). These reactions catalyse the formation of metabolites such as

15-keto-PG and 13,14-dihydro-15-keto PGs that have significantly reduced bioactivities (Tai et al. 2002). 15-PGDH exhibits a rather broad substrate specificity and can deactivate a range of PGs. The PG deactivation system has been identified in epidermal keratinocytes, whole human skin and hair follicles (Rhodes et al. 2009; Judson et al. 2010; Michelet et al. 2008).

Lipoxygenase-Derived Lipid Mediators

Lipoxygenases (LOX) mediate the addition of molecular oxygen in a stereoselective manner, generating a range of hydroperoxy fatty acyls that are reduced to hydroxy fatty acids or further metabolised to leukotrienes (LT; Fig. 4.2). Although LOX isozymes are classified based on their phylogenetic profile, they are conventionally defined by their positional selectivity when oxygenating AA (Brash 1999; Kuhn and O'Donnell 2006). Human and animal skin have been shown to express 5-, 8-, 12- and 15-LOX activities that can catalyse the oxygenation of many PUFA including LA, AA, EPA, DGLA and DHA to produce and array of mono- and polyhydroxy fatty acids (Miller et al. 1991; Lu et al. 2010; Rhodes et al. 2009; Nicolaou et al. 2012; Mayer et al. 1984).

The enzyme 5-LOX requires 5-lipoxygenase activating protein (FLAP) for activation; it converts AA to the unstable 5-hydroperoxy eicosatetraenoic acid (HPETE) that is subsequently reduced to 5-hydroxy eicosatetraenoic acid (HETE; Brash 1999). Oxidation of 5-HETE by the enzyme 5-hydroxyeicosanoid dehydrogenase (5-HEDH) produces 5-oxo-eicosatetraenoic acid (ETE; Grant et al. 2009), while dehydration of 5-HETE can generate LTA_4 precursor to LTB_4 and the cysteinyl leukotrienes LTC_4, LTD_4 and LTE_4 (Brain and Williams 1990). Studies exploring the prevalence of 5-LOX in the skin have suggested that its activity is rather low and related primarily to epidermal keratinocytes in particular during their differentiation (Ziboh et al. 1984; Grabbe et al. 1984a; Janssen-Timmen et al. 1995), Langerhans cells (Doepping et al. 2007; Spanbroek et al. 1998) and infiltrating leukocytes (Serhan et al. 2003), while the formation of cutaneous cysteinyl-leukotrienes has been attributed to infiltrating and not resident skin cells (Church et al. 2002).

12-LOX activity is mediated by a number of isozymes: the cytosolic leukocyte-type 12-LOX and microsomal platelet-type 12-LOX are found in a number of cell types, while the 12*R*-LOX are unique mammalian skin cells and appear to be important in inflammatory conditions such as psoriasis (Boeglin et al. 1998; Kuhn and O'Donnell 2006). The epidermis-type lipoxygenase-3 (eLOX-3) contributes to the terminal differentiation of keratinocytes and, together with 12*R*-LOX, is important for the integrity of the epidermal barrier through formation of the epidermis-specific O-linoleoyl-omega-hydroxyceramides and hepoxilins, another AA-derived group of bioactive lipids (Fig. 4.2) (Epp et al. 2007; Brash et al. 2007; Nigam et al. 2007; Zheng et al. 2011).

The activity of 15-LOX is attributed to two isoforms: the reticulocyte-type 15-LOX-1 and the epidermis-type 15-LOX-2. It is noteworthy that 15-LOX-1 has

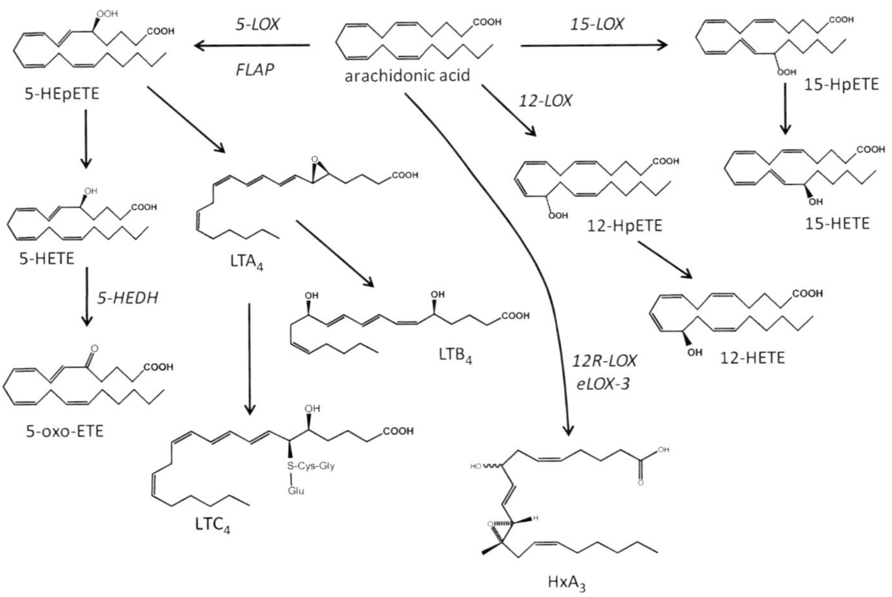

Fig. 4.2 Schematic representation of lipoxygenase-mediated pathways leading to the biosynthesis of hydroxy fatty acid species from arachidonic acid. *LOX* lipoxygenase, *LT* leukotriene, *HxA_3* hepoxilin A_3

high homology to the leukocyte-type 12-LOX, an isoform presenting both 12- and 15-HETE activities (referred to as 12/15-LOX; Kuhn and O'Donnell 2006). When the substrate is LA, the main product of 15-LOX-1 is 13*S*-HODE, a metabolite found in high abundance in the skin (Rhodes et al. 2009; Nicolaou et al. 2012). Although mammalian 12- and 15-LOX isozymes oxygenate, primarily, AA and LA, they can also utilise other PUFA including DGLA, EPA and DHA, found in either free or esterified in phospholipids and lipoproteins (Kuhn and O'Donnell 2006; Gron et al. 1993; Heitmann et al. 1995). Recent reports have revealed the presence of a large number of PUFA-derived hydroxy fatty acids in the skin confirming the presence of many cutaneous LOX isozymes, although the exact cellular origin and specific bioactivities of all these mediators are still to be explored (Nicolaou et al. 2012; Rhodes et al. 2009; Lu et al. 2010; McDaniel et al. 2011; Ziboh et al. 2000b; Pilkington et al. 2013; Shappell et al. 2001). Finally, murine epidermis is expressing 8-LOX, a homologue of the inducible 15-LOX-2 that has not yet been found in human skin (Jisaka et al. 1997; Krieg et al. 1998).

LOX reactions are also important in transcellular metabolism generating polyhydroxylated PUFA species with, principally, anti-inflammatory protective biological profiles (Sala et al. 2010; Calder 2009). These LOX-LOX or acetylated-COX-2-LOX or CYP-LOX mediated reactions generate lipoxins (LX) from AA, resolvins (Rv) and protectins (PD) from both EPA and DHA, and maresins (MaR) from DHA (Serhan et al. 2008a). To date, there are no reports on the actual formation of these species in the skin, however, the biological precursors of some of these mediators have been identified (e.g. 15-HETE for LX, 18-HEPE for RvE1 and 17-HDHA for RvD1; McDaniel et al. 2011; Nicolaou et al. 2012; Pilkington et al. 2013).

Cytochrome P450-Derived Lipid Mediators

Although the family of cytochrome P450 (CYP) monooxygenases are better known for their role in detoxification of drugs and xenobiotics, their prevalence in skin suggests potential contribution of PUFA oxygenation (Baron et al. 2008; Neis et al. 2010; Rolsted et al. 2008). CYP isozymes relevant to the formation of eicosanoid-like species are epoxygenases and hydrolases (Oliw 1994): when AA is the substrate, CYP-catalysed epoxygenation can occur in every double bond resulting in four regioisomeric *cis*-epoxyeicosatrienoic acids (EET), namely: 5,6-, 8,9- 11,12- and 14,15-EET, that can be formed as either *R, S* or *S, R* enantiomers. These epoxides are then transformed to biologically inactive dihydroxyeicosatetraenoic acids (DHET) by the enzyme epoxide hydrolase. Mid-chain oxidation of AA can produce a range of HETE including 5-, 8-, 9-, 11-, 12- and 15-HETE similar to the ones formed by LOX, while CYP-catalysed omega-hydroxylations can form 16-, 17-, 18-, 19- and 20-HETE (Spector 2009; Zeldin 2001; Konkel and Schunck 2011; Fig. 4.3). CYP can utilise a range of PUFA including LA, EPA and DHA, to form LA-derived epoxides such as the 9,10- and 12,13-epoxyoctadecamonoenoic acids (EpOME) and HODE, EPA-derived HEPE and five epoxyeicosatetraenoic acids (EEQ), and DHA-derived HDHA and six regioisomeric epoxydocosapentaenoic acids (EDP). Although some of these products could contribute to transcellular biosynthesis of poly-hydroxy PUFA mediators such as RvE and RvD, the presence of the latter has not yet been confirmed in skin (Serhan et al. 2008b; Konkel and Schunck 2011).

Non-enzymatically Produced Oxidised Lipid Mediators

Free radical catalysed oxygenation of PUFA results in lipid mediators with structures similar to the ones produced via the COX-, LOX- or CYP- mediated reactions (Fig. 4.4). However, these oxidised lipids are regio- and/or stereo-isomers of the enzymatically-produced ones, exhibit different and/or less potent bioactivities or are inert, and, in many cases, their presence has been used as marker of local or systemic oxidative stress (Salomon 2005; Grundmann et al. 2004; Niki 2008). One of the most interesting families of oxidised lipids is the group of isoprostanes (isoP) that are racemic diastereoisomers of PGs (Milne et al. 2008). They are formed in situ via the oxidation of PUFA already esterified in phospholipids are released by PLA_2. The AA-derived F_2-isoP are chemically stable and the plasma and/or urine levels of 8-iso $PGF_{2\alpha}$ are frequently used as biomarker of oxidative stress. Increased

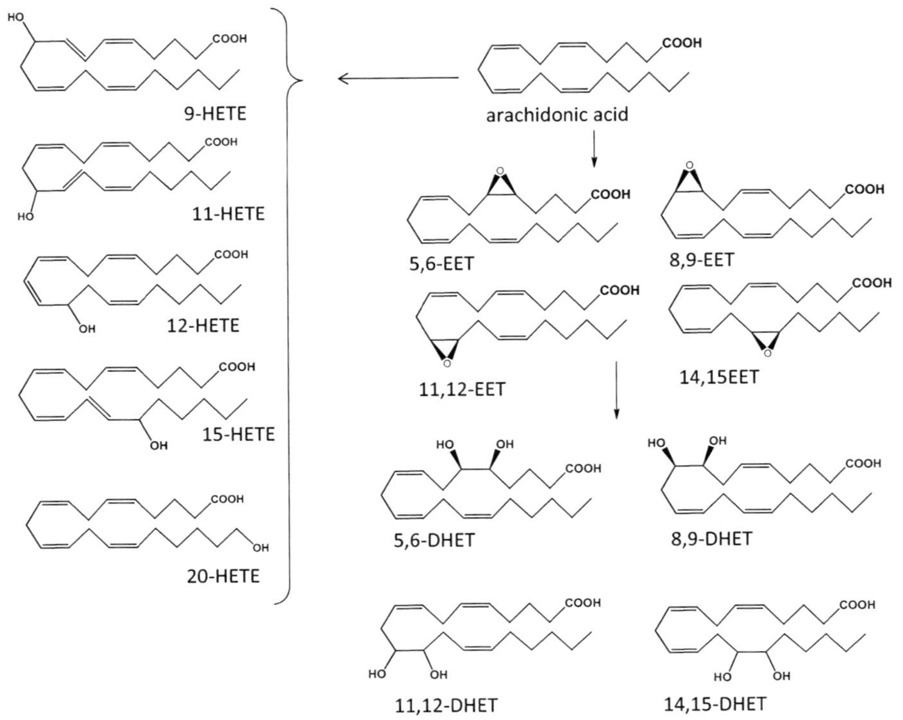

Fig. 4.3 Schematic representation of main cytochrome P450 (CYP) derived mediators of arachidonic acid. *HETE* hydroxy eicosatetraenoic acid, *EET* epoxy eicosatetraenoic acid, *DHET* dihydro eicosatetraenoic acid

levels of this metabolite in UVR-irradiated skin microdialysates and psoriasis patients further support the role of oxidation in cutaneous inflammatory conditions (Grundmann et al. 2004; Wiswedel 2009). Recently, the formation of IsoP products from EPA (F_3-isoP; Song et al. 2009) and DHA (F_4-isoP or F_4-neuroprostanes; Yin et al. 2005) cyclopentenone isoP (Musiek et al. 2006) has been reported, however their prevalence in human skin has not yet been explored.

Other oxidised PUFA derivatives produced by non-enzymatic reactions include the levuglandins (LG) and isolevuglandins (isoLG), a group of highly reactive keto-aldehydes, which can form toxic protein and nucleic acid adducts, although, to date, there are no reports on their involvement in skin oxidation (Salomon 2005). Lipid peroxidation, a feature of many skin disorders, can also generate highly reactive toxic aldehydes, including 4-hydroxy-2-nonenal (HNE), the main metabolite formed upon peroxidation of n-6PUFA and 4-hydoxy-2-hexanal (HHE), that is produced from n-3PUFA; both HNE and HHE have been used as indices of PUFA peroxidation (Catala 2010; Briganti and Picardo 2003). Studies suggest that HNE can bind and modify proteins in aged skin fibroblasts, while keratinocytes have the ability to process and detoxify this cytotoxic aldehyde (Aldini et al. 2003;

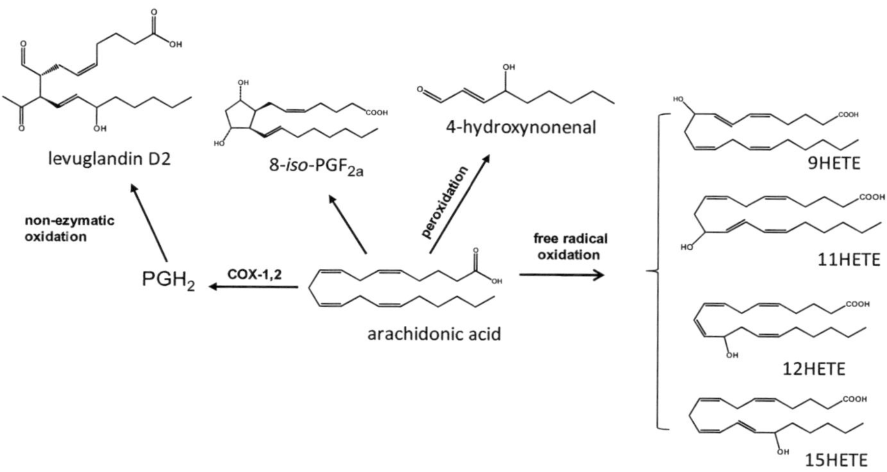

Fig. 4.4 Examples of oxygenated lipid mediators produced during the non-enzymatic oxidation of arachidonic acid. *HETE* eicosatetraenoic acid, *iso-PG* isoprostane

Jorgensen et al. 2014). Finally, lipid peroxidation can generate a wide range of HODE, HETE, HEPE an HDHA regio- and stereo-isomers that are formed in a non-selective random manner and may be used as markers of cutaneous disease (Baer et al. 1991; Grundmann et al. 2004).

Bioactivities of Oxidised Lipid Mediators and Role in Inflammatory Skin Disease

Bioactive lipid mediators are produced by skin cells at relatively low concentrations and are used to maintain critical functions related to physiology and homeostasis. However, when produced at higher concentrations and in response to various stimuli, lipid mediators can modulate cell proliferation, inflammation and tumorigenesis, and this is of particular interest to acute and chronic cutaneous inflammatory diseases including atopic dermatitis, psoriasis, photosensitivity disorders, photoageing, sunburn and skin cancer. Nutritional supplementation with PUFA has the potential to alter the cutaneous fatty acid profile and this has been explored as means of developing chemoprotective and therapeutic interventions (Ziboh et al. 2000a; Lee et al. 2003; Nicolaou 2013; Kendall and Nicolaou 2013; De Spirt et al. 2009). These approaches were based on the ability of the fatty acid treatment to reduce the prevalence of AA-derived pro-inflammatory HETE and increase the concentration of less inflammatory derivatives such as 15-HEPE and 15-HETrE (Calder 2006). Examples include animal and human studies using the n-3PUFA ALA, EPA and DHA that reported decreased sensitivity to sunburn and increased immune protection (Orengo et al. 1992; Rhodes et al. 1994, 2003; Pilkington et al. 2013;

Takemura et al. 2002). A similar rationale has been used to investigate the role of n-3PUFA-derived mediators in wound healing where the profile of lipid chemo-attractants can influence the leukocytic infiltrate and may affect the degree of epithelization (Lu et al. 2010; McDaniel et al. 2011; Martin and Leibovich 2005; Miller et al. 1991). Furthermore, GLA, DGLA and DHA have showed promising results in improving atopic dermatitis (Gueck et al. 2004; Kawashima et al. 2008), while n-3PUFA supplementation during pregnancy results in fewer children with atopic eczema (Makrides et al. 2013). Finally, GLA, DGLA and EPA supplements, as well as local application and intravenous administration of EPA, have been considered in for the treatment of psoriasis (Zulfakar et al. 2007; Mayser et al. 2002; Ricketts et al. 2011; Miller et al. 1988).

Bioactivity of Cutaneous Prostanoids

Prostaglandins, especially the AA-derived ones, are some of the most abundant cutaneous lipid mediators. PGE_2 is produced by both keratinocytes and fibroblasts, and can act as a keratinocyte chemo-attractant with possible involvement in wound healing (Parekh et al. 2009; Ruzicka 1992). Additionally, it contributes to the cross-talk and biochemical support between skin layers (Leong et al. 1996; Sato et al. 1997), while its role on keratinocyte proliferation and differentiation makes it important for the epidermal barrier (Honma et al. 2005; Konger et al. 1998; Pentland and Needleman 1986). Furthermore, PGE_2 is involved in epidermal melanocyte biology mediating post-inflammatory pigmentary responses and melanocyte dendricity (Scott et al. 2004, 2007; Gledhill et al. 2010). Overall, PGE_2 exhibits pro-inflammatory and vasodilatory activities, induces cellular proliferation, modulates immune suppression and is a strong tumour promoter (Harris et al. 2002; Larsson et al. 2004; Narayanan et al. 2011). Its concentration is increased following UVR exposure, following upregulation of cutaneous PLA_2 and COX-2, a major event in photo-carcinogenesis (Tober et al. 2006; Kabashima et al. 2007; Rundhaug and Fischer 2008; Rhodes et al. 2009). UVR appears to also inhibit 15-PGDH in human skin allowing for a short-term increase in PGE_2 levels suggesting that higher levels of this eicosanoid may be of importance in the orchestration of cutaneous response to UVR-induced injury (Judson et al. 2010). This role of PGE2 is further supported by studies showing that it can play a part in the resolution of inflammation and tissue repair (Levy et al. 2001), while continuous production of PGE_2 in unresolved inflammation may contribute to the immune suppression observed in photo-carcinogenesis (Narayanan et al. 2011).

PGD_2 is one of the principle prostanoids involved in atopic dermatitis, and this is linked to the involvement of Langerhans cells and mast cells, that are considered to be the main cutaneous courses of PGD_2 (Sugimoto et al. 2006; Matsushima et al. 2011; Kolgen et al. 2002). The vasoactive PGI_2 and immunosuppressive PGE_2 have also been suggested to be actively involved in this cutaneous condition (Nakajima et al. 2010; Honda et al. 2009). Reduced levels of PGD_2 have been reported in animal and human skin post UVR. This has also been linked to the UVR-mediated Langerhans cells migration out of the epidermis that is part of the solar radiation-induced immunosuppression (Ikai et al. 1989; Rhodes et al. 2009; Meunier et al. 1995). Finally, PGD_2 is precursor to the anti-inflammatory cyclopentanone prostaglandins PGJ_2, Δ^{12}-PGJ_2 and 15d-PGJ_2 that are formed through non-enzymatic hydrolysis (Straus and Glass 2001; Scher and Pillinger 2009). Although Δ^{12}-PGJ_2, a PPARγ agonist, has been shown to exhibit anti-proliferative effects in epidermal cells in vitro, the formation of such potent electrophiles by skin cells has not yet been well documented (Ikai et al. 1995).

Bioactivity of Cutaneous Hydroxy and Epoxy Fatty Acids

Leukotrienes and 5-oxo-ETE are potent chemo-attractants that can contribute to cutaneous inflammation and allergy (Grant et al. 2009; Boyce 2007). However, although the involvement of LTB_4 has been shown in animal models of atopic dermatitis (Andoh et al. 2012), it is likely that the origin of 5-LOX-derived mediators can be attributed to dermal neutrophilic infiltrates and not the resident skin cells (Church et al. 2002).

12-HETE is a potent leukocyte chemo-attractant that can also stimulate cell proliferation and enhance tumour cell survival (Dailey and Imming 1999; Honn et al. 1994). Binding sites for 12-HETE have been identified in human keratinocytes and Langerhans cells suggesting its potential involvement in wound healing and cutaneous inflammation (Arenberger et al. 1992, 1993; Ruzicka 1992). Conversely, 15-HETE is considered, primarily, as an anti-inflammatory mediator that can counteract some of the pro-inflammatory properties of 12-HETE and PGE_2, and may contribute to the resolution of cutaneous inflammation (Serhan et al. 2003; Vachier et al. 2002; Rhodes et al. 2009). Interestingly, there is a degree of reciprocal regulation between dermal and epidermal 12- and 15-LOX activities and related metabolites, alluding to the complex relationship, cross-talk and biochemical interactions occurring between these skin areas (Kragballe et al. 1986b; Yoo et al. 2008). 15-LOX and its metabolites may also mediate anti-cancer and protective effects similarly to its homologous murine 8-LOX (Schweiger et al. 2007), while 15-HETE may act as metabolic precursor of the anti-inflammatory lipoxins, although the latter have not yet been identified in skin (Serhan et al. 2003). The same anti-inflammatory profile is shared by the 15-LOX-derived LA and DGLA metabolites, 13-HODE and 15-HETrE (Nicolaou et al. 2012; Ziboh et al. 2000b; Xi et al. 2000), while the DHA-derived 14,21-dihydroxy-docosahexaenoic acid (HDHA) has recently been reported to mediate wound healing (Lu et al. 2010). 13-HODE, is one of the most abundant hydroxy fatty acid in the skin and has been shown to reverse epidermal hyperproliferation (Cho and Ziboh 1994). Recent reports have shown a possible involvement of HODE and oxo-ODE in pain, although this has not yet been explored in skin (Ruparel et al. 2012; Sisignano et al. 2013).

UVR upregulates cutaneous 12- and 15-LOX, with concomitant increase of the related hydroxy fatty acids (Ziboh et al. 2000b; Yoo et al. 2008; Rhodes et al. 2009). Interestingly, LOX-derived metabolites appear to peak later than COX-derived prostanoids, suggesting that these two groups of lipids may be important for events related to different phases of the inflammatory response, including initiation and resolution (Rhodes et al. 2009). LOX-mediators are also important in psoriasis that is characterised by high levels of 12-HETE, while chiral analysis revealed the prevalence of 12*R*-HETE in psoriatic scales pointing at the involvement of 12*R*-LOX (Boeglin et al. 1998). Furthermore, increased production of PGE_2 and $PGF_{2\alpha}$ are less prominent in psoriatic tissue, and 15-HETE appears decreased in uninvolved psoriatic skin compared to healthy epidermis (Grabbe et al. 1984b; Ikai 1999; Kragballe et al. 1986a). Finally, 8, 9- and 11-HETE have also been detected in human skin, although their exact origin has not yet been fully elucidated and may be products of CYP-mediated reactions or non-enzymatic oxidation of AA (Rhodes et al. 2009; Nicolaou et al. 2012). The same applies to the prevalence of 13-(*R,S*)-HODE in psoriatic skin that is characterised by increased oxidative stress, although the contribution of *R*-hydroxy-fatty acid producing CYP isoforms cannot be excluded (Zhou et al. 2009; Baer et al. 1990).

Although some AA-specific mono-oxygenases have been identified in epidermal keratinocytes (Keeney et al. 1998a, 1998b), the overall contribution of CYP in cutaneous eicosanoid production remains to be explored. CYP-derived eicosanoids include various EET that are involved in vascular relaxation and angiogenesis, their biologically inactive DHET metabolites, various mid-chain HETE (e.g. 8-,9-,11-, 12-,15-HETE), and the omega-hydroxylated 20-HETE, a potent vasoconstrictor (Spector 2009; Kroetz and Xu 2005; Panigrahy et al. 2010; Iliff et al. 2010). EET have been identified in mouse skin cells and 11,12- and 14,15-EET have been shown to play a part in the cornification of human and mouse keratinocytes (Du et al. 2005; Ladd et al. 2003). It has been reported that UVR upregulates CYP4A11 in human keratinocytes, however, the effect of UVR on the activity and expression of cutaneous CYP450 isoforms is not well documented (Gonzalez et al. 2001). Finally, it is of interest to note that CYP isozymes can mediate the oxidation of eicosanoids, as suggested by the co-localisation of COX-2, mPGES-1 and CYP4F8 in psoriatic lesions leading to oxidation of PGE_2 (Stark et al. 2006).

Acknowledgements Research support provided by the The Wellcome Trust (WT094028) is gratefully acknowledged. We thank Dr. Paula Urquhart for her help to compose the figures shown in the manuscript.

References

Aldini G, Granata P, Orioli M, Santaniello E, Carini M. Detoxification of 4-hydroxynonenal (HNE) in keratinocytes: characterization of conjugated metabolites by liquid chromatography/electrospray ionization tandem mass spectrometry. J Mass Spectrom. 2003;38:1160–8.

Andersen S, Sjursen W, Laegreid A, Volden G, Johansen B. Elevated expression of human nonpancreatic phospholipase A2 in psoriatic tissue. Inflammation. 1994;18:1–12.
Andoh T, Nishikawa Y, Yamaguchi-Miyamoto T, Nojima H, Narumiya S, Kuraishi Y. Thromboxane A2 induces itch-associated responses through TP receptors in the skin in mice. J Invest Dermatol. 2007;127:2042–7.
Andoh T, Haza S, Saito A, Kuraishi Y. Involvement of leukotriene B4 in spontaneous itch-related behaviour in NC mice with atopic dermatitis-like skin lesions. Exp Dermatol. 2012;20:894–8.
Arenberger P, Kemeny L, Rupec R, Bieber T, Ruzicka T. Langerhans cells of the human skin possess high-affinity 12(S)-hydroxyeicosa tetraenoic acid receptors. Eur J Immunol. 1992;22:2469–72.
Arenberger P, Kemeny L, Ruzicka T. Characterization of high-affinity 12(S)-hydroxyeicosatetraenoic acid (12(S)-HETE) binding sites on normal human keratinocytes. Epithelial Cell Biol. 1993;2:1–6.
Baer AN, Costello PB, Green FA. Free and esterified 13(R, S)-hydroxyoctadecadienoic acids: principal oxygenase products in psoriatic skin scales. J Lipid Res. 1990;31:125–30.
Baer AN, Costello PB, Green FA. Stereospecificity of the products of the fatty acid oxygenases derived from psoriatic scales. J Lipid Res. 1991;32:341–7.
Baron JM, Wiederholt T, Heise R, Merk HF, Bickers DR. Expression and function of cytochrome p450-dependent enzymes in human skin cells. Curr Med Chem. 2008;15:2258–64.
Bergers M, Van de Kerkhof PC, Happle R, Mier PD. Membrane-bound phospholipase C activity in normal and psoriatic epidermis. Acta Derm Venereol. 1990;70:57–9.
Bito T, Nishigori C. Impact of reactive oxygen species on keratinocyte signaling pathways. J Dermatol Sci. 2012;68:3–8.
Black AK, Greaves MW, Hensby CN, Plummer NA. Increased prostaglandins E2 and F2alpha in human skin at 6 and 24 h after ultraviolet B irradiation (290–320 nm). Br J Clin Pharmacol. 1978;5:431–6.
Black AT, Gray JP, Shakarjian MP, Mishin V, Laskin DL, Heck DE, Laskin JD. UVB light upregulates prostaglandin synthases and prostaglandin receptors in mouse keratinocytes. Toxicol Appl Pharmacol. 2008;232:14–24.
Boeglin WE, Kim RB, Brash AR. A 12R-lipoxygenase in human skin: mechanistic evidence, molecular cloning, and expression. Proc Natl Acad Sci U S A. 1998;95:6744–9.
Boyce JA. Mast cells and eicosanoid mediators: a system of reciprocal paracrine and autocrine regulation. Immunol Rev. 2007;217:168–85.
Brain SD, Williams TJ. Leukotrienes and inflammation. Pharmacol Ther. 1990;46:57–66.
Brash AR. Lipoxygenases: occurrence, functions, catalysis, and acquisition of substrate. J Biol Chem. 1999;274:23679–82.
Brash AR, Yu Z, Boeglin WE, Schneider C. The hepoxilin connection in the epidermis. FEBS J. 2007;274:3494–502.
Breyer RM, Bagdassarian CK, Myers SA, Breyer MD. Prostanoid receptors: subtypes and signaling. Annu Rev Pharmacol Toxicol. 2001;41:661–90.
Briganti S, Picardo M. Antioxidant activity, lipid peroxidation and skin diseases. What's new. J Eur Acad Dermatol Venereol. 2003;17:663–9.
Bryniarski K, Biedron R, Jakubowski A, Chlopicki S, Marcinkiewicz J. Anti-inflammatory effect of 1-methylnicotinamide in contact hypersensitivity to oxazolone in mice; involvement of prostacyclin. Eur J Pharmacol. 2008;578:332–8.
Burr GO, Burr MM. A new deficiency disease produced by the rigid exclusion of fat from the diet. J Biol Chem. 1929;82:345–67.
Calder PC. Polyunsaturated fatty acids and inflammation. Prostaglandins Leukot Essent Fatty Acids. 2006;75:197–202.
Calder PC. Polyunsaturated fatty acids and inflammatory processes: new twists in an old tale. Biochimie. 2009;91:791–5.
Carsberg CJ, Ohanian J, Friedmann PS. Ultraviolet radiation stimulates a biphasic pattern of 1,2-diacylglycerol formation in cultured human melanocytes and keratinocytes by activation of phospholipases C and D. Biochem J. 1995;305(Pt 2):471–7.

Catala A. A synopsis of the process of lipid peroxidation since the discovery of the essential fatty acids. Biochem Biophys Res Commun. 2010;399:318–23.

Cathcart MC, Reynolds JV, O'byrne KJ, Pidgeon GP. The role of prostacyclin synthase and thromboxane synthase signaling in the development and progression of cancer. Biochimi Biophys Acta. 2010;1805:153–66.

Chan A, Mauro T. Acidification in the epidermis and the role of secretory phospholipases. Dermatoendocrinology. 2011;3:84–90.

Chapkin RS, Ziboh VA. Inability of skin enzyme preparations to biosynthesize arachidonic acid from linoleic acid. Biochem Biophys Res Commun. 1984;124:784–92.

Chen X, Gresham A, Morrison A, Pentland AP. Oxidative stress mediates synthesis of cytosolic phospholipase A2 after UVB injury. Biochim Biophys Acta. 1996;1299:23–33.

Cho Y, Ziboh VA. 13-Hydroxyoctadecadienoic acid reverses epidermal hyperproliferation via selective inhibition of protein kinase C-beta activity. Biochem Biophys Res Commun. 1994;201:257–65.

Cho JW, Park K, Kweon GR, Jang BC, Baek WK, Suh MH, Kim CW, Lee KS, Suh SI. Curcumin inhibits the expression of COX-2 in UVB-irradiated human keratinocytes (HaCaT) by inhibiting activation of AP_1: p38 MAP kinase and JNK as potential upstream targets. Exp Mol Med. 2005;37:186–92.

Church MK, Griffiths TJ, Jeffery S, Ravell LC, Cowburn AS, Sampson AP, Clough GF. Are cysteinyl leukotrienes involved in allergic responses in human skin? Clin Exp Allergy. 2002;32:1013–9.

Colombe L, Vindrios A, Michelet JF, Bernard BA. Prostaglandin metabolism in human hair follicle. Exp Dermatol. 2007;16:762–9.

Dailey LA, Imming P. 12-Lipoxygenase: classification, possible therapeutic benefits from inhibition, and inhibitors. Curr Med Chem. 1999;6:389–98.

De spirt S, Stahl W, Tronnier H, Sies H, Bejot M, Maurette JM, Heinrich U. Intervention with flaxseed and borage oil supplements modulates skin condition in women. Br J Nutr. 2009;101:440–5.

Dennis EA, Cao J, Hsu YH, Magrioti V, Kokotos G. Phospholipase A2 enzymes: physical structure, biological function, disease implication, chemical inhibition, and therapeutic intervention. Chem Rev. 2011;111:6130–85.

Doepping S, Funk CD, Habenicht AJ, Spanbroek R. Selective 5-lipoxygenase expression in Langerhans cells and impaired dendritic cell migration in 5-LO-deficient mice reveal leukotriene action in skin. J Invest Dermatol. 2007;127:1692–700.

Du L, Yermalitsky V, Ladd PA, Capdevila JH, Mernaugh R, Keeney DS. Evidence that cytochrome P450 CYP2B19 is the major source of epoxyeicosatrienoic acids in mouse skin. Arch Biochem Biophys. 2005;435:125–33.

Epp N, Furstenberger G, Muller K, De juanes S, Leitges M, Hausser I, Thieme F, Liebisch G, Schmitz G, Krieg P. 12R-lipoxygenase deficiency disrupts epidermal barrier function. J Cell Biol. 2007;177:173–82.

Fisher GJ, Talwar HS, Baldassare JJ, Henderson PA, Voorhees JJ. Increased phospholipase C-catalyzed hydrolysis of phosphatidylinositol-4,5-bisphosphate and 1,2-sn-diacylglycerol content in psoriatic involved compared to uninvolved and normal epidermis. J Invest Dermatol. 1990;95:428–35.

Gledhill K, Rhodes LE, Brownrigg M, Haylett AK, Masoodi M, Thody AJ, Nicolaou A, Tobin DJ. Prostaglandin-E2 is produced by adult human epidermal melanocytes in response to UVB in a melanogenesis-independent manner. Pigment Cell Melanoma Res. 2010;23:394–403.

Gonzalez MC, Marteau C, Franchi J, Migliore-Samour D. Cytochrome P450 4A11 expression in human keratinocytes: effects of ultraviolet irradiation. Br J Dermatol. 2001;145:749–57.

Grabbe J, Czarnetzki BM, Mardin M. Release of lipoxygenase products of arachidonic acid from freshly isolated human keratinocytes. Arch Dermatol Res. 1984a;276:128–30.

Grabbe J, Czarnetzki BM, Rosenbach T, Mardin M. Identification of chemotactic lipoxygenase products of arachidonate metabolism in psoriatic skin. J Invest Dermatol. 1984b;82:477–9.

Grant GE, Rokach J, Powell WS. 5-Oxo-ETE and the OXE receptor. Prostaglandins Other Lipid Mediat. 2009;89:98–104.

Gresham A, Masferrer J, Chen X, Leal-Khouri S, Pentland AP. Increased synthesis of high-molecular-weight cPLA2 mediates early UV-induced PGE2 in human skin. Am J Physiol. 1996;270:C1037–50.

Gron B, Iversen L, Ziboh V, Kragballe K. Distribution of monohydroxy fatty acids in specific human epidermal phospholipids. Exp Dermatol. 1993;2:38–44.

Grundmann JU, Wiswedel I, Hirsch D, Gollnick HP. Detection of monohydroxyeicosatetraenoic acids and F2-isoprostanes in microdialysis samples of human UV-irradiated skin by gas chromatography-mass spectrometry. Skin Pharmacol Physiol. 2004;17:37–41.

Gueck T, Seidel A, Baumann D, Meister A, Fuhrmann H. Alterations of mast cell mediator production and release by gamma-linolenic and docosahexaenoic acid. Vet Dermatol. 2004;15:309–14.

Hansen HS, Jensen B. Essential function of linoleic acid esterified in acylglucosylceramide and acylceramide in maintaining the epidermal water permeability barrier. Evidence from feeding studies with oleate, linoleate, arachidonate, columbinate and alpha-linolenate. Biochim Biophys Acta. 1985;834:357–63.

Hanson D, Deleo V. Long-wave ultraviolet light induces phospholipase activation in cultured human epidermal keratinocytes. J Invest Dermatol. 1990;95:158–63.

Harris SG, Padilla J, Koumas L, Ray D, Phipps RP. Prostaglandins as modulators of immunity. Trends Immunol. 2002;23:144–50.

Heitmann J, Iversen L, Kragballe K, Ziboh VA. Incorporation of 15-hydroxyeicosatrienoic acid in specific phospholipids of cultured human keratinocytes and psoriatic plaques. Exp Dermatol. 1995;4:74–8.

Honda T, Matsuoka T, Ueta M, Kabashima K, Miyachi Y, Narumiya S. Prostaglandin E(2)-EP(3) signaling suppresses skin inflammation in murine contact hypersensitivity. J Allergy Clin Immunol. 2009;124:809–18e2.

Honma Y, Arai I, Hashimoto Y, Futaki N, Sugimoto M, Tanaka M, Nakaike S. Prostaglandin D2 and prostaglandin E2 accelerate the recovery of cutaneous barrier disruption induced by mechanical scratching in mice. Eur J Pharmacol. 2005;518:56–62.

Honn KV, Tang DG, Gao X, Butovich IA, Liu B, Timar J, Hagmann W. 12-lipoxygenases and 12(S)-HETE: role in cancer metastasis. Cancer Metastasis Rev. 1994;13:365–96.

Ikai K. Psoriasis and the arachidonic acid cascade. J Dermatol Sci. 1999;21:135–46.

Ikai K, Imamura S. Prostaglandin D2 in the skin. Int J Dermatol. 1988;27:141–9.

Ikai K, Ujihara M, Kanauchi H, Urade Y. Effect of ultraviolet irradiation on the activity of rat skin prostaglandin D synthetase. J Invest Dermatol. 1989;93:345–8.

Ikai K, Yamamoto M, Matsuyoshi N, Fukushima M. Effect of cytotoxic prostaglandin, delta 12-prostaglandin J2 on E-cadherin expression in transformed epidermal cells in culture. Prostaglandins Leukot Essent Fatty Acids. 1995;52;303–7.

Iliff JJ, Jia J, Nelson J, Goyagi T, Klaus J, Alkayed NJ. Epoxyeicosanoid signaling in CNS function and disease. Prostaglandins Other Lipid Mediat. 2010;91:68–84.

Janssen-Timmen U, Vickers PJ, Wittig U, Lehmann WD, Stark HJ, Fusenig NE, Rosenbach T, Radmark O, Samuelsson B, Habenicht AJ. Expression of 5-lipoxygenase in differentiating human skin keratinocytes. Proc Natl Acad Sci U S A. 1995;92:6966–70.

Jisaka M, Kim RB, Boeglin WE, Nanney LB, Brash AR. Molecular cloning and functional expression of a phorbol ester-inducible 8S-lipoxygenase from mouse skin. J Biol Chem. 1997;272:24410–6.

Jorgensen P, Milkovic L, Zarkovic N, Waeg G, Rattan SI. Lipid peroxidation-derived 4-hydroxynonenal-modified proteins accumulate in human facial skin fibroblasts during ageing in vitro. Biogerontology. 2014;15(1):105–10.

Judson BL, Miyaki A, Kekatpure VD, Du B, Gilleaudeau P, Sullivan-Whalen M, Mohebati A, Nair S, Boyle JO, Granstein RD, Subbaramaiah K, Krueger JG, Dannenberg AJ. UV radiation inhibits 15-hydroxyprostaglandin dehydrogenase levels in human skin: evidence of transcriptional suppression. Cancer Prev Res (Phila). 2010;3:1104–11.

Kabashima K, Nagamachi M, Honda T, Nishigori C, Miyachi Y, Tokura Y, Narumiya S. Prostaglandin E2 is required for ultraviolet B-induced skin inflammation via EP2 and EP4 receptors. Lab Invest. 2007;87:49–55.
Kanda N, Ishikawa T, Watanabe S. Prostaglandin D2 induces the production of human beta-defensin-3 in human keratinocytes. Biochem Pharmacol. 2010;79:982–9.
Kawashima H, Tateishi N, Shiraishi A, Teraoka N, Tanaka T, Tanaka A, Matsuda H, Kiso Y. Oral administration of dihomo-gamma-linolenic acid prevents development of atopic dermatitis in NC/Nga mice. Lipids. 2008;43:37–43.
Keeney DS, Skinner C, Travers JB, Capdevila JH, Nanney LB, King LE Jr, Waterman MR. Differentiating keratinocytes express a novel cytochrome P450 enzyme, CYP2B19, having arachidonate monooxygenase activity. J Biol Chem. 1998a;273:32071–9.
Keeney DS, Skinner C, Wei S, Friedberg T, Waterman MR. A keratinocyte-specific epoxygenase, CYP2B12, metabolizes arachidonic acid with unusual selectivity, producing a single major epoxyeicosatrienoic acid. J Biol Chem. 1998b;273:9279–84.
Kendall AC, Nicolaou A. Bioactive lipid mediators in skin inflammation and immunity. Prog Lipid Res. 2013;52:141–64.
Kolgen W, Both H, Van Weelden H, Guikers KL, Bruijnzeel-Koomen CA, Knol EF, Van Vloten WA, De Gruijl FR. Epidermal langerhans cell depletion after artificial ultraviolet B irradiation of human skin in vivo: apoptosis versus migration. J Invest Dermatol. 2002;118:812–7.
Konger RL, Malaviya R, Pentland AP. Growth regulation of primary human keratinocytes by prostaglandin E receptor EP2 and EP3 subtypes. Biochim Biophys Acta. 1998;1401:221–34.
Konger RL, Billings SD, Thompson AB, Morimiya A, Ladenson JH, Landt Y, Pentland AP, Badve S. Immunolocalization of low-affinity prostaglandin E receptors, EP and EP, in adult human epidermis. J Invest Dermatol. 2005;124:965–70.
Konkel A, Schunck WH. Role of cytochrome P450 enzymes in the bioactivation of polyunsaturated fatty acids. Biochim Biophys Acta. 2011;1814:210–22.
Kragballe K, Duell EA, Voorhees JJ. Selective decrease of 15-hydroxyeicosatetraenoic acid (15-HETE) formation in uninvolved psoriatic dermis. Arch Dermatol. 1986a;122:877–80.
Kragballe K, Pinnamaneni G, Desjarlais L, Duell EA, Voorhees JJ. Dermis-derived 15-hydroxy-eicosatetraenoic acid inhibits epidermal 12-lipoxygenase activity. J Invest Dermatol. 1986b;87:494–8.
Krieg P, Kinzig A, Heidt M, Marks F, Furstenberger G. cDNA cloning of a 8-lipoxygenase and a novel epidermis-type lipoxygenase from phorbol ester-treated mouse skin. Biochim Biophys Acta. 1998;1391:7–12.
Kroetz DL, Xu F. Regulation and inhibition of arachidonic acid omega-hydroxylases and 20-HETE formation. Annu Rev Pharmacol Toxicol. 2005;45:413–38.
Kudo I, Murakami M. Prostaglandin E synthase, a terminal enzyme for prostaglandin E2 biosynthesis. J Biochem Mol Biol. 2005;38:633–8.
Kuhn H, O'Donnell VB. Inflammation and immune regulation by 12/15-lipoxygenases. Prog Lipid Res. 2006;45:334–56.
Ladd PA, Du L, Capdevila JH, Mernaugh R, Keeney DS. Epoxyeicosatrienoic acids activate transglutaminases in situ and induce cornification of epidermal keratinocytes. J Biol Chem. 2003;278:35184–92.
Larsson SC, Kumlin M, Ingelman-Sundberg M, Wolk A. Dietary long-chain n-3 fatty acids for the prevention of cancer: a review of potential mechanisms. Am J Clin Nutr. 2004;79:935–45.
Lee JL, Mukhtar H, Bickers DR, Kopelovich L, Athar M. Cyclooxygenases in the skin: pharmacological and toxicological implications. Toxicol Appl Pharmacol. 2003;192:294–306.
Leong J, Hughes-Fulford M, Rakhlin N, Habib A, Maclouf J, Goldyne ME. Cyclooxygenases in human and mouse skin and cultured human keratinocytes: association of COX-2 expression with human keratinocyte differentiation. Exp Cell Res. 1996;224:79–87.
Levy BD, Clish CB, Schmidt B, Gronert K, Serhan CN. Lipid mediator class switching during acute inflammation: signals in resolution. Nat Immunol. 2001;2:612–9.
Lu Y, Tian H, Hong S. Novel 14,21-dihydroxy-docosahexaenoic acids: structures, formation pathways, and enhancement of wound healing. J Lipid Res. 2010;51:923–32.

Makrides M, Gunaratne AW, Collins CT. Dietary n-3 LC-PUFA during the perinatal period as a strategy to minimize childhood allergic disease. Nestle Nutr Inst Workshop Ser. 2013;77:155–62.

Martin P, Leibovich SJ. Inflammatory cells during wound repair: the good, the bad and the ugly. Trends Cell Biol. 2005;15:599–607.

Masoodi M, Nicolaou A, Gledhill K, Rhodes LE, Tobin DJ, Thody AJ. Prostaglandin D production in FM55 melanoma cells is regulated by alpha-melanocyte-stimulating hormone and is not related to melanin production. Exp Dermatol. 2010;19:751–3.

Massey KA, Nicolaou A. Lipidomics of polyunsaturated-fatty-acid-derived oxygenated metabolites. Biochem Soc Trans. 2011;39:1240–6.

Matsushima Y, Satoh T, Yamamoto Y, Nakamura M, Yokozeki H. Distinct roles of prostaglandin D2 receptors in chronic skin inflammation. Mol Immunol. 2011;49:304–10.

Maury E, Prevost M-C, Simon M-F, Redoules D, Ceruti I, Tarroux R, Charveron M, Chap H. Identification of two secreted phospholipases A2 in human epidermis. J Invest Dermatol. 2000;114:960–6.

Mayer B, Rauter L, Zenzmaier E, Gleispach H, Esterbauer H. Characterization of lipoxygenase metabolites of arachidonic acid in cultured human skin fibroblasts. Biochim Biophys Acta. 1984;795:151–61.

Mayser P, Grimm H, Grimminger F. n-3 fatty acids in psoriasis. Br J Nutr. 2002;87(Suppl 1):S77–82.

Mccusker MM, Grant-Kels JM. Healing fats of the skin: the structural and immunologic roles of the omega-6 and omega-3 fatty acids. Clin Dermatol. 2010;28:440–51.

Mcdaniel JC, Massey K, Nicolaou A. Fish oil supplementation alters levels of lipid mediators of inflammation in microenvironment of acute human wounds. Wound Repair Regen. 2011;19:189–200.

Meunier L, Bata-Csorgo Z, Cooper KD. In human dermis, ultraviolet radiation induces expansion of a CD36 + CD11b + CD1-macrophage subset by infiltration and proliferation; CD1 + Langerhans-like dendritic antigen-presenting cells are concomitantly depleted. J Invest Dermatol. 1995;105:782–8.

Michelet JF, Colombe L, Gautier B, Gaillard O, Benech F, Pereira R, Boulle C, Dalko-Csiba M, Rozot R, Neuwels M, Bernard BA. Expression of NAD + dependent 15-hydroxyprostaglandin dehydrogenase and protection of prostaglandins in human hair follicle. Exp Dermatol. 2008;17:821–8.

Miller CC, Mccreedy CA, Jones AD, Ziboh VA. Oxidative metabolism of dihomogammalinolenic acid by guinea pig epidermis: evidence of generation of anti-inflammatory products. Prostaglandins. 1988;35:917–38.

Miller CC, Tang W, Ziboh VA, Fletcher MP. Dietary supplementation with ethyl ester concentrates of fish oil (n-3) and borage oil (n-6) polyunsaturated fatty acids induces epidermal generation of local putative anti-inflammatory metabolites. J Invest Dermatol. 1991;96:98–103.

Milne GL, Yin H, Morrow JD. Human biochemistry of the isoprostane pathway. J Biol Chem. 2008;283:15533–7.

Muller K, Krieg P, Marks F, Furstenberger G. Expression of PGF(2alpha) receptor mRNA in normal, hyperplastic and neoplastic skin. Carcinogenesis. 2000;21:1063–6.

Murakami M, Kudo I. Prostaglandin E synthase: a novel drug target for inflammation and cancer. Curr Pharm Des. 2006;12:943–54.

Musiek ES, Breeding RS, Milne GL, Zanoni G, Morrow JD, Mclaughlin B. Cyclopentenone isoprostanes are novel bioactive products of lipid oxidation which enhance neurodegeneration. J Neurochem. 2006;97:1301–13.

Nakajima S, Honda T, Sakata D, Egawa G, Tanizaki H, Otsuka A, Moniaga CS, Watanabe T, Miyachi Y, Narumiya S, Kabashima K. Prostaglandin I2-IP signaling promotes Th1 differentiation in a mouse model of contact hypersensitivity. J Immunol. 2010;184:5595–603.

Narayanan DL, Saladi RN, Fox JL. Ultraviolet radiation and skin cancer. Int J Dermatol. 2011;49:978–86.

Neis MM, Wendel A, Wiederholt T, Marquardt Y, Joussen S, Baron JM, Merk HF. Expression and induction of cytochrome p450 isoenzymes in human skin equivalents. Skin Pharmacol Physiol. 2010;23:29–39.

Nicolaou A. Eicosanoids in skin inflammation. Prostaglandins Leukot Essent Fatty Acids. 2013;88:131–8.

Nicolaou A, Masoodi M, Gledhill K, Haylett AK, Thody AJ, Tobin DJ, Rhodes LE. The eicosanoid response to high dose UVR exposure of individuals prone and resistant to sunburn. Photochem Photobiol Sci. 2012;11(2):371–80

Nigam S, Zafiriou MP, Deva R, Ciccoli R, Roux-Van Der Merwe R. Structure, biochemistry and biology of hepoxilins: an update. FEBS J. 2007;274:3503–12.

Niki E. Lipid peroxidation products as oxidative stress biomarkers. Biofactors. 2008;34:171–80.

Oka M, Edamatsu H, Kunisada M, Hu L, Takenaka N, Dien S, Sakaguchi M, Kitazawa R, Norose K, Kataoka T, Nishigori C. Enhancement of ultraviolet B-induced skin tumor development in phospholipase Cepsilon-knockout mice is associated with decreased cell death. Carcinogenesis. 2011;31:1897–902.

Oliw EH. Oxygenation of polyunsaturated fatty acids by cytochrome P450 monooxygenases. Prog Lipid Res. 1994;33:329–54.

Orengo IF, Black HS, Wolf JE Jr. Influence of fish oil supplementation on the minimal erythema dose in humans. Arch Dermatol Res. 1992;284:219–21.

Panigrahy D, Kaipainen A, Greene ER, Huang S. Cytochrome P450-derived eicosanoids: the neglected pathway in cancer. Cancer Metastasis Rev. 2010;29:723–35.

Parekh A, Sandulache VC, Singh T, Cetin S, Sacks MS, Dohar JE, Hebda PA. Prostaglandin E2 differentially regulates contraction and structural reorganization of anchored collagen gels by human adult and fetal dermal fibroblasts. Wound Repair Regen. 2009;17:88–98.

Pentland AP, Needleman P. Modulation of keratinocyte proliferation in vitro by endogenous prostaglandin synthesis. J Clin Invest. 1986;77:246–51.

Pilkington SM, Rhodes LE, Al-Aasswad NM, Massey KA, Nicolaou A. Impact of EPA ingestion on COX- and LOX-mediated eicosanoid synthesis in skin with and without a pro-inflammatory UVR challenge—report of a randomised controlled study in humans. Mol Nutr Food Res. 2013;58:580–90.

Rhodes LE, O'Farrell S, Jackson MJ, Friedmann PS. Dietary fish-oil supplementation in humans reduces UVB-erythemal sensitivity but increases epidermal lipid peroxidation. J Invest Dermatol. 1994;103:151–4.

Rhodes LE, Shahbakhti H, Azurdia RM, Moison RM, Steenwinkel MJ, Homburg MI, Dean MP, Mcardle F, Beijersbergen Van Henegouwen GM, Epe B, Vink AA. Effect of eicosapentaenoic acid, an omega-3 polyunsaturated fatty acid, on UVR-related cancer risk in humans. An assessment of early genotoxic markers. Carcinogenesis. 2003;24:919–25.

Rhodes LE, Gledhill K, Masoodi M, Haylett AK, Brownrigg M, Thody AJ, Tobin DJ, Nicolaou A. The sunburn response in human skin is characterized by sequential eicosanoid profiles that may mediate its early and late phases. FASEB J. 2009;23:3947–56.

Ricketts JR, Rothe MJ, Grant-Kels JM. Nutrition and psoriasis. Clin Dermatol. 2011;28:615–26.

Rolsted K, Kissmeyer AM, Rist GM, Hansen SH. Evaluation of cytochrome P450 activity in vitro, using dermal and hepatic microsomes from four species and two keratinocyte cell lines in culture. Arch Dermatol Res. 2008;300:11–8.

Rundhaug JE, Fischer SM. Cyclo-oxygenase-2 plays a critical role in UV-induced skin carcinogenesis. Photochem Photobiol. 2008;84:322–9.

Ruparel S, Green D, Chen P, Hargreaves KM. The cytochrome P450 inhibitor, ketoconazole, inhibits oxidized linoleic acid metabolite-mediated peripheral inflammatory pain. Mol Pain. 2012;8:73.

Ruzicka T. The role of the epidermal 12-hydroxyeicosatetraenoic acid receptor in the skin. Eicosanoids. 1992;5 Suppl:S63–5.

Sala A, Folco G, Murphy RC. Transcellular biosynthesis of eicosanoids. Pharmacol Rep. 2010;62:503–10.

Salomon RG. Levuglandins and isolevuglandins: stealthy toxins of oxidative injury. Antioxid Redox Signal. 2005;7:185–201.

Sato T, Kirimura Y, Mori Y. The co-culture of dermal fibroblasts with human epidermal keratinocytes induces increased prostaglandin E2 production and cyclooxygenase 2 activity in fibroblasts. J Invest Dermatol. 1997;109:334–9.

Satoh T, Moroi R, Aritake K, Urade Y, Kanai Y, Sumi K, Yokozeki H, Hirai H, Nagata K, Hara T, Utsuyama M, Hirokawa K, Sugamura K, Nishioka K, Nakamura M. Prostaglandin D2 plays an essential role in chronic allergic inflammation of the skin via CRTH2 receptor. J Immunol. 2006;177:2621–9.

Schafer L, Kragballe K. Abnormalities in epidermal lipid metabolism in patients with atopic dermatitis. J Invest Dermatol. 1991;96:10–5.

Scher JU, Pillinger MH. 15d-PGJ2: the anti-inflammatory prostaglandin? Clin Immunol. 2005;114:100–9.

Scher JU, Pillinger MH. The anti-inflammatory effects of prostaglandins. J Invest Med. 2009;57:703–8.

Schurer NY, Stremmel W, Grundmann JU, Schliep V, Kleinert H, Bass NM, Williams ML. Evidence for a novel keratinocyte fatty acid uptake mechanism with preference for linoleic acid: comparison of oleic and linoleic acid uptake by cultured human keratinocytes, fibroblasts and a human hepatoma cell line. Biochim Biophys Acta. 1994;1211:51–60.

Schuster VL. Prostaglandin transport. Prostaglandins Other Lipid Mediat. 2002;68–69:633–47.

Schweiger D, Furstenberger G, Krieg P. Inducible expression of 15-lipoxygenase-2 and 8-lipoxygenase inhibits cell growth via common signaling pathways. J Lipid Res. 2007;48:553–64.

Scott G, Leopardi S, Printup S, Malhi N, Seiberg M, Lapoint R. Proteinase-activated receptor-2 stimulates prostaglandin production in keratinocytes: analysis of prostaglandin receptors on human melanocytes and effects of PGE2 and PGF2alpha on melanocyte dendricity. J Invest Dermatol. 2004;122:1214–24.

Scott G, Jacobs S, Leopardi S, Anthony FA, Learn D, Malaviya R, Pentland A. Effects of PGF2alpha on human melanocytes and regulation of the FP receptor by ultraviolet radiation. Exp Cell Res. 2005;304:407–16.

Scott G, Fricke A, Fender A, Mcclelland L, Jacobs S. Prostaglandin E2 regulates melanocyte dendrite formation through activation of PKCzeta. Exp Cell Res. 2007;313:3840–50.

Serhan CN, Jain A, Marleau S, Clish C, Kantarci A, Behbehani B, Colgan SP, Stahl GL, Merched A, Petasis NA, Chan L, Van Dyke TE. Reduced inflammation and tissue damage in transgenic rabbits overexpressing 15-lipoxygenase and endogenous anti-inflammatory lipid mediators. J Immunol. 2003;171:6856–65.

Serhan CN, Chiang N, Van Dyke TE. Resolving inflammation: dual anti-inflammatory and pro-resolution lipid mediators. Nat Rev Immunol. 2008a;8:349–61.

Serhan CN, Yacoubian S, Yang R. Anti-inflammatory and proresolving lipid mediators. Annu Rev Pathol. 2008b;3:279–312.

Shappell SB, Keeney DS, Zhang J, Page R, Olson SJ, Brash AR. 15-Lipoxygenase-2 expression in benign and neoplastic sebaceous glands and other cutaneous adnexa. J Invest Dermatol. 2001;117:36–43.

Shimura C, Satoh T, Igawa K, Aritake K, Urade Y, Nakamura M, Yokozeki H. Dendritic cells express hematopoietic prostaglandin D synthase and function as a source of prostaglandin D2 in the skin. Am J Pathol. 2010;176:227–37.

Sisignano M, Angioni C, Ferreiros N, Schuh CD, Suo J, Schreiber Y, Dawes JM, Antunes-Martins A, Bennett DL, Mcmahon SB, Geisslinger G, Scholich K. Synthesis of lipid mediators during UVB-induced inflammatory hyperalgesia in rats and mice. PLoS ONE. 2013;8:e81228.

Sjursen W, Brekke OL, Johansen B. Secretory and cytosolic phospholipase A(2)regulate the long-term cytokine-induced eicosanoid production in human keratinocytes. Cytokine. 2000;12:1189–94.

Smith WL, Dewitt DL, Garavito RM. Cyclooxygenases: structural, cellular, and molecular biology. Annu Rev Biochem. 2000;69:145–82.

Smith WL, Urade Y, Jakobsson PJ. Enzymes of the cyclooxygenase pathways of prostanoid biosynthesis. Chem Rev. 2012;111:5821–65.
Song WL, Paschos G, Fries S, Reilly MP, Yu Y, Rokach J, Chang CT, Patel P, Lawson JA, Fitzgerald GA. Novel eicosapentaenoic acid-derived F3-isoprostanes as biomarkers of lipid peroxidation. J Biol Chem. 2009;284:23636–43.
Spanbroek R, Stark HJ, Janssen-Timmen U, Kraft S, Hildner M, Andl T, Bosch FX, Fusenig NE, Bieber T, Radmark O, Samuelsson B, Habenicht AJ. 5-Lipoxygenase expression in Langerhans cells of normal human epidermis. Proc Natl Acad Sci U S A. 1998;95:663–8.
Spector AA. Arachidonic acid cytochrome P450 epoxygenase pathway. J Lipid Res. 2009;50 Suppl:S52–6.
Stark K, Torma H, Oliw EH. Co-localization of COX-2, CYP4F8, and mPGES-1 in epidermis with prominent expression of CYP4F8 mRNA in psoriatic lesions. Prostaglandins Other Lipid Mediat. 2006;79:114–25.
Straus DS, Glass CK. Cyclopentenone prostaglandins: new insights on biological activities and cellular targets. Med Res Rev. 2001;21:185–210.
Sugimoto M, Arai I, Futaki N, Hashimoto Y, Honma Y, Nakaike S. Role of COX-1 and COX-2 on skin PGs biosynthesis by mechanical scratching in mice. Prostaglandins Leukot Essent Fatty Acids. 2006;75:1–8.
Surh YJ, Na HK, Park JM, Lee HN, Kim W, Yoon IS, Kim DD. 15-Deoxy-Delta(1)(2),(1)(4)-prostaglandin J(2), an electrophilic lipid mediator of anti-inflammatory and pro-resolving signaling. Biochem Pharmacol. 2011;82:1335–51.
Tai HH, Ensor CM, Tong M, Zhou H, Yan F. Prostaglandin catabolizing enzymes. Prostaglandins Other Lipid Mediat. 2002;68–69:483–93.
Takeda K, Yokoyama S, Aburatani H, Masuda T, Han F, Yoshizawa M, Yamaki N, Yamamoto H, Eguchi N, Urade Y, Shibahara S. Lipocalin-type prostaglandin D synthase as a melanocyte marker regulated by MITF. Biochem Biophys Res Commun. 2006;339:1098–106.
Takemura N, Takahashi K, Tanaka H, Ihara Y, Ikemoto A, Fujii Y, Okuyama H. Dietary, but not topical, alpha-linolenic acid suppresses UVB-induced skin injury in hairless mice when compared with linoleic acids. Photochem Photobiol. 2002;76:657–63.
Tober KL, Wilgus TA, Kusewitt DF, Thomas-Ahner JM, Maruyama T, Oberyszyn TM. Importance of the EP(1) receptor in cutaneous UVB-induced inflammation and tumor development. J Invest Dermatol. 2006;126:205–11.
Ueno N, Takegoshi Y, Kamei D, Kudo I, Murakami M. Coupling between cyclooxygenases and terminal prostanoid synthases. Biochem Biophys Res Commun. 2005;338:70–6.
Ujihara M, Horiguchi Y, Ikai K, Urade Y. Characterization and distribution of prostaglandin D synthetase in rat skin. J Invest Dermatol. 1988:90:448–51.
Uozumi N, Kume K, Nagase T, Nakatani N, Ishii S, Tashiro F, Komagata Y, Maki K, Ikuta K, Ouchi Y, Miyazaki J-I, Shimizu T. Role of cytosolic phospholipase A2 in allergic response and parturition. Nature. 1997;390:618–622.
Urade Y, Eguchi N. Lipocalin-type and hematopoietic prostaglandin D synthases as a novel example of functional convergence. Prostaglandins Other Lipid Mediat. 2002;68–69:375–82.
Vachier I, Chanez P, Bonnans C, Godard P, Bousquet J, Chavis C. Endogenous anti-inflammatory mediators from arachidonate in human neutrophils. Biochem Biophys Res Commun. 2002;290:219–24.
Watanabe K. Prostaglandin F synthase. Prostaglandins Other Lipid Mediat. 2002;68–69:401–7.
Watanabe K. Recent reports about enzymes related to the synthesis of prostaglandin (PG) F(2) (PGF(2alpha) and 9alpha, 11beta-PGF(2)). J Biochem. 2012;150:593–6.
Weber TJ, Markillie LM, Chrisler WB, Vielhauer GA, Regan JW. Modulation of JB6 mouse epidermal cell transformation response by the prostaglandin F2alpha receptor. Mol Carcinog. 2002;35:163–72.
Wiswedel I. F(2)-isoprostanes: sensitive biomarkers of oxidative stress in vitro and in vivo: a gas chromatography-mass spectrometric approach. Methods Mol Biol. 2009;580:3–16.

Woodward DF, Jones RL, Narumiya S. International Union of Basic and Clinical Pharmacology. LXXXIII: classification of prostanoid receptors, updating 15 years of progress. Pharmacol Rev. 2012;63:471–538.
Xi S, Pham H, Ziboh VA. Suppression of proto-oncogene (AP-1) in a model of skin epidermal hyperproliferation is reversed by topical application of 13-hydroxyoctadecadienoic acid and 15-hydroxyeicosatrienoic acid. Prostaglandins Leukot Essent Fatty Acids. 2000;62:13–9.
Yin H, Musiek ES, Gao L, Porter NA, Morrow JD. Regiochemistry of neuroprostanes generated from the peroxidation of docosahexaenoic acid in vitro and in vivo. J Biol Chem. 2005;280:26600–11.
Yoo H, Jeon B, Jeon MS, Lee H, Kim TY. Reciprocal regulation of 12- and 15-lipoxygenases by UV-irradiation in human keratinocytes. FEBS Lett. 2008;582:3249–53.
Zeldin DC. Epoxygenase pathways of arachidonic acid metabolism. J Biol Chem. 2001;276:36059–62.
Zheng Y, Yin H, Boeglin WE, Elias PM, Crumrine D, Beier DR, Brash AR. Lipoxygenases mediate the effect of essential fatty acid in skin barrier formation: a proposed role in releasing omega-hydroxyceramide for construction of the corneocyte lipid envelope. J Biol Chem. 2011;286:24046–56.
Zhou Q, Mrowietz U, Rostami-Yazdi M. Oxidative stress in the pathogenesis of psoriasis. Free Radic Biol Med. 2009;47:891–905.
Ziboh VA. Prostaglandins, leukotrienes, and hydroxy fatty acids in epidermis. Semin Dermatol. 1992;11:114–20.
Ziboh VA, Chapkin RS. Metabolism and function of skin lipids. Prog Lipid Res. 1988;27:81–105.
Ziboh VA, Lord JT, Penneys NS. Alterations of prostaglandin E2-9-ketoreductase activity in proliferating skin. J Lipid Res. 1977;18:37–43.
Ziboh VA, Casebolt TL, Marcelo CL, Voorhees JJ. Biosynthesis of lipoxygenase products by enzyme preparations from normal and psoriatic skin. J Invest Dermatol. 1984;83:426–30.
Ziboh VA, Miller CC, Cho Y. Metabolism of polyunsaturated fatty acids by skin epidermal enzymes: generation of antiinflammatory and antiproliferative metabolites. Am J Clin Nutr. 2000a;71:361S–6S.
Ziboh VA, Miller CC, Cho Y. Significance of lipoxygenase-derived monohydroxy fatty acids in cutaneous biology. Prostaglandins Other Lipid Mediat. 2000b;63:3–13.
Ziboh VA, Cho Y, Mani I, Xi S. Biological significance of essential fatty acids/prostanoids/lipoxygenase-derived monohydroxy fatty acids in the skin. Arch Pharm Res. 2002;25:747–58.
Zulfakar MH, Edwards M, Heard CM. Is there a role for topically delivered eicosapentaenoic acid in the treatment of psoriasis? Eur J Dermatol. 2007;17:284–91.
Zuo Y, Zhuang DZ, Han R, Isaac G, Tobin JJ, Mckee M, Welti R, Brissette JL, Fitzgerald ML, Freeman MW. ABCA12 maintains the epidermal lipid permeability barrier by facilitating formation of ceramide linoleic esters. J Biol Chem. 2008;283:36624–35.

Part III
Epidermal Lipids

Chapter 5
Skin Ceramides

Philip W. Wertz

Core Messages

- Ceramides are the most abundant group of lipids in the stratum corneum.
- Stratum corneum ceramides are structurally diverse.
- Ceramides are synthesized in the viable epidermis and glycosylated.
- Ceramides are transported to the bottom of the stratum corneum in lamellar granules.
- Both free and covalently bound ceramides are important for permeability barrier function.

Introduction

An early investigation by Kooyman using human foot callus (Kooyman 1932) demonstrated that as epidermal differentiation progresses to produce the stratum corneum, there is a loss of phospholipids and accumulation of a different polar lipid. In 1965, Nicolaides identified this neutral polar stratum skin lipid class as ceramide (Nicolaides 1965). This finding was based on infrared spectral data and was added as a footnote to the proof of the manuscript (personal communication). Long (1970) confirmed the earlier findings using cow snout, but was not aware of the report by Nicolaides. Further progress on the chemistry of the skin ceramides was not made until the studies in the mid- to late-1970s by G. Maurice Gray and Harold Yardley and their associates (Gray and Yardley 1975a, b; Gray et al. 1978a, b; Gray and White 1978; Yardley and Summerly 1981). This series of studies established that ceramide content of human, pig, and rat epidermal cells increases with increasing differentiation, and ceramides are major lipids in the stratum corneum. Overall, the lipid composition of the human and pig epidermis were very

P. W. Wertz (✉)
University of Iowa, Iowa City, IA 52242, USA
E-mail: Philip-wertz@uiowa.edu

A. Pappas (ed.), *Lipids and Skin Health,* DOI 10.1007/978-3-319-09943-9_5

close, while that of the rat was somewhat similar. The main building blocks of the epidermal ceramides in pig and human were shown to include sphingosine and dihydrosphingosine and phytosphingosine as the base components, and normal fatty acids and α-hydroxyacids as the amide-linked fatty acids. In addition, the major glucosylceramides found in the viable portion of the epidermis were isolated, and the amide-linked fatty acid was shown to be an unusual, very long hydroxyacid.

In 1983, a complete set of structures for the ceramides from pig epidermis was published (Wertz and Downing 1983). This included six chromatographically separable fractions, which were labeled ceramide 1–ceramide 6, going from least polar to most polar. The least polar of these consisted of 30- to 34-carbon-long ω-hydroxyacids, amide-linked to a mixture of sphingosines and dihydrosphingosines with ester-linked linoleate on the ω-hydroxyl group. The unusual hydroxyacid noted earlier by Gray et al. proved to be the ω-hydroxyacid found in this acylceramide and the analogous acylglucosylceramide found in the viable epidermis. Ceramide 2 contained normal fatty acids, mostly C20:0–C28:0, amide-linked to sphingosines and dihydrosphingosines. Ceramide 3 contained the same fatty acids found in ceramide 2 but now amide-linked to phytosphingosines. Ceramides 4 and 5 contained α-hydroxyacids amide-linked to sphingosines and dihydrosphingosines. They differed in that ceramide 4 contained mainly C24:0–C26:0 hydroxyacids, while ceramide 5 contained C16:0 α-hydroxyacid. The simple nomenclature used for the pig epidermal ceramides was adequate for this set, but attempts to use a numeric nomenclature system for human epidermal ceramides became very confusing. To alleviate this problem, Motta et al. introduced a new system of ceramide nomenclature (Motta et al. 1993). In this system, the amide-linked fatty acid is designated as N for normal fatty acid, A for α-hydroxyacid and O for ω-hydroxyacid, and the base component is designated as S for sphingosine/dihydrosphingosine, P for phytosphingosine, and H for 6-hydroxysphingosine. The presence of an ester-linked fatty acid is designated with a prefix E. Thus, pig ceramide 1 is ceramide EOS. Pig ceramide 2 is ceramide NS and so on.

There were several partial characterizations of human epidermal ceramides in the 1980s and 1990s, but the first reasonably complete descriptions of the human epidermal ceramides came in 2003 (Ponec et al. 2003). The ceramides reported included ceramides EOS, EOP, EOH, NS, NP, NH, AS, AP, and AH, as shown in Fig. 5.1. In each ceramide-containing sphingosine, dihydrosphingosine was also reported. More recently, it has become possible to examine molecular species of ceramides by means of LC/MS (Smeden et al. 2011). This analysis revealed all of the individual ceramide species in the types reported by Ponec et al. In addition, there were several minor unidentified components.

In addition to the free ceramides, it was recognized that ω-hydroxyceramides become covalently attached to the outer surface of the cornified envelope (Wertz and Downing 1986, 1987; Swartzendruber et al. 1987). In addition to ceramide OS, the covalently bound lipid in porcine stratum corneum generally included small proportions of ω-hydroxyacid and normal fatty acid.

In humans, there are additionally covalently bound ceramides OP and OH (Wertz et al. 1989; Robson et al.1994; Hill et al. 2006). The ω-hydroxyacids and

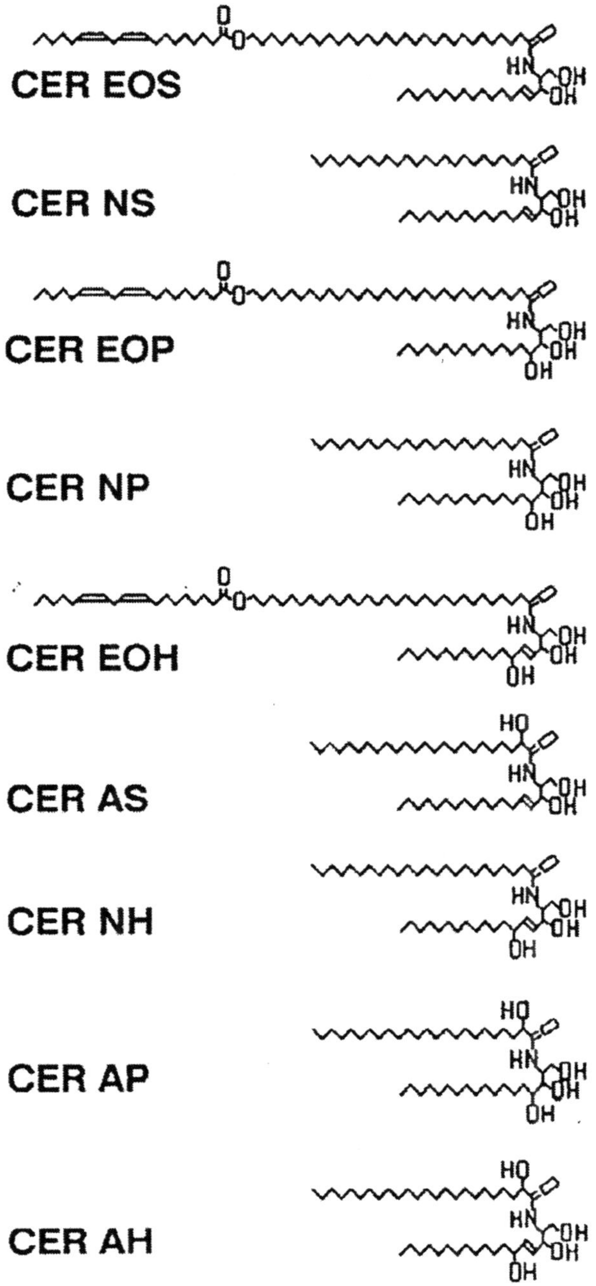

Fig. 5.1 Structures of human stratum corneum ceramides (Ponec et al. 2003) accompanied by shorthand nomenclature (Motta et al. 1993). Ceramides are arranged from least polar at the *top* of the display to most polar at the *bottom*

ω-hydroxyceramides are ester-linked to acidic groups on the outer surface of the cornified envelope through the ω-hydroxyl group (Stewart and Downing 2001).

Biosynthetic Pathways

The rate limiting enzyme in sphingolipids biosynthesis is serine palmitoyl transferase (Radin 1984). This enzyme condenses serine with palmitoyl CoA in an NADPH-dependent reaction with loss of a CO_2 and formation of 3-ketodihydrosphingosine. An NADPH-requiring reductase then reduces the ketone to produce dihydrosphingosine, and N-acylation then produces a simple ceramide. Various modifications of the base component of this initial ceramide include hydroxylations on carbon-4 of the base component to produce phytosphingosine, introduction of a trans double bond between carbon-4 and carbon-5 to produce sphingosine, and introduction of the double bond followed by hydroxylation on carbon-6 to produce 6-hydroxysphingosine. The fatty acid component of the initial ceramide can be hydroxylated on either carbon-2 to produce an α-hydroxyacid or on the terminal carbon to produce an ω-hydroxyacid. At least some of the hydroxylation reactions require vitamin C (Ponec et al. 1997). Ceramides are produced in the viable epidermis and are almost immediately converted to glucosylceramides. The sequence of hydroxylation reactions, double bond introduction, and glycosylation is not established. Most of the glucosylceramides are incorporated into lamellar granules. The lamellar granules also contain glucocerebrosidase (Freinkel and Traczyk 1985; Grayson et al. 1985; Madison et al. 1998), and at the time when lamellar granule contents are extruded into the intercellular spaces, this enzyme converts the glucosylceramides to ceramides. This lipid processing appears to be necessary for formation of mature intercellular lamellae (Holleran et al. 1993).

A special case that should be noted involves the linoleate-containing glucosylceramides and their analogous linoleate-containing ceramides. Some of the linoleate-containing acylglucosylceramide is thought to be associated with the lamellar structures within the granules (Wertz and Downing 1982), whereas most of the lamellar granule-associated acylglucosylceramides are in the bounding membrane of the granule with the glucosyl moiety on the inside (Wertz 2008). The acylglucosylceramide inside the lamellar granules is extruded into the intercellular space, after which it is deglycosylated to yield ceramide EOS. The acylglucosylceramide in the bounding membrane of the lamellar granule is the precursor of ω-hydroxyceramides that become covalently attached to the outer surface of the cornified envelope (Wertz 2000). The acylglucosylceramide is moved into position when the bounding membrane of the lamellar granule fuses into the cell plasma membrane. The process of attachment involves two stereoselective lipoxygenase attacks on the linoleate (Zheng et al. 2011) followed by possible transglutaminase-mediated formation of the ester linkage (Nemes et al. 1999) coupled with or following removal of the oxidized linoleate. This process leads to attachment of glucosyl-ω-hydroxyceramides (Doering et al. 1999). A glucocerebrosidase then removes the glucosyl moiety.

One or more ceramidases may act upon the covalently bound ω-hydroxyceramidases to release the long-chain base component, leaving behind covalently bound ω-hydroxyacid. The free long-chain bases are potent antimicrobials, and this may be an important part of the innate immune system of the skin (Drake et al. 2008).

Structures, Proportions, and Properties

As previously mentioned, the first complete set of epidermal ceramide structures was that from pig (Wertz and Downing 1983). It contained six chromatographically distinct fractions corresponding to ceramides EOS (7.7 wt.%), NS (42.4 wt.%), NP (10.2 wt.%), AS long (12.1 wt.%), AS short (10.5 wt.%), and AP (17.2 wt.%). The ω-hydroxyacids in ceramide EOS were 30- to 34-carbons long, and the sphingoid bases were 16- to 22-carbons long and included sphingosines and dihydrosphingosines. The fatty acids in ceramide NS were mainly saturated entities 24- to 28-carbons in length, and the bases were similar to those found in ceramide EOS. Ceramide NP contained fatty acid components similar to ceramide NS, but the phytosphingosines ranged from 16- to 24-carbons in length. Ceramides AS long and AS short bothcontain long-chain bases similar to those in ceramides EOS and NS. They differ in that ceramide AS long contains mainly 24- to 28-carbon α-hydroxyacids, while ceramide AS short contains almost exclusively α-hydroxypalmitic acid. Finally, ceramide AP contains phytosphingosines similar to those in ceramide NP and α-hydroxyacids similar to those in ceramide AS long.

In humans, the situation is more complicated. There are three ceramides that contain 6-hydroxysphingosine, which is not present in pig, and there is a linoleate-containing ceramide EOP, which is not present in pig (Ponec et al. 2003). The nine ceramide types found in humans are EOS (8.3 wt.%), NS (20.6 wt.%), EOP (6.4 wt.%), NP (18.1 wt.%), EOH (5.0 wt.%), AS(4.4 wt.%), NH (15.4 wt.%), AP (8.7 wt.%), and AH (13.0 wt.%). Chain-length distributions were generally similar to those found in the pig ceramides.

In accord with the finding of three linoleate-containing acylceramides in human epidermis, there are three corresponding covalently bound ω-hydroxyceramides (OS, OP, and OH) in human stratum corneum (Wertz et al. 1989; Robson et al. 1994; Hill et al. 2006).

When isolated, the natural epidermal ceramides are white amorphous solids. Synthetic ceramides, where there is no chain length variation, can be crystalline. It is noteworthy that since ceramides do not have bulky polar headgroups, such as those found in phospholipids, they are cylindrical in shape. This makes them ideal for formation of highly ordered, and thereby impermeable, membranes. In addition, except for the linoleate in the acylceramides, the epidermal ceramides contain very few cis double bonds. This is favorable for resistance of oxidative damage on exposure to air.

Catabolism

Once ceramides reach the stratum corneum, they are relatively stable. There are, however, several ceramidases that can hydrolyze ceramides to release fatty acid and long-chain base. The extent of this hydrolysis under normal circumstances in minimal, probably due to the scarcity of free water in the intercellular spaces. It is, nevertheless, thought that the release of free long-chain bases is biologically significant in that the long-chain bases are potent antimicrobials and may function in the innate immunity of the skin (Drake et al. 2008). There is some evidence that free long-chain bases may also be anti-inflammatory.

References

Doering T, Proia RL, Sandhoff K. Accumulation of protein bound epidermal glucosylceramides in beta-glucocerebrosidase deficient type 2 Gaucher mice. FEBS Lett. 1999;447:167–70.

Drake DR, Brogden KA, Dawson DV, Wertz PW. Antimicrobial lipids at the skin surface. J Lipid Res. 2008;49:4–11.

Freinkel RK, Traczyk TN. Lipid composition and acid hydrolase content of lamellar granules of fetal rat epidermis. J Invest Dermatol. 1985;85:295–8.

Gray GM, Yardley HJ. Lipid composition of cells isolated from pig, human and rat epidermis. J Lipid Res. 1975a;16:434–40.

Gray GM, Yardley HJ. Different populations of pig epidermal cells: isolation and lipid composition. J Lipid Res. 1975b;16:441–7.

Gray GM, White RJ. Glycosphingolipids and ceramides in human and pig epidermis. J Invest Dermatol. 1978;70:336–41.

Gray GM, White RJ, Majer JR. 1-(3'-O-acyl)-beta-glucosyl-N-dihydroxypentatriacontadienoyl-sphingosine, a major component of the glucosylceramides of pig and human epidermis. Biochim Biophys Acta. 1978a;528:127–37.

Gray GM, King IA, Yardley HJ. The plasma membrane of granular cells from pig epidermis: isolation and lipid and protein composition. J Invest Dermatol. 1978b;71:131–35.

Grayson S, Johnson-Winegar AG, Wintroub BU, Isseroff RR, Epstein EH Jr., Elias PM. Lamellar body-enriched fractions from neonatal mice: preparative techniques and partial characterization. J Invest Dermatol. 1985;85:289–94.

Hill J, Paslin D, Wertz PW. A new covalently bound ceramide from human stratum corneum—omega-hydroxyacylphytosphingosine. Int J Cosmet Sci. 2006;28:225–30.

Holleran WM, Takaqi Y, Menon GK, Lesler G, Feingold KR, Elias PM. Processing of epidermal glucosylceramides is required for optimal mammalian cutaneous permeability barrier function. J Clin Invest. 1993;9:1656–64.

Kooyman DJ: Lipid in the skin. Some changes in the lipids of the epidermis during the process of keratinization. Arch Derm Syph. 1932;25:444–50.

Long VJW. Variations in lipid composition at different depths in cow snout epidermis. J Invest Dermatol. 1970;55:269–73.

Madison KC, Sando GN, Howard EJ, True CA, Gilbert D, Swartzendruber DC, Wertz PW. Lamellar granule biogenesis: a role for ceramide glucosyltransferase, lysosomal enzyme transport and the Golgi. J Invest Dermatol Symp Proc. 1998;3:80–6.

Motta S, Monti M, Sesana S, Caputo R, Carelli S, Ghidoni R. Ceramide composition of the psoriatic scale. Biochim Biophys Acta. 1993;1182:147–51.

Nemes Z, Marekov LN, Fesus L, Steinert PM. A novel function for transglutaminase 1: attachment of long-chain omega-hydroxyceramides to involucrin by ester bond formation. Proc Natl Acad Sci U S A. 1999;96:8402–7.

Nicolaides N. Skin lipids II. Lipid class composition of sample from various species and anatomic sites. J Am Oil Chem Soc. 1965;42:691–702.

Ponec M, Weerheim A, Kempenaar J, Mulder A, Gooris GS, Bouwstra JA, Mommas AM. The formation of competent barrier lipids in reconstructed human epidermis requires the presence of vitamin C. J Invest Dermatol. 1997;109:348–55.

Ponec M, Weerheim A, Lankhorst P, Wertz P. New acylceramide in native and reconstructed epidermis. J Invest Dermatol. 2003;120:581–5.

Radin NS. Biosynthesis of the sphingoid bases: a provocation. J Lipid Res. 1984;25:1536–40.

Robson KJ, Stewart ME, Michelsen S, Lazo ND, Downing DT. 6-Hydroxy-4-sphingenine in human epidermal ceramides. J Lipid Res. 1994;35:2060–9.

Stewart ME, Downing DT. The omega-hydroxyceramides of pig epidermis are attached to corneocytes solely through omega-hydroxyl groups. J Lipid Res. 2001;42:1105–10.

Swartzendruber DC, Wertz PW, Madison KC, Downing DT. Evidence that the corneocyte has a chemically bound lipid envelope. J Invest Dermatol. 1987;88:709–13.

van Smeden J, Hoppel L, van der Heijden R, Hankemeier T, Vreeken RJ, Bouwstra JA. LC/MS analysis of stratum corneum lipids: ceramide profiling and discovery. J Lipid Res. 2011;52:1211–21.

Wertz PW. Lipids and barrier function of the skin. Acta Derm Venereol Suppl. 2000;208:7–11.

Wertz PW. Changes in epidermal lipids and sebum secretion with aging. In: Dayan N, editor. Skin aging handbook. Norwich: William Andrew; 2008. pp. 91–104.

Wertz PW, Downing DT. Glycolipids in mammalian epidermis: structure and function in the water barrier. Science. 1982;217:1261–2.

Wertz PW, Downing DT. Ceramides of pig epidermis: structure determination. J Lipid Res. 1983;24:759–65.

Wertz PW, Downing DT. Covalent attachment of omega-hydroxyacid derivatives to epidermal macromolecules: a preliminary characterization. Biochem Biophys Res Commun. 1986;137:992–7.

Wertz PW, Downing DT. Covalently bound omega-hydroxyacylsphingosine in the stratum corneum. Biochim Biophys Acta. 1987;917:108–11.

Wertz PW, Madison KC, Downing DT. Covalently bound lipids in human stratum corneum. J Invest Dermatol. 1989;92:109–11.

Yardley HJ, Summerly R. Lipid composition and metabolism in normal and diseased epidermis. Pharmacol Ther. 1981;13:357–83.

Zheng Y, Yin H, Boeglin WE, Elias PM, Crumrine D, Beier DR, Brash AR. Lipoxygenases mediate the effect of essential fatty acid in skin barrier formation: a proposed role in releasing omega-hydroxyceramide for construction of the corneocyte lipid envelope. J Biol Chem. 2011;286:24046–56.

Chapter 6
Cholesterol Metabolism in the Epidermis

G. K. Menon and K. R. Feingold

Core Messages

- Epidermal cholesterol synthesis is largely autonomous.
- Cholesterol as well as several compounds in its synthetic pathway is crucial for normal skin functions.
- Lack of cholesterol or alterations in cholesterol synthesis would lead to greatly compromised epidermal barrier functions.
- Cholesterol sulfate levels in epidermis are significant in comparison with other tissues, and the "cholesterol sulfate cycle" is important for epidermal differentiation, as well as barrier formation and orderly desquamation of the stratum corneum.

Introduction

"Cholesterol is good for your skin" may be an improbable headline that you would see in a health magazine—but it is the reality. Cholesterol is required to form membranes in cells and is also crucial for the formation of the extracellular lamellae in the stratum corneum that provides the permeability barrier for the skin. Keratinocytes in the epidermis, as they stratify and differentiate, synthesize a variety of structural proteins (primarily keratins but also the proteins that form the cornified envelope such as involucrin and loricrin) as well as lipids which are packaged in secretory organelles termed as the epidermal lamellar bodies (LBs). With transmission electron microscopy, these organelles are seen as ovoid to spherical 0.2–0.5 μm structures containing tightly packed flattened discs enclosed in a membrane (Fig. 6.1).

G. K. Menon (✉)
ASI Corporation, 1005 Route 202/206 Bridgewater, NJ 08807, USA
e-mail: Gmenon@ashland.com

K. R. Feingold
UCSF, San Francisco, USA

A. Pappas (ed.), *Lipids and Skin Health*, DOI 10.1007/978-3-319-09943-9_6

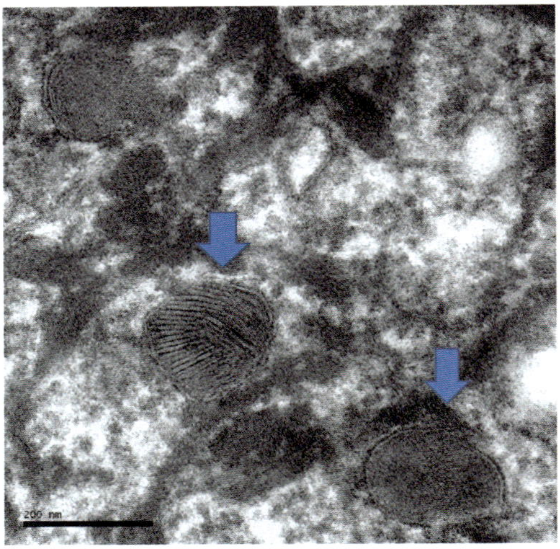

Fig. 6.1 Epidermal LBs

They predominantly contain cholesterol, glucosylceramides, and phospholipids in addition to a variety of proteins including proteolytic and lipolytic enzymes and antimicrobial peptides. At terminal differentiation into corneocytes, these LBs are secreted from the keratinocytes into the extracellular space at the stratum granulosum–stratum corneum interface. Following secretion the disc-like contents of the LBs fuse end-to-end, forming longer pleated sheets that are further processed by a battery of lipid catabolizing enzymes, including beta glucocerebrosidase which cleaves off the sugar moiety from glucosylceramides converting them to ceramides and phospholipases which metabolize the phospholipids resulting in free fatty acids. The resulting mixture of cholesterol, ceramides, and free fatty acids is organized into multiple lamellae that occupy the extracellular spaces of the stratum corneum, occluding these domains to create a hydrophobic matrix (the physical basis of the permeability barrier) preventing the movement of water across the skin. This organization of the stratum corneum has been likened to a "brick and mortar" wall, an appropriate, if far too simplistic description to convey a composite material that has made life possible on land. It is often said that this roughly 10-μm-thick tissue is what stands between life and certain death in a desiccating environment.

A paucity of cholesterol adversely affects the assembly of epidermal lamellar granules, which originate from the Golgi complex. This results in LBs with an abnormal morphology, domain separations within the stratum corneum lipids, and an increased efflux of water, quantifiable by increased transepidermal water loss (TEWL) values. It is also known that irrespective of blood cholesterol levels, the epidermis makes its own cholesterol, acting as an autonomous organ. The epidermis ramps up cholesterol synthesis when the permeability barrier is challenged, progressively returning to basal levels along with the restoration of permeability

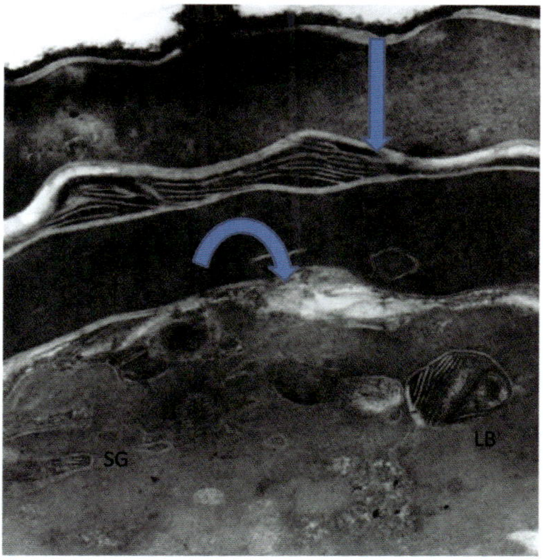

Fig. 6.2 Abnormal SC lipid organization

barrier function (Menon et al. 1985). Although aged skin maintains normal permeability barrier function in the basal state its ability to up-regulate cholesterol synthesis decreases in an age-dependent manner resulting in an impairment in the ability to restore permeability barrier function to normal after perturbations (Ghadially et al. 1995).

Although the autonomous nature of epidermal cholesterol synthesis was considered as one of the reasons that orally taken statins do not induce changes in skin excepting for a very small percentage of patients (who get perioral dermatitis or pigmentation abnormalities in this area). Nevertheless, several reports claim that the prolonged statin therapy could have undesirable side effects on several tissues including skin (Kiortsis et al. 2007; Hydzik and Szpak 2011) (Indeed topically applied statins in animal models, adversely affect epidermal cholesterol synthesis leading to scaly skin conditions and ultrastructural correlates of defective stratum corneum lipid organization (Fig. 6.2) with increased rates of TEWL (Feingold et al. 1991)). Again, initially an inhibition of cholesterol synthesis plays a role in the abnormal structure and function of stratum corneum, but at later time points, due to a compensatory increase in epidermal HMG-CoA reductase activity, cholesterol synthesis returns to normal. However, a concomitant increase in fatty acid synthesis results in an excess of fatty acids that alter the molar ratio of the stratum corneum lipids resulting in barrier abnormalities.

The authors have done some of the early and basic studies on impact of epidermal cholesterol metabolism on skin barrier function. Presently, we summarize the vast amount of work that our group and others have done in the last couple of decades that has contributed to the current state of understanding of epidermal cholesterol metabolism and skin barrier.

Unraveling the Significance of Cholesterol for Skin Permeability Barrier Function

Considered on a weight basis, mammalian epidermis is a very active site of cholesterol synthesis (Feingold et al. 1983). Within the cell, cholesterol is synthesized in the endoplasmic reticulum that contains the rate limiting enzyme for sterologenesis—HMG-CoA reductase. The significance of cholesterol for the stratum corneum permeability barrier is quite evident from the fact that the "mortar lipids" are made up of equimolar ratios of cholesterol, ceramides, and free fatty acids (Elias and Menon 1991). Studies on rodents, following acute permeability barrier disruption, demonstrated a marked and rapid increase in cholesterol synthesis (Feingold 1991). Additionally, there was an upregulation of mRNA and protein level as well as the activity of HMG-CoA reductase, a key rate-limiting enzyme in the cholesterol synthetic pathway (Jackson et al. 1992; Proksch et al. 1992, 1995). Barrier disruption also results in a notable increase in the percentage of HMG-CoA reductase in the active dephosphorylated form and this increase in activity is seen as early as 15 min after permeability barrier disruption (Proksch et al. 1990). It was also found that the extent of barrier disruption (reflected in the TEWL values) required to activate the enzyme by dephosphorylation is less than that required to increase the enzyme protein levels. The increase in HMG-CoA reductase occurs in both the upper and lower epidermal cell layers (Proksch et al. 1990). Expression of other key enzymes in the cholesterol synthetic pathway, such as HMG-CoA synthase, farnesyl diphosphate synthase, and squalene synthase, also increase after acute permeability barrier perturbation (Harris et al. 1997). However, when an occlusive membrane experimentally "restored" the permeability barrier towards normal following barrier disruption, it inhibits the increase in epidermal cholesterol synthesis as well as the mRNA levels of the cholesterol synthetic enzymes (Menon et al. 1985; Proksch et al. 1990). However, when a semipermeable membrane replaces the occlusive cover, the cholesterol synthetic response goes on unabated. These results indicate that the increase in cholesterol synthesis is an adaptive response to barrier disruption, rather than a nonspecific injury response.

Experiments with enzyme inhibitors that inhibited cholesterol synthesis showed the significance of epidermal cholesterol synthesis in permeability barrier homeostasis. Topical application of statins (inhibiting HMG-CoA reductase activity) blocked the characteristic upregulation in epidermal cholesterol synthesis that followed acute barrier disruption resulting in a delay in the recovery of permeability barrier function (Feingold et al. 1990). Interestingly, topical application of statins did not block the first wave of secretion of the nascent LBs. However, the reappearance of LBs in stratum granulosum cells (i.e., the synthesis of new LBs) is delayed and the newly synthesized LBs display abnormal internal contents (Zetterstein et al. 1998). In addition, the lamellar lipid organization of the stratum corneum barrier is also altered in appearance, indicating that an abnormality of mortar lipids is the underlying cause of the permeability barrier defect (Fig. 6.2). Topical application of cholesterol or its intermediary compound mevalonate (formed by HMG-CoA reductase) overcomes these effects demonstrating that the statin-induced defects are not nonspecific toxic effects of topical inhibitors but rather are due to a decrease in cholesterol (Feingold et al. 1990). These studies demonstrate the crucial role of epidermal cholesterol synthesis in providing cholesterol for permeability barrier homeostasis.

Further demonstrating the importance of epidermal cholesterol synthesis for permeability barrier function are studies in mice, which are deficient in 3β-hydroxysterol-Δ24, the enzyme that catalyzes the conversion of desmosterol to cholesterol. These mice fail to generate cholesterol in the epidermis, but they have an abundance of desmosterol. However, they die soon after birth due to lack of a functional permeability barrier (Mirza et al. 2006).

The potential role of extracutaneous cholesterol is suggested by the observation that inhibiting de novo cholesterol synthesis in the epidermis results in only a modest delay in the restoration of barrier function to normal. This suggests that extracutaneous cholesterol could provide the cholesterol needed for barrier restoration. Both LDL receptors and SR-B1 receptors are expressed in the epidermis and these receptors would allow for uptake of lipoproteins that contain cholesterol. Notably, the expression of both LDL receptors and SR-B1 receptors increases with permeability barrier disruption. Moreover, this increase can be prevented by occlusion that artificially restores barrier function to normal indicating that the increase in LDL receptor and SR-B1 receptor expression is linked to permeability barrier function.

Another aspect of cholesterol metabolism concerns membrane transporters such as ABCA1, which are responsible for cholesterol efflux from cells. ABCA1 plays a major role in cholesterol efflux providing a mechanism for regulating cellular cholesterol levels. Expression of ABCA1 has been documented in both undifferentiated and differentiated keratinocytes as well as the upper and lower layers of epidermis (Jiang et al. 2010). The authors also found that ABCA1 expression is decreased following acute barrier disruption, perhaps leading to reduced cholesterol efflux, allowing the cells to utilize cholesterol to meet the demands of increased LB synthesis and recovery of permeability barrier function. In addition to ABCA1, ABCG1 is also expressed in cultured human keratinocytes and murine epidermis, and it is known to be induced during the process of keratinocyte differentiation, with increasing levels towards the outer epidermis. It is regulated by liver X receptor (LXR) and PPAR activators as well as by cellular sterol levels. Moreover, acute permeability barrier disruption increases the expression of ABCG1 (Jiang et al. 2010). While ABCG1 had been thought to act at the plasma membrane to efflux cholesterol, Tarling and Edwards (2011) showed that ABCG1 also localizes to the intracellular endosomes. The transport of cholesterol to epidermal LBs may be partially mediated by ABCG1, as ABCG1 KO mice display defective (empty appearing) LBs, and a reduction in SC lamellar membranes (Jiang et al. 2010). However, permeability barrier function is normal in these animals indicating that ABCG1 is not essential for the formation of LBs suggesting that there may be multiple mechanisms for cholesterol transport into LBs.

Other barriers present in the epidermis are also influenced by the cholesterol synthetic pathway. For example, CoQ 10 is an important antioxidant that functions in the oxidant barrier of skin and its role in skin health is very well appreciated (Shindo et al. 1994). CoQ 10 is derived from the same pathway that leads to cholesterol and drugs that inhibit HMG-CoA reductase activity would also adversely affect CoQ 10 production (Mabuchi et al. 2005). In the skin, the reduced form of CoQ10 (ubiquinol) acts as an antioxidant with tenfold higher levels in the epidermis than in the dermis (Shindo et al. 1994). Inoue et al. (2008) found that normal human keratinocytes in vitro, in the presence of CoQ 10 showed decreased IL-6 production following UVB exposure. Additionally, when fibroblasts that were pretreated with CoQ 10 were exposed to conditioned media from UVB exposed keratinocytes, the characteristic response of MMP-1 production was also downregulated. In a clinical study, daily applications of a 1 % CoQ10 cream for 5 months were also reported to reduce the wrinkle score grade (Inoue et al. 2008), which they attributed to a reduction in Matrix Metaloprotease (MMP) production (Inoue et al. 2008). Such studies underlie the significance of the cholesterol synthetic pathway.

Additionally, within the epidermal compartment that houses melanocytes, cholesterol also has an important role in another facet of the barrier—namely melanogenesis. Schallrueter et al. (2009) showed that human epidermal melanocytes have the capacity of cholesterol signaling via Apo-B/LDL receptor and that cholesterol under in vitro conditions increases melanogenesis. Although the significance of melanin as a UV barrier has often been disputed, the microparsol of melanosomes that cap the basal keratinocyte nuclei has long been considered to be crucial for DNA protection in these cells. Active melanocytes no doubt have a role in epidermal barrier, as permeability barrier recovery under experimental conditions in mice and humans is more efficient in darkly pigmented skin (Reed et al. 1995).

Additionally, dehydrocholesterol is a precursor for vitamin D, whose importance in health is very well recognized. UV induced synthesis of vitamin D makes a major contribution to total body vitamin D stores. In fact if exposed to sufficient sunlight one does not have to ingest vitamin D as the skin is capable of making all the vitamin D that is required. In the epidermis 1, 25 dihydroxy vitamin D binds to the vitamin D receptor, which heterodimerizes with RXR to control the transcription of several genes that regulate the differentiation of keratinocytes and permeability barrier function. Considering all of the important roles that the cholesterol biosynthetic pathway assumes in keratinocytes the predominantly autonomous nature of epidermal cholesterol synthesis appears to be a huge selective advantage.

Effects of Aging

Like many other functions that decline with age, epidermal cholesterol synthesis is decreased in the elderly and this decrease is associated with a decrease in HMG-CoA activity. Permeability barrier function (measured by TEWL) appears normal in aging or actually better than normal, often attributed to the decreased microcirculation of the aged skin. Whatever be the case for normal TEWL in aged skin, there is a delay in the recovery of permeability barrier function following acute perturbation, in both aged mice as well as humans (around 75 years of age). This is associated with a decrease in extracellular lipids of the stratum corneum, which is due to a decrease in epidermal cholesterol synthesis (Ghadially et al. 1995). The LB synthetic and secretory responses are blunted in aged skin, but topical application of mevalonic acid or cholesterol can correct the abnormality demonstrating the key defect is the inability of aged skin to make sufficient amounts of cholesterol (Haratake et al. 2000).

Yet another cholesterol connection with ageing changes in skin could be related to its role in modulating metalloproteinases—which play a significant role in dermal changes manifested as wrinkles, especially in photoaging. Quan et al. (2009) have shown that following experimental UV irradiation of human skin, all three *UV* irradiation inducible *MMPs* were produced primarily in the *epidermis*. Thus, in photoaging, the epidermis plays a significant role in inducing dermal changes. Additionally, cholesterol treatment of fibroblasts has been reported to increase expression of tissue inhibitor of metalloproteinases (TIMP) (Tyagi et al. 1997), and a declining cholesterol synthesis in aging skin could also modulate the activity of metalloproteinases, enhancing the dermal changes. A recent paper (Kim et al. 2010) suggests that cholesterol depletion leads to inhibition of TIMP-2 which in turn leads to the conversion of proMMP2 into active MMP2 in human dermal fibroblasts. Therefore, it appears that the presence of cholesterol may also be fundamental for healthy extracellular matrix of the dermis.

Cholesterol Sulfate

While cholesterol sulfate is a minor sterol constituent in blood and nonkeratinizing gastrointestinal epithelia, with a low cholesterol–cholesterol sulfate ratio (about 500:1), in the stratum corneum of normal epidermis the cholesterol–cholesterol sulfate ratio is 10:1–5:1 (Ponec and Williams 1986), indicative of it having an important role in SC structure and function. Cholesterol sulfate is a normal constituent of hair and nails (Wertz and Downing 1988, and in skin, it is primarily located in the epidermis, where it is formed through enzymatic sulfoconjugation of cholesterol which is mediated by sulfotranferases. Cytosolic sulfotransferases (SULTs) are a superfamily of enzymes that catalyze the sulfoconjugation of drugs, neurotransmitters, xenobiotics, hormones, and most importantly for epidermal biology, sterols. The SULT2 family is primarily responsible for the sulfation of neutral steroids and sterols. SULT2A1 catalyzes the conversion of DHEA to DHEA sulfate while the SULT2B1a isoform preferentially sulfonates pregnenolone. SULT2B1a and SULT2A1 are not expressed in keratinocytes or epidermis. SULT2B1b, which is expressed in keratinocytes and epidermis, preferentially catalyzes the conversion of cholesterol to cholesterol sulfate and thus, the SULT2B1b isoform accounts for the majority of cholesterol SULT activity in the epidermis. SULT2B1b expression and cholesterol SULT activity is increased in keratinocytes during calcium induced differentiation and SULT2B1b is observed in both the basal and granular layer of the epidermis (Higashi et al. 2004). PPARα, PPARβ/δ, PPARγ, and LXR activators, which stimulate differentiation, increase SULT2Blb expression and cholesterol SULT activity while retinoic acid, which inhibits differentiation, inhibits cholesterol SULT activity (Jiang et al. 2005; Jetten et al. 1989).

SULT2B1b generates cholesterol sulfate in the lower nucleated cell layers of the epidermis while the activity of steroid sulfatase (SSase), which catalyzes the conversion of cholesterol sulfate to cholesterol and sulfate, peaks in the outer epidermis (stratum corneum). Based on these observations, Epstein et al. (1984) proposed that an "epidermal cholesterol sulfate cycle" exists in the epidermis in which cholesterol sulfate is synthesized in the lower epidermis, and then metabolized back to cholesterol in the outer epidermis. Thus, the content of cholesterol sulfate in the epidermis increases from 1 to 5 % of total lipid as nucleated epidermal cells migrate from the basal to the granular layer and then declines again to 1 % as corneocytes migrate from the inner to outer stratum corneum (Long et al. 1985; Rearick et al. 1987). Abnormalities in this cholesterol sulfate cycle lead to abnormal desquamation and the permeability barrier abnormality that occurs in recessive X linked ichthyosis (RXLI) (see below). The significance of epidermal cholesterol sulfate cycle becomes obvious when the multiple important roles this compound plays in the epidermis are examined.

Role of Cholesterol Sulfate in Keratinocyte Differentiation

Both the content of cholesterol sulfate and the activity of cholesterol sulfate transferase (SULT2B1b) increase with keratinocyte differentiation. The synthesis of cholesterol sulfate is thus closely linked to epidermal differentiation, including normal cornification (Jetten et al. 1989; Higashi et al. 2004). This relationship is apparent from the much higher levels of cholesterol sulfate in keratinizing than in mucosal epithelia. Conversely, reversal of keratinization by the application of exogenous retinoids, which induces mucous metaplasia in keratinizing epithelia, markedly reduces the levels of tissue cholesterol sulfate (Jetten et al. 1989). In addition, concomitant with the formation of a functionally competent stratum corneum (Komuves et al. 1999), both cholesterol sulfate levels and SULT2B1b expression increase late in epidermal development in utero.

The rise in cholesterol sulfate concentrations that occurs in conjunction with keratinocyte differentiation may not be just a marker of differentiation but may in fact reflect a signaling role for the molecule in inducing keratinocyte differentiation. The addition of cholesterol sulfate to keratinocytes in culture or overexpressing SULT2B1b in keratinocytes, which leads to an increase in cholesterol sulfate synthesis, stimulates keratinocyte differentiation (Hanley et al. 2001; Ikuta et al. 1994). Furthermore, cholesterol sulfate is a potent transcriptional regulator of keratinocyte gene expression (Strott and Higashi 2003), stimulating epidermal differentiation by several mechanisms: (1) It activates the η isoform of protein kinase C (PKC) (Kashiwagi et al. 2002), which stimulates the phosphorylation of epidermal structural proteins and increases cornified envelope formation. (2) It regulates the transcription of several proteins involved in cornified envelope formation, including transglutaminase 1 (TGM1) and involucrin. The increase in transcription is mediated by an AP-1 binding site in the promoter region of these proteins (Hanley et al. 2001). Cholesterol sulfate increases the expression of Fra-1, Fra-2, and Jun D, members of the AP-1 family of transcription factors. These two mechanisms are likely linked, because PKC activation by cholesterol sulfate could also phosphorylate AP-1 transcription factors thereby enhancing the transcriptional regulation of differentiation-linked proteins, such as TG-1 and involucrin (Kuroki et al. 2000). (3) Finally, recent studies have shown that cholesterol sulfate increases profilaggrin expression through the induction of RORα (Hanyu et al. 2012). Cholesterol sulfate induces profilaggrin expression in a dose-dependent fashion when applied exogenously, but not when RORα is silenced with siRNA. RORα and RORγ play an important role in the regulation of genes encoding several phase I and phase II metabolic enzymes, including several 3β-hydroxysteroid dehydrogenases, cytochrome P450 enzymes, and SULTs (Kang et al. 2007) and thus cholesterol sulfate may play a role in regulating drug metabolism in the epidermis. Cholesterol sulfate is a ligand for RORα and stimulates the expression of RORα which together would increase the transcriptional activity of RORα. These observations provide biochemical and molecular mechanisms whereby cholesterol sulfate could stimulate keratinocyte differentiation.

While the studies described above demonstrate that the addition of cholesterol sulfate stimulates keratinocyte differentiation recent studies have importantly demonstrated a physiologic role for cholesterol sulfate formation in regulating keratinocyte differentiation. Using shRNA, to inhibit the expression of SULT2B1b in keratinocytes in culture, Shimada and colleagues have demonstrated that concomitant with a decrease in cholesterol sulfate formation there was a decrease in the expression of involucrin. Moreover, using shRNA they were able to show that inhibiting the expression of SULT2B1b in mouse epidermis similarly reduced the expression of involucrin (Shimada et al. 2008). TPA treatment of keratinocytes in culture or mouse skin in vivo increases the expression of both SULT2B1b and involucrin. Treatment of keratinocytes in culture or mouse skin with shRNA to prevent the TPA induced increase in SULT2B1b inhibits the ability of TPA to increase involucrin expression (Matsuda et al. 2011). These observations strongly suggest that the synthesis of cholesterol sulfate in the epidermis is not only a marker of differentiation but may also play a key role in inducing keratinocyte differentiation.

Steroid Sulfatase

The breakdown of cholesterol sulfate is catalyzed by the enzyme SSase and mutations in this gene lead to X-linked ichthyosis. The gene that encodes SSase is located distally on the short arm of the Y chromosome (Reed et al. 2005). Retinoids and 1,25(OH)$_2$ vitamin D3 induce SSase activity and expression (Hughes et al. 2001). Additionally, exogenous cholesterol sulfate (Maloney et al. 1984) induces SSase enzyme activity. In normal epidermis, SSase protein and enzyme activity are low in the basal and spinous layers, but increase in the outer epidermis, peaking in the granular layer (Elias et al. 1984). SSase enzyme activity persists into the stratum corneum, where it desulfates cholesterol sulfate, allowing for normal desquamation and contributing to the pool of cholesterol available to from lamellar bilayer formation. In ultrastructural cytochemical studies, SSase activity localizes in the cytosol of keratinocytes, within LBs, and in the extracellular space of the lower stratum corneum, following delivery by LB exocytosis (Elias et al. 2004). Thus, SSase, similar to other lipid hydrolases that are involved in the extracellular processing of secreted lipids, utilizes the LB secretory system for delivery to the extracellular spaces of the stratum corneum where it participates in the regulation of permeability barrier homeostasis, as well as regulating desquamation (Elias and Menon 1991).

Role of Cholesterol Sulfate in Desquamation

In normal epidermis, cholesterol sulfate levels decline from about 5 % of lipid mass in the stratum granulosum to about 1 % of lipid mass in the outer stratum corneum, through ongoing hydrolysis by SSase (Elias et al. 1984; Long et al. 1985; Ranasinghe et al. 1986) (Fig. 6.3). In contrast, the stratum corneum in patients with XLI typically contains 10–12 % cholesterol sulfate (Williams and Elias 1981; Elias et al. 1984). As mentioned above mutations in SSase lead to XLI and the marked inhibition of desquamation is the major clinical abnormality in these patients.

Cholesterol sulfate is not concentrated in LBs (Grayson et al. 1985; Elias et al. 2004) and the pathway by which it is delivered to the extracellular spaces of the stratum corneum is unknown. Cholesterol sulfate is extremely amphiphilic, which may allow it to traverse cell membranes in search of the highly hydrophobic, extracellular domains of the stratum corneum (Ponec and Williams 1986). Alternatively, cholesterol sulfate could be actively transported from the stratum granulosum cells into the extracellular space of the stratum corneum by specific transporters that transport sulfated conjugates of lipophilic compounds (ABCC1, ABCC3, and/or ABCC4 (MRP1, MRP3, and MRP4)) (Konig et al. 1999). These transporters are expressed in keratinocytes and the level of expression increases with differentiation (Kielar et al. 2003).

Cholesterol sulfate is a serine protease (SP) inhibitor and thereby prevents the degradation of CDs, which are required for normal desquamation (Williams 1991;

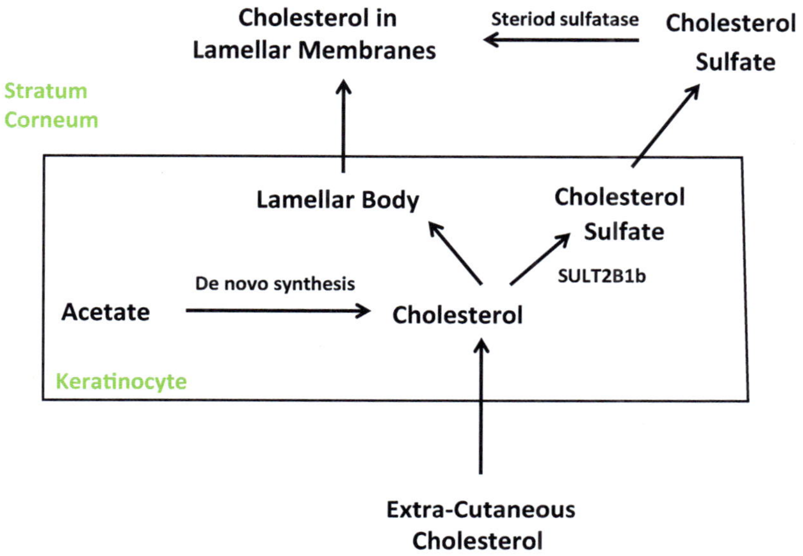

Fig. 6.3 Cholestrol pathways in skin

Sato et al. 1998; Elias et al. 2004). Additionally, the acidic pH of normal stratum corneum inhibits Klk5 and Klk7 activities (Hachem et al. 2002, 2003, 2009), and the pH of the stratum corneum in XLI is even more acidic than normal (Ohman and Vahlquist 1998), which would further inhibit Klk activity in the stratum corneum of patients with XLI (Elias et al. 2004).

Role of Cholesterol Sulfate in Permeability Function

XLI patients display only a minimal barrier abnormality under basal conditions (Lavrijsen et al. 1993; Johansen et al. 1995; Hoppe et al. 2012), but following acute barrier disruption, the recovery kinetics are significantly delayed (Zettersten et al. 1998), suggesting that excess cholesterol sulfate adversely effects permeability barrier homeostasis. Excess cholesterol sulfate forms nonlamellar domains in both model lipid mixtures and in XLI scale (see review by Elias et al. 2014). However, the barrier abnormality in XLI could also be due to decreased cholesterol (Elias et al. 2004) in the stratum corneum from the failure to breakdown cholesterol sulfate to cholesterol and/or to cholesterol sulfate-mediated inhibition of HMG-CoA reductase leading to decreased cholesterol synthesis in the viable epidermis (Williams and Elias 1981). A similar decrease in cholesterol results in a barrier abnormality in association with the formation of abnormal extracellular lamellar membranes in experimental animals (Feingold et al. 1990). It is noteworthy that barrier abnormalities typically result in epidermal hyperplasia and inflammation in other ichthyoses, but XLI provokes little epidermal hyperplasia or inflammation. This minimal pathology may be accounted for by gene expression studies that show very few genes are altered in the epidermis of patients with XLI (Hoppe et al. 2012).

Conclusion

It should be clear to the reader that cholesterol and the products of cholesterol synthesis play crucial roles in the epidermis. Whereas in many situations, for example, atherosclerosis, cholesterol induces disease, in the epidermis cholesterol is required for normal function and without cholesterol and the products of cholesterol synthesis cutaneous function would be greatly compromised.

References

Elias PM, Menon GK. Structural and lipid biochemical correlates of the epidermal permeability barrier. Adv Lipid Res. 1991;24:1–26.

Elias PM, Williams ML, Maloney ME, Bonifas JA, Brown BE, Grayson S, Epstein EH Jr. Stratum corneum lipids in disorders of cornification. Steroid sulfatase and cholesterol sulfate in normal desquamation and the pathogenesis of recessive X-linked ichthyosis. J Clin Invest. 1984;74:1414–21.

Elias PM, Crumrine D, Rassner U, Hachem J-P, Menon GK, Man AW, Choi MH, Leypold L, Feingold KR, Williams ML. Basic of abnormal desquamation and permeability barrier dysfunction in RXLI. J. Invest. Dermatol. 2004;122:314–319.

Elias PM, Williams ML, Choi E-U, Feingold KR. Role of cholesterol sulfate in epidermal structure and function: lessons from X-linked ichthyosis. Biochem Biophys Acta Mol Cell Biol Lipids. 2014;1841(3):353–61.

Epstein EH, Williams ML, Elias PM. The epidermal cholesterol sulfate cycle. J Am Acad Dermatol. 1984;10:866–8.

Feingold KR, Wiley MH, MacRae G, Lear S, Moser AH, Zsigmond G, Siperstein MD. Denovo sterologenesis in the intact rat. Metabolism. 1983;32:75–81.

Feingold KR. The regulation and role of epidermal lipid synthesis. Adv Lipid Res. 1991;24:57–82.

Feingold KRE, Man MQ, Menon GK, Cho SS, Brown BE, Elias PM. Cholesterolsynthesis is required for cutaneous barrier function in mice. J Clin Invest. 1990;86:1738–45.

Feingold KR, Man MQ, Proksch E, Menon GK, Brown BE, Elias PM. The lovastatin-treated rodent: a new model of barrier disruption and epidermal hyperplasia. J Invest Dermatol. 1991;96:201–9.

Ghadially R, Brown BE, Sequeira-Martin SM, Feingold KR, Elias PM. The aged epidermal permeability barrier. Structural, functional, and lipid biochemical abnormalities in humans and a senescent murine model. J Clin Invest. 1995;95:2281–90.

Grayson S, Johnson-Vinegar AG, Wintraub BU, Isserhoff RR, Epstein EH, Jr, Elias PM. Lamellar body-enriched fractions from neonatal mice: preparative techniques and partial characterization. J. Invest. Dermatol. 1985;85:289–294.

Hachem JP, Crumrine D, Fluhr J, Brown BE, Feingold KR, Elias PM. pH directly regulates epidermal permeability barrier homeostasis, and stratum corneum integrity/cohesion. J Invest Dermatol. 2003;121:345–53.

Hachem JP, Wagberg F, Schmuth M, et al. Serine protease activity and residual LEKTI expression determine phenotype in Netherton Syndrome. J. Invest. Dermatol. 2006;126:1609–1621.

Hanley K, Wood L, Ng DC, et al. Cholesterol sulfate stimulates involucrin transcription in keratinocytes by increasing Fra-1, Fra-2, and Jun D. J Lipid Res. 2001;42:390–8.

Hanyu O, Nakae H, Miida T, Higashi Y, Fuda H, Endo M, Kohjitani A, Sone H, Strott CA. Cholesterol sulfate induces expression of the skin barrier protein filaggrin in normal human epidermal keratinocytes through induction of RORα. Biochem Biophys Res Commun. 2012;428(1):99–104.

Haratake A, Ikenaga K, Katoh N, Uchiwa H, Hirano S, Yasuno H. Topical mevalonic acid stimulates de novo cholesterol synthesis and epidermal permeability barrier homeostasis in aged mice. J Invest Dermatol. 2000;114:247–52.

Harris IR, Farrel AM, Grunfeld C, Holleran WM, Elias PM, Feingold KR. Permeability barrier disruption coordinately regulates mRNA levels for key enzymes of cholesterol, fatty acids and ceramide synthesis in the epidermis. J Invest Dermatol. 1997;109:783–7.

Higashi Y, Fuda H, Yanai H, et al. Expression of cholesterol sulfotransferase (SULT2B1b) in human skin and primary cultures of human epidermal keratinocytes. J Invest Dermatol. 2004;122:1207–13.

Hoppe T, Winge MC, Bradly M, Nordenskjold M, Vahlquist A, et al. X-linked recessive ichthyosis: an impaired barrier function evokes limited gene responses before and after moisturizing treatments. Br. J. Dermatol. 2012;167:514–522.

Hydzik P, Szpak D. Side effects of the HMG-CoA reductase inhibitors (statins). Lupus erythematosus induced by atorvastatin therapy. Przegl Lek. 2011;68:495–8.

Ikuta T, Chida K, Tajima O, et al. Cholesterol sulfate, a novel activator for the eta isoform of protein kinase C. Cell Growth Differ. 1994;5:943–7.

Inoue M, Ooe M, Fuji K, Matsunaka M, Ichihashi M. Mechanisms of inhibitory effects of Co Q(10) on UVB-induced wrinkle formation in vitro and in vivo. Biofactors. 2008;32:237–43.

Jackson SM, Wood LC, Lauer S, Taylor JM, Cooper AD, Elias PM, Feingold KR. Effect of cutaneous permeability barrier disruption on HMG-CoA reductase, LDL receptor, and apolipoprotein E mRNA levels in the epidermis of hairless mice. J Lipid Res. 1992;33:1307–14.

Jetten AM, George MA, Pettit GR, et al. Action of phorbol esters, bryostatins, and retinoic acid on cholesterol sulfate synthesis: relation to the multistep process of differentiation in human epidermal keratinocytes. J Invest Dermatol. 1989;93:108–15.

Jiang YJ, Lu B, Tarling E, Kim P, Man M-Q, Crumrine D, Edwards P, Elias PM, Feingold KR. Regulation of ABCG1 expression in human keratincoytes and murine epidermis. J Lipid Res. 2010;51:3185–95.

Kang HS, Angers M, Beak JY, Wu X, Gimble JM, Wada T, Xie W, Collins JB, Grissom SF, Jetten AM. Gene expression profiling reveals a regulatory role for RORα and RORγ in phase I and phase II metabolism. Physiol Genomics. 2007;31:281–94.

Kashiwagi M, Ohba M, Chida K, et al. Protein kinase C eta (PKC eta): its involvement in keratinocyte differentiation. J Biochem (Tokyo). 2002;132:853–7.

Kielar D, Kaminski WE, Liebisch G, Piehler A, Wenzel JJ, Mohle C, Heilerl S, Langman T, Friedrich SO, Bottcher A, Barlage S, Drobnik W, Schmitz G. Adenosine triphosphate binding cassette (ABC) transporters are expressed and regulated during terminal keratinocyte differentiation: a potential role for ABCA7 in epidermal lipid reorganization. J Invest Dermatol. 2003;121:465–74.

Kim S, Hee-Oh J, Lee Y, Le J, Cho KH, Chung JHO. Induction of tissue inhibitor of matrix metalloprotease-2 by cholesterol depletion leads to the conversion of pro MMP-2 into active MMP-2 in human dermal fibriblasts. Exp Mol Med. 2010;42:38–46.

Kiortsis DN, et al. Statin-associated adverse effects beyond muscle and liver toxicity. Atherosclerosis. 2007;195:7–16.

Komuves LG, Hanley K, Jiang Y, et al. Induction of selected lipid metabolic enzymes and differentiation-linked structural proteins by air exposure in fetal rat skin explants. J Invest Dermatol. 1999;112:303–9.
Konig J, Nies AT, Cui Y, Leier I, Keppler D. Conjugate export pumps of the multidrug resistance protein (MRP) family: localization, substrate specificity, and MRP2-mediated drug resistance. Biochim Biophys Acta Biomembr. 1999;1461:377–94.
Kuroki T, Ikuta T, Kashiwagi M, et al. Cholesterol sulfate, an activator of protein kinase C mediating squamous cell differentiation: a review. Mutat Res. 2000;462:189–95.
Lavrijsen APM, Oestmens E, Hermans J, Bodde HE, Vermeer BJ, Ponec M. Barrier function parameters in various keratinization disorders: transepidermal water loss and vascular response to hexyl nicotinate. Br. J. Dermatol. 1993;129:547–554.
Long SA, Wertz PW, Strauss JS, et al. Human stratum corneum polar lipids and desquamation. Arch Dermatol Res. 1985;277:284–7.
Mabuchi H, Higashikata T, Kawashiri M, Katsuda S, Mizuno M, Nohara A, Inazu A, Koizumi J, Kobayashi J. Reduction of serum ubiquinol-10 and ubiquinone-10 levels by atorvastatin in hypercholesterolemic patients. J Atheroscler Thromb. 2005;12:111–9.
Maloney ME, Williams ML, Epstein EH Jr, Law MY, Fritsch PO, Elias PM. Lipids in the pathogenesis of ichthyosis: topical cholesterol sulfate-induced scaling in hairless mice. J Invest Dermatol. 1984;83:252–6.
Matsuda T, Shimada M, Sato A, Akase T, Yoshinari K, Nagata K, Yamazoe Y. Tumor necrosis Factor-alpha-Nuclear Factor-Kappa B signaling enhances St2b2 expression during 12-O-tetradecanoylphorbol-13-acetate-induced epidermal hyperplasia. Biol. Pharm. Bull. 2011;34:183–190.
Menon GK, Feingold KR, Moser AH, Brown BE, Elias PM. De novo sterologenesis in the skin II. Regulation by cutaneous barrier requirements. J Lipid Res. 1985;26:418–27.
Mirza R, Hayasaka S, Takagishi Y, Kambe F, Ohmori S, Maki M, Yamamoto K, Murakami T, Kaji D, Zadworny D, Murata Y, Seo H. DHCR24 gene knockout mice demonstrate lethal dermopathy with differentiation and maturation defects in the epidermis. J Invest Dermatol. 2006;126:638–47.
Ohman H, Vahlquist A. The pH gradient over the stratum corneum differs in X-linked recessive and autosomal dominant Ichthyosis: a clue to the molecular origin of the "acid skin mantle"? J Invest dermatol. 1998;111:674–7.
Ponec M, Williams ML. Cholesterol sulfate uptake and outflux in cultured human keratinocytes. Arch Dermatol Res. 1986;279:32–6.
Proksch E, Elias PM, Feingold KR. Regulation of 3-hydroxy-3-methylglutaryl-coenzyme A reductase activity in murine epidermis. Modulation of enzyme content and activation state by barrier requirements. J Clin Invest. 1990;85:874–82.
Proksch E, Feingold KR, Elias PM. Epidermal HMG CoA reductase activity in essential fatty acid deficiency: barrier requirements rather than eicosanoid generation regulate cholesterol synthesis. J Invest Dermatol. 1992;99:216–20.
Proksch E, Elias PM, Feingold KR. Localization and regulation of epidermal 3-hydroxy-3-methylglutaryl-coenzyme A reductase activity by barrier requirement. J Clin Invest. 1995;95:2281–90.
Quan T, Qui Z, Xia W, Shao Y, Voorhees JV, Fisher GJ. Matrix degrading metalloproteinases in photoaging. J Invest Dermatol Symp Proc. 2009;14:20–4.
Ranasinghe AW, Wertz PW, Downing DT, Mackenzie IC. Lipid composition of cohesive and desquamated corneocytes from mouse ear skin. J Invest Dermatol. 1986;86:187–90.
Rearick JI, Albro PW, Jetten AM. Increase in cholesterol sulfotransferase activity during in vitro squamous differentiation of rabbit tracheal epithelial cells and its inhibition by retinoic acid. J Biol Chem. 1987a;262:13069–74.
Reed T, Ghadially R, Elias PM. Skin type, but neither race or gender, influence epidermal permeability function. Arch Dermatol. 1995;131:1134–8.
Reed MJ, Purohit A, Woo LW, et al. Steroid sulfatase: molecular biology, regulation, and inhibition. Endocr Rev. 2005;26:171–202.
Sato J, Denda M, Nakanishi J, Nomura J, Koyama J. Cholesterol sulfate inhibits proteases that are involved in desquamation of stratum corneum. J Invest Dermatol. 1998;111:1523–747.

Schallreuter KU, Hasse S, Rokos H, Chavan B, Shalbaf M, Spencer JD, Wood JM. Cholesterol regulates melanogenesis in human epidermal melanocytes and melanoma cells. Exp Dermatol. 2009;18:680–8.

Shimada M, Matsuda T, Sato A, Akase T, Matsubara T, Nagata K, Yamazoe Y. Expression of a skin cholesterol sulfotransferase, St2b2, is a trigger of epidermal cell differentiation. Xenobiotica. 2008;38:1487–1499.

Shindo Y, Wit E, Han E, Epstein W, Packer L. Enzymic and non-enzymic antioxidants in epidermis and dermis of human skin. J Invest Dermatol. 1994;102:122–4.

Strott CA, Higashi Y. Cholesterol sulfate in human physiology: what's it all about? J Lipid Res. 2003;44:1268–78.

Tarling EJ, Edwards PA. ATP binding cassette transporter G1 (ABCG1) is an intracellular sterol transporter. PNAS. 2011;108:19719–24.

Tyagi SC, Kumar S, Katwa L. Differential regulation of extracellular matrix metalloproteinase and tissue inhibitor by heparin and cholesterol in fibroblast cells. J Mol Cell Cardiol. 1997;29:391–404.

Wertz PW, Downing DT. Integral lipids of human hair. Lipids. 1988;23:878–81.

Williams ML, Elias PM. Stratum corneum lipids in disorders of cornification: increased cholesterol sulfate content of stratum corneum in recessive X-linked ichthyosis. J Clin Invest. 1981;68(6):1404–10.

Zetterstein E, Man M-Q, Sato J, Denda M, Farrell A, Ghadially R, Williams ML, Feingold KR, Elias PM. Recessive x-Linked Ichthyosis: role of cholesterol-sulfate accumulation in the barrier abnormality. J Invest Dermatol. 1998;111:784–90.

Chapter 7
Nuclear Hormone Receptors and Epidermal Differentiation

Nguan Soon Tan and Walter Wahli

Core Messages

- Nuclear hormone receptors are involved in skin epidermal differentiation and dictate the formation of epidermal barrier.
- Vitamin D receptor (VDR) null mice exhibit defects in epidermal differentiation and develop alopecia, which is also found in many patients with mutations in this receptor.
- Suprabasal expression of functional retinoic acid receptors/retinoid X receptors (RAR/RXR) is required for retinoic-mediated epidermal hyperplasia.
- Peroxisome proliferator-activated receptors (PPAR) and liver X receptor (LXR) likely mediate the expression of keratinocyte differentiation-associated genes through a common pathway.
- Drugs (agonists and antagonists) that target the nuclear hormone receptors constitute one of the largest and most potent groups of pharmaceuticals currently in use and thus hold promising potential for the development of improved skin care.

N. S. Tan (✉)
School of Biological Sciences, Nanyang Technological University,
60 Nanyang Drive, Singapore 637551, Singapore
e-mail: nstan@ntu.edu.sg

Institute of Molecular and Cell Biology, 61 Biopolis Drive,
Proteos, Singapore 138673, Singapore
e-mail: nstan@imcb.a-star.edu.sg

W. Wahli
Lee Kong Chian School of Medicine, Nanyang Technological University,
50 Nanyang Drive, Singapore 637553, Singapore
e-mail: walter.wahli@ntu.edu.sg

Center for Integrative Genomics, National Research Center Frontier in Genetics,
University of Lausanne, Bâtiment Génopode, 1015 Lausanne, Switzerland
e-mail: walter.wahli@unil.ch

A. Pappas (ed.), *Lipids and Skin Health*, DOI 10.1007/978-3-319-09943-9_7

Introduction

The skin is the human organ with the largest surface area, and it provides the body's first line of defense against dehydration, injury, infection, and physical and environmental challenges. Among the numerous cells present in the skin epidermis, keratinocytes are the predominant cell type and are also the major contributor to the protective functions of this organ. Keratinocytes undergo an elaborate terminal differentiation process to form a constantly renewing mechanical and chemical barrier between the body and its immediate environment (Tobin 2006; Segre 2006; Madison 2003). Therefore, keratinocyte differentiation is an essential process in the production of the epidermal barrier. Four cell layers can be distinguished morphologically in the healthy epidermis: the stratum basale, stratum spinosum, stratum granulosum, and stratum corneum. The stratum basale is the proliferative cell compartment, and it keeps the epidermis firmly anchored to the basal lamina. The differentiation and keratinization processes begin in the stratum spinosum through the production of fibrillar proteins that aggregate into tonofibrils and are involved in desmosome formation, which generate strong connections between adjacent keratinocytes. Cells of the stratum granulosum are characterized by keratohyalin granules, which contain proteins that promote the hydration and cross-linking of keratin. Cells close to the stratum corneum secrete lamellar bodies into the extracellular space; lamellar bodies are composed of proteins and lipids and participate in the formation of the hydrophobic lipid envelope, which is important for barrier properties. These cells lose their organelles and become dead corneocytes of the stratum corneum (Lippens et al. 2005; Radoja et al. 2006). The formation of the epidermal barrier relies on controlled transcriptional changes that lead the cells through a gradual terminal differentiation process. In this chapter, we discuss the existing knowledge concerning the differentiation-dependent regulation of gene expression in keratinocytes by nuclear hormone receptors.

Keratinocyte Differentiation

As nuclear receptors affect all steps of keratinocyte differentiation, these steps are summarized below to provide a better understanding of the rest of the chapter. In healthy epidermis, only the basal layer contains mitotic cells. The keratinocytes of the basal layer are tightly packed and form a cell palisade attached to the basal lamina. The cytoskeleton in basal layer keratinocytes contains keratins K5 and K14 and small amounts of K15, which anchor the epidermis firmly to its substratum (Porter and Lane 2003). These keratins attach to hemidesmosomes through the laminin V receptor $\alpha6\beta4$ integrin. Additional integrins, $\alpha2\beta1$, $\alpha3\beta1$, and $\alpha5\beta1$, also bind to the matrix proteins of the basal lamina and provide the basal keratinocytes with essential information about their physical location (Adams and Watt 1989, 1990). The detachment of integrins from their extracellular matrix ligands is involved in initiating differentiation (Watt et al. 1993; Adams and Watt 1993). The junction proteins in the basal keratinocytes include hemidesmosomal proteins, as well as desmogleins 2 and 3, desmocollin 3, E- and P-cadherins, and α1-connexin (Tsuruta et al. 2002; Godsel et al. 2004; Dusek et al. 2006; Livshits et al. 2012; Tinkle et al. 2008; Perez-Moreno and Fuchs 2006; Jamora and Fuchs 2002). Furthermore, the cell surface of the basal keratinocytes is populated with many receptors, thus providing responsiveness to signals from the surrounding cells, including dermal fibroblasts. Conversely, basal keratinocytes can produce a large variety of secreted peptides, growth factors, cytokines, and chemokines (Gröne 2002; auf demKeller et al. 2004; Piepkorn et al. 1998). Therefore, the keratinocytes in the basal layer do not passively respond to signals from other cell types, but can also initiate and modulate signals in the surrounding tissue (Radoja et al. 2006). In this communication network, the epithelial–mesenchymal interaction is recognized as having a significant impact on epidermal differentiation (Chong et al. 2009; Smola et al. 1993; Maas-Szabowski et al. 1999). The keratinocytes in the suprabasal, spinosum, and granulosum layers do not divide under normal conditions; however, they can be induced to do so under specific conditions, such as wound healing (Garlick and Taichman 1994).

The transit amplifying cells from the basal layer detach from the basement membrane after several rounds of cell division and migrate upward toward the surface of the skin, giving rise to the stratum spinosum (Blanpain and Fuchs 2006; Cangkrama et al. 2013). As the keratinocytes leave the basal compartment, a significant rearrangement of the intermediate filament networks occurs to increase the tensile strength of the cells. These cells synthesize large amounts of K1 and K10, accompanied by the downregulation of K5 and K14. Thus, K1 and K10 are biomarkers of early differentiation (Porter and Lane 2003). The intermediate filaments anchor themselves to the desmosomes to enable the distribution of tensile strain throughout the entire skin tissue. Although desmosomes are present in cells from the basal to the granular layer, specific desmosomal proteins are expressed in a differentiation-dependent manner, similar to intermediate filaments. For example, the desmosomal cadherins have gradients of expression in the layers of the skin: the desmoglein 1 and desmocollin 1 levels are highest in the superficial layers, while the gradients of desmoglein 3 and desmocollin 3 have the opposite configuration (i.e., they are expressed at the highest levels in the basal layer; Chidgey et al. 1997; Chidgey 1997). As epidermal differentiation progresses to the advanced stage, the keratinocytes express filaggrin and trichohyalin, which interact with the keratin filaments. These two proteins initially accumulate in cytoplasmic keratohyalin granules, which are prominent morphological hallmarks in cells of the stratum granulosum (Manabe and O'Guin 1992). As keratinocytes enter the final phase of differentiation, they undergo metabolic and morphological alterations that culminate in the accumulation of structural proteins, precursor forms of enzymes, and large amount of lipids in their storage organelles (keratohyalin granules and lamellar bodies). The accumulation of these components is essential for the creation of the outermost stratum corneum (Kessner et al. 2008; Proksch et al. 2008; Kypriotou et al. 2012). The immediate availability of these components enables keratinocyte cornification and the formation of extracellular lipid-enriched "lamellar membranes" to take place rapidly in the upper granular and stratum corneum layers (Feingold and Jiang 2011).

During keratinocyte differentiation, an increase in the cellular content of highly saturated very-long-chain fatty acids, cholesterol sulfate, unesterified cholesterol, and sphingolipids takes place in parallel with a decrease of glycerophospholipids, triglycerides, and polyunsaturated very-long-chain fatty acids (Schurer and Elias 1991; Schürer et al. 1991). Sphingolipids, predominantly ceramides, contribute to ~50% of the stratum corneum lipids, which are the building blocks of the epidermal barrier (Schurer and Elias 1991; Elias and Menon 1991). A lipid-enriched extracellular milieu in the stratum corneum is formed by the exocytosis of lamellar bodies from the upper stratum granulosum cells. In conjunction with the deposition of the lipid matrix, extensive cross-linking of loricin, involucrin, and other structural proteins by transglutaminases leads to the formation of the cornified envelope (Lorand and Graham 2003; Nemes et al. 1999; Nemes and Steinert 1999). This structure is essential for the mechanical strength and waterproofing of the skin (Fig. 7.1).

Transcriptional Regulation of Keratinocyte Differentiation: Nuclear Hormone Receptors

The regulation of keratinocyte differentiation has been widely studied, and transcriptional regulation is known to play an important role in epidermal differentiation. Although numerous transcription factors, including activator protein 1 (AP-1), p63, Notch, and NF-κB (Eckert and Welter 1996; Angel et al. 2001; Koster and Roop 2004; Lefort and Dotto 2004; McKenzie and Sabin 2003), among others, have been implicated in the terminal differentiation process; however, the identification of the roles of nuclear hormone receptors has opened a new level of understanding with respect to keratinocyte differentiation in association with lipid metabolism (Feingold 2007). Nuclear hormone receptors, one of the largest known classes of transcription factors (coded by 48 genes in humans), have been implicated in skin epidermal differentiation. Because of their direct effect on transcription and hormone responsiveness and the availability of synthetic ligands, nuclear hormone receptors undoubtedly comprise one of the most exciting families of regulatory proteins from the clinical perspective. In fact, drugs that target nuclear hormone receptors constitute one of the largest and most potent groups of pharmaceuticals currently in use (Berkenstam and Gustafsson 2005). Thus, nuclear hormone receptors have attracted increasing attention as potential drug targets owing to their ability to stimulate different aspects of epidermal differentiation and lipid homeostasis. The mechanistic actions of nuclear hormone receptors are well reviewed elsewhere and will not be discussed here (Robyr et al. 2000; Nagy and Schwabe 2004; Aranda and Pascual 2001). Although the expression of several nuclear hormone receptors in the epidermis has been described (Table 7.1), only a small number of these receptors have been recognized as important regulators of epidermal differentiation and lipid homeostasis. Here, we discuss the role of the VDR, RARs, RXRs, and PPARs.

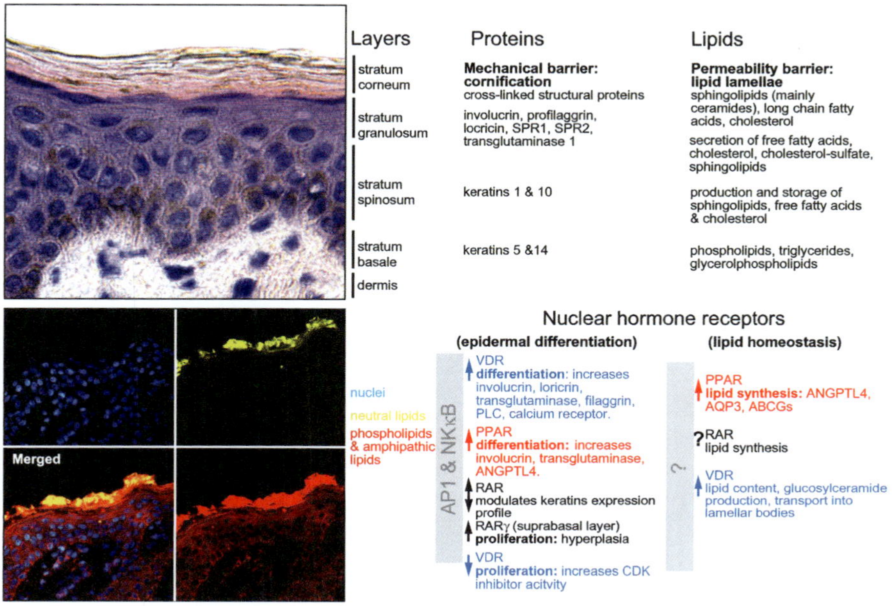

Fig. 7.1 Keratinocyte differentiation involves changes in the protein and lipid composition of different epidermal layers. Epidermal proliferation and differentiation are essential for the formation of the cornified layer and permeability barrier in the stratum corneum of the skin. During differentiation, epidermal lipids are synthesized by keratinocytes and stored in epidermal lamellar bodies containing cholesterol, phospholipids, ceramides, and hydrolytic enzymes. Hydrolytic enzymes are delivered to stratum corneum interstices, where they convert secreted ceramides and phospholipids, including sphingomyelin, into ceramides and free fatty acids. Although the effect of nuclear hormone receptors, such as peroxisome proliferator-activated receptors (PPAR) and vitamin D receptor (VDR), on skin lipid synthesis has been observed, the mechanistic details remain unclear. Keratinocytes synthesize specific cytoskeletal proteins: basal (K5 and K14) and suprabasal (K1 and K10) keratins, as well as cornified envelope-associated proteins, particularly involucrin and loricrin. Transglutaminase 1 enzyme is responsible for convalently interconnecting structural proteins, thus conferring mechanical strength to the skin. Nuclear hormone receptors have been shown to modulate keratinocyte differentiation either directly through their cognate responsive elements or indirectly through functional interactions with common transcription factors, such AP-1 and NFκB

Vitamin D Receptor (VDR)

In addition to providing a protective shield against external insults and dehydration, the epidermis is a major source of vitamin D for the body. Although much of the provitamin D in our bodies is activated by hydroxylation in the liver and kidney, the epidermis is also capable of producing active vitamin D3, which may play a role locally. Under the influence of ultraviolet exposure, the photolysis of 7-dehydrocholesterol results in the endogenous production of vitamin D. In addition, keratinocytes also possess the necessary enzymes (e.g., CYP27A1, CYP27B1,

Table 7.1 Epidermal expression of ligand-activated nuclear hormone receptors

Nuclear hormone receptors	Nomenclature	References
Glucocorticoid receptor	NR3C1	(Donet et al. 2008)
Estrogen receptor β	NR3A	(Pelletier and Ren 2004)
Retinoic acid receptor (α, γ)	NR1B1, NR1B3	(Elder et al. 1991)
Retinoid X receptor (α, β)	NR2B1, NR2B2	(Thacher et al. 2000)
Vitamin D receptor	NR1I1	(Bikle et al. 2004)
Peroxisome proliferator-activated receptor (α, β, γ)	NR1C1, NR1C2, NR1C3	(Michalik and Wahli 2007)
Liver X receptor (α, β)	NR1H3, NR1H2	(Kömüves et al. 2002)
Thyroid receptor	NR1A1, NR1A2	(Ohtsuki et al. 1992)
Farnesoid X receptor	NR1H4	(Higashiyama et al. 2008)

25-hydroxylase, and 1-hydroxylase) to further metabolize vitamin D to its active form, 1,25-dihydroxyvitamin D (1,25$(OH)_2$D) (Bikle 2011, 2012).

The most striking feature of VDR-knockout mice is the development of alopecia, which is also found in many patients with mutations in the VDR, a condition referred to as hereditary vitamin D resistance (Malloy et al. 1999; Li et al. 1997; Bikle et al. 2006). These mutant mice also exhibit a defect in epidermal differentiation, as exhibited by reduced levels of involucrin and loricrin, as well as a loss of keratohyalin granules (Xie et al. 2002). They also exhibit a reduction in the lipid content of the lamellar bodies concomitant with a reduction in glucosylceramide production and transport to the lamellar bodies, which leads to a defective permeability barrier. The activation of VDR by 1,25$(OH)_2$D exerts an antiproliferative effect by increasing cyclin-dependent kinase inhibitor activity, reducing proliferation in the basal layer of the epidermis, and promoting the sequential differentiation of keratinocytes to form the upper layers of the epidermis. 1,25$(OH)_2$D increases the expression of involucrin, transglutaminase, loricrin, filaggrin, phospholipase C (PLC), and the calcium receptor at sub-nanomolar concentrations (Smith et al. 1986; Hosomi et al. 1983; Bikle and Pillai 1993; Pillai and Bikle 1991). The pro-differentiation effect of the VDR is thought to be partly due to its involvement in the regulation of intracellular calcium concentrations in addition to its effect on the promoter activity of numerous keratinocyte differentiation genes (Bikle et al. 2001, 2002). Increased expression of the calcium receptor renders the keratinocytes more sensitive to the prodifferentiating action of calcium. All of the PLC family members are induced by 1,25$(OH)_2$D and calcium, and blocking PLC-γ1 expression prevents both 1,25$(OH)_2$D-stimulated and calcium-stimulated differentiation. Calcium and 1,25$(OH)_2$D also functionally interact to induce involucrin and transgluataminase (Bikle et al. 2004; Bikle 2004). A possible mechanism for this synergy regarding the induction of the involucrin gene is that the calcium response element and vitamin D response element in the involucrin promoter are in close proximity. A recurring mechanistic action for VDR and other nuclear hormone receptors is the pivotal involvement of transcription factor AP-1. Mutations in the AP-1 site of the calcium response element block both calcium and 1,25$(OH)_2$D induction of the involucrin genes; however, mutations of the vitamin D response element only block its response to 1,25$(OH)_2$D. The specificity of VDR action within the skin is attributed, at least in part, to the involvement of different coregulators (Oda et al. 2004, 2009). For the proliferation of keratinocytes in the basal layer of the epidermis, the mediator complex Vitamin D receptor interacting protein (DRIP) is the dominant coregulator. In the more differentiated keratinocytes of the epidermis, the steroid receptor coactivator (SRC) complexes, which share similar nuclear receptor binding determinants with DRIP but constitute functionally distinct complexes, dominate the VDR functions. At a high-nanomolar range of $1,25(OH)_2D$, the antiproliferative effect of VDR is accompanied by reduced c-myc and cyclin D1 expression and increases in the cell cycle inhibitors p15, p21, and p27. In addition, agonist-activated VDR regulates the processing of long-chain glycosylceramides, which are critical for the formation of the permeability barrier (Bikle 2011). Thus, owing to its clear prodifferentiation effects, $1,25(OH)_2D$ is used in the treatment of psoriasis (Samarasekera et al. 2013). Prescription cream or solution containing calcipotriene (Dovonex, Sorilux) is used to treat mild to moderate psoriasis, but this treatment cause irritation on sensitive area of the skin. Calcitriol ($1,25(OH)_2D$), marketed under various trade names including Rocaltrol (Roche), Calcijex (Abbott), Decostriol (Mibe, Jesalis), and Vectical (Galderma), is equally effective and possibly less irritating than calcipotriene.

Retinoic Acid Receptors (RARs) and Retinoid X Receptors (RXRs)

The dermatological use of retinoids (vitamins A derivatives) preceded the description of their mechanism of action and the discovery of their receptors (Rees 1975). There are two major groups of retinoid-responsive nuclear hormone receptors: the RARs and the RXRs. Each of these groups of retinoid receptors has three subtypes (α, β, γ). RARs and RXRs are activated by 9-cis-retinoic acid, while all-trans-retinoic acid only activates RARs (Fisher et al. 1995; Fisher and Voorhees 1996). The functional retinoid receptor is a heterodimer of one RAR and one RXR molecule. RXR is also the obligate dimerization partner of other nuclear hormone receptors, such as PPAR, VDR, and LXR; thus, the effects of retinoids are expected to be wide-ranging. In the epidermis, retinoid signaling is carried out by RARα, RARγ, RXRα, and RXRβ. The protein expression levels of RXRs far exceed those of RARs, suggesting that RXRs can also partner with other nuclear hormone receptors present in the same cell. RARγ and RXRα are the most abundantly expressed retinoid-responsive nuclear hormone receptors in both human (Fisher et al. 1994) and mouse epidermis (Darwiche et al. 1995).

When topically applied to adult skin, retinoic acid induces epidermal hyperplasia that results from hyperproliferation of basal keratinocytes, which leads to thickening of the differentiated spinous and granular layers (Fisher and Voorhees 1996). Retinoic acid treatment also decreases the integrity of the stratum corneum, impairing the adult skin permeability barrier and increasing trans-epidermal water loss (Elias et al. 1981). Despite the well-known effect of retinoic acid in induced epidermal hyperplasia, the skin of RARα- and RARβ-null mice appeared normal, while the skin of RARγ-null mice displayed minor defects of granular keratinocyte differentiation (Chapellier et al. 2002; Lohnes et al. 1993; Lufkin et al. 1993;

Ghyselinck et al. 1997). The skin of the mouse devoid of all three RARs exhibited an epidermal phenotype similar to that of the RARγ-null mouse, suggesting little functional redundancy between RARα and RARγ for the proliferation and differentiation of adult keratinocytes. Importantly, it also indicated that RARs in basal keratinocytes do not perturb homeostatic epidermal self-renewal, suggesting that RAR-dependent signaling is dispensable for homeostasis. As expected, retinoic acid treatment of control mice led to marked epidermal thickening associated with increased proliferation. However, this hyperproliferation was not observed in retinoic acid-treated skin that was devoid of RARγ specifically only in the suprabasal layer, or in RARγ-null mice. Thus, RARγ is required in suprabasal keratinocytes for RA-induced epidermal hyperplasia, partly through increased paracrine signaling involving heparin-binding epidermal growth factor (EGF)-like growth factor (HB-EGF) (Li et al. 2000; Chapellier et al. 2002). Although RXRα and RXRβ were detected in epidermal keratinocytes, genetic evidence indicated that RXRα has a clearly dominant role. Indeed, the skin of adult RXRβ-null mice appeared normal. Interestingly, the mice with an epidermal-specific deletion of RXRα developed progressive alopecia with typical features of degenerated hair follicles in addition to utriculi and dermal cysts, which can all be attributed to defects in hair cycles (Li et al. 2000). With respect to hair follicle development, the epidermal-specific RXRα knockout mice phenocopied the VDR-null mice (Li et al. 2000, 2001). In contrast to VDR-null mice, RXRα mutant mice also exhibited interfollicular keratinocyte hyperproliferation, as well as abnormal terminal differentiation and increased dermal cellularity associated with a skin inflammatory reaction (Li et al. 1997, 2000; Yoshizawa et al. 1997). These features most likely reflect the involvement of other heterodimeric partners of RXRα, such as the RARs, thyroid hormone receptors, and the PPARs.

The in vitro and in vivo effects of retinoic acid on keratinocytes differ (Törmä 2011). For example, retinoic acid treatment induced keratinocyte growth arrest in in vitro monolayer keratinocyte culture, while it stimulated keratinocyte proliferation in reconstructed three-dimensional (3D) human skin, similar to when topically applied to the skin in vivo. Furthermore, retinoids also induced different keratin expression profiles in cultured monolayer keratinocytes compared with 3D skin equivalents (Törmä 2011). Keratin gene expression is affected by retinoids in many complex ways, and it is obvious that classical activation of retinoic acid response elements (RAREs) by RAR/RXR heterodimers is not always involved. For example, a number of keratin genes (e.g., genes encoding keratins 5, 6, 14, and 17) reportedly carry positive or negative RAREs, while no RAREs have been identified in keratin 2 or keratin 4 genes (Kerns et al. 2010; Ma et al. 1997). Thus, it is likely that these latter two retinoid-regulated keratins are among the many genes that are indirectly influenced by retinoids through mechanisms that do not require RAREs (Lu et al. 1994; Balmer and Blomhoff 2002; Radoja et al. 1997). Among the indirect mechanisms through which retinoic acid and its receptors regulate the differentiation and proliferation of epidermal keratinocytes is the antagonistic activity of AP-1 and NF-κB (Törmä 2011).

Peroxisome Proliferator-Activated Receptors (PPARs)

PPARs are among the most recently identified nuclear hormone receptors. They have attracted attention due to their crucial role in lipid homeostasis and their potential to regulate cell differentiation (Desvergne and Wahli 1999; Icre et al. 2006; Michalik and Wahli 2007). PPARs form permissive heterodimers with RXR, thus, allowing both the ligands of PPAR and RXR to regulate the transcription of PPAR target genes (Tan et al. 2005). None of the three PPARs (α, γ, and β/δ) are detectable in the adult mouse interfollicular epidermis (Michalik et al. 2001). The expression of PPARα and PPARβ/δ were transiently upregulated at wound edges after injury. Epidermal PPARβ/δ expression is also stimulated by phorbol esters, hair plucking, and epidermal inflammation (Tan et al. 2001, 2003, 2004b; Michalik et al. 2001). Although all three PPARs are detectable in the human epidermis, PPARβ/δ is the most abundant subtype, followed by PPARα and PPARγ.

Macroscopic examination of the skin of adult PPARα-knockout mice showed normal skin architecture (Lee et al. 1995). However, a detailed microscopic analysis of these mutant mice revealed a thin stratum granulosum with focal parakeratosis, suggesting impaired keratinocyte differentiation. Fetal epidermal development in PPARα-knockout mice was delayed, with defects in the formation of the stratum corneum. The overexpression of PPARα in the epidermis of transgenic mice resulted in impaired embryonic epidermal development associated with a thinner epidermis and fewer hair follicles than normal mice; however, no abnormality was detected in the adult epidermis (Yang et al. 2006). These observations suggest that PPARα may be important for embryonic epidermal development, but dispensable for adult epidermal homeostasis. Consistent with the very low level of PPARγ expression in the epidermis, no defect in epidermal maturation was observed in PPARγ-heterozygous or PPARγ-knockout mice derived from placental rescue, demonstrating that PPARγ is not required for epidermal differentiation (Rosen et al. 1999; Barak et al. 1999). Histological analysis of PPARβ/δ-knockout mice did not reveal any defect in skin epidermal architecture during fetal development or adulthood (Michalik et al. 2001; Peters et al. 2000). However, PPARβ/δ-knockout mice exhibited retardation of postnatal hair follicle morphogenesis, particularly at the hair peg stage (Di-Poï et al. 2005). This finding was attributed to the PPARβ/δ-mediated temporal activation of the antiapoptotic Akt1 pathway in vivo, which protects keratinocytes in hair pegs from apoptosis and is required for normal hair follicle development (Di-Poï et al. 2005).

PPARα, PPARβ/δ, and PPARγ are ligand-activated nuclear hormone receptors, and their agonists have been shown to lead to the stimulation of the expression of numerous genes necessary for the formation of the cornified envelope and lamellar membrane barrier. Such genes include involucrin, filaggrin, loricrin, transglutaminase 1, aquaporin 3 (AQP3), angiopoetin-like 4 protein (ANGPTL4), and the ATP-binding cassette subfamily G members 1 and 12 (ABCG1, ABCG12) (Pal et al. 2011; Hanley et al. 1997; Mao-Qiang et al. 2004; Jiang et al. 2008, 2010, 2011). AQP3, a member of the aquaglyceroporin family, which transports water and glycerol, is robustly expressed in the epidermis and plays important roles in stratum corneum hydration, permeability barrier function, and wound healing (Jiang et al.

2008). ABCG1 is expressed in cultured human keratinocytes and murine epidermis and is induced during keratinocyte differentiation, with increased levels observed in the outer epidermis (Jiang et al. 2010). ABCG1-null mice display abnormal lamellar body contents and secretion leading to impaired lamellar bilayer formation, indicating a potential role for ABCG1 in normal lamellar body formation and secretion (Jiang et al. 2010). Using 3D skin substitutes, ligand-activated PPARβ/δ indirectly stimulated keratinocyte differentiation, as is required for de novo gene transcription and protein translation (Pal et al. 2011). PPARβ/δ stimulates the expression of ANGPTL4; deficiency of ANGPTL4 in human cultured keratinocytes and the skin of mice results in diminished expression of various protein kinase C isotypes and phosphorylated AP-1, whose roles in keratinocyte differentiation are well established (Pal et al. 2011).

Interestingly, many effects of PPAR agonists on epidermal differentiation and lipid synthesis have also been observed with respect to agonists of another nuclear hormone receptor, LXR. Such findings suggest that both PPAR and LXR nuclear hormone receptors likely mediate the expression of differentiation-associated genes through a common pathway (Feingold and Jiang 2011; Demerjian et al. 2009; Man et al. 2006; Feingold 2007; Schmuth et al. 2008). The promoter regions of many genes whose expression levels increased during differentiation, including involucrin, loricin, and transglutaminase 1, have AP-1 binding sites. ANGPTL4, encoded by a PPAR target gene, mediates the activation and binding of JUNB and c-JUN to the promoter region of human involucrin and transglutaminase type 1 genes, respectively (Pal et al. 2011). Various studies have shown that the deletion or mutation of such AP-1 sites abolishes the stimulation of those genes by PPARs and LXR activators (Kömüves et al. 2000; Hanley et al. 2000).

Wound healing occurs as a high-priority survival response to skin injuries. The expression of PPARβ/δ is upregulated in adult epidermis by inflammatory stimuli during skin injury, which also provokes keratinocyte activation (Michalik et al. 2001; Tan et al. 2001, 2007). The activated PPARβ/δ has an antiapoptotic effect on the keratinocytes. Hence, it is protecting them from cytokine-induced apoptosis during the inflammatory phase of wound repair, which maintains a sufficient number of viable migratory keratinocytes for the reepithelialization phase of the healing process (Di-Poi et al. 2002). PPARβ/δ in wound fibroblasts plays a regulatory role in controlling keratinocytes proliferation (Chong et al. 2009). Thus, our studies and that of others have provided compelling evidence for a role of PPARβ/δ as a valuable pharmacologic wound healing target, impacting numerous events essential for wound healing (Tan et al. 2003, 2004a, b; Michalik and Wahli 2006, 2007).

Additional Remarks

Many of the genes expressed during keratinocyte differentiation must be expressed in a coordinated manner. A variety of transcription factors (AP-1, AP-2, POU domain, Sp1, and STAT factors) are expressed in the epidermis, and multiple members of the same family are present in many cases (e.g., AP-1 and POU domain factors). AP-1 plays a pivotal role in nuclear hormone receptor-mediated keratinocyte differentiation, cornification, and lamellar body formation. It is a heterodimeric protein composed of proteins belonging to the c-Fos, c-JUN, activating transcription factor (ATF), and Jun dimerization protein (JDP) families. The simultaneous expression of multiple members of a single transcription factor family provides numerous opportunities for complex regulation. Several studies suggest that specific members of these families interact with specific keratinocyte genes. All of the known AP-1 factors are expressed in the epidermis, each in a specific spatial pattern that suggests the potential to regulate multiple genes. It will be important to determine the role of each of these members in nuclear hormone receptor-mediated keratinocyte gene expression and maturation. Eventually, using the diverse tools that are currently available, we can expect the elucidation of all of the steps between the interaction of the ligands with their receptors and the activation of target gene expression.

Summary

The ultrastructure of the epidermis and the well-orchestrated expression of its structural elements are of paramount importance in the distribution of the mechanical forces throughout large areas of the skin and greatly contribute to its tensile integrity. A complex and gradual change of gene expression that allows the compositional modification of this mechanical scaffold while maintaining its integrity is required for the continuous renewal of the epidermis, particularly its waterproof barrier. Although many aspects of the structural organization of the skin and keratinocyte differentiation have been studied, our mechanistic understanding of these processes as regulated by nuclear hormone receptors in a temporal and spatial fashion remains incomplete. Agonist and antagonist drugs that target the nuclear hormone receptors constitute one of the largest and most potent groups of pharmaceuticals currently in use, and therefore hold great potential for applications to improve skin barrier function. Hence, there is a need for further exploration of the mechanisms by which nuclear hormone receptor signaling contributes to epidermal maturation.

Acknowledgements Work in the authors' laboratories was supported by grants from the National Medical Research Council, Singapore (NMRC/1280/2010) and the Ministry of Education, Singapore (MOE2010-T2-2) to N.S.T., the Swiss National Science Foundation (including personal grants to W.W., support from the National Research Center Frontiers in Genetics, and support from the Sinergia program). We also acknowledge support from the 7th Framework Programme of the European Commission (TORNADO), the Bonizzi Theler Foundation, the Roche Foundation, and the Etat de Vaud.

References

Adams JC, Watt FM. Fibronectin inhibits the terminal differentiation of human keratinocytes. Nature. 1989;340:307–9.

Adams JC, Watt FM. Changes in keratinocyte adhesion during terminal differentiation: reduction in fibronectin binding precedes alpha 5 beta 1 integrin loss from the cell surface. Cell. 1990;63:425–35.

Adams JC, Watt FM. Regulation of development and differentiation by the extracellular matrix. Development. 1993;117:1183–98.

Angel P, Szabowski A, Schorpp-Kistner M, et al. Function and regulation of AP-1 subunits in skin physiology and pathology. Oncogene. 2001;20:2413–23.

Aranda A, Pascual A. Nuclear hormone receptors and gene expression. Physiol Rev. 2001;81:1269–304.

auf demKeller U, Krampert M, Kümin A, et al. Keratinocyte growth factor: effects on keratinocytes and mechanisms of action. Eur J Cell Biol. 2004;83:607–12.

Balmer JE, Blomhoff R. Gene expression regulation by retinoic acid. J Lipid Res. 2002;43:1773–808.

Barak Y, Nelson MC, Ong ES, et al. PPAR gamma is required for placental, cardiac, and adipose tissue development. Mol Cell. 1999;4:585–95.

Berkenstam A, Gustafsson JA. Nuclear receptors and their relevance to diseases related to lipid metabolism. Curr Opin Pharmacol. 2005;5:171–6.

Bikle DD. Vitamin D regulated keratinocyte differentiation. J Cell Biochem. 2004;92:436–44.

Bikle DD. Vitamin D metabolism and function in the skin. Mol Cell Endocrinol. 2011;347:80–9.

Bikle DD. Vitamin D and the skin: physiology and pathophysiology. Rev Endocr Metab Disord. 2012;13:3–19.

Bikle DD, Pillai S. Vitamin D, calcium, and epidermal differentiation. Endocr Rev. 1993;14:3–19.

Bikle DD, Ng D, Tu CL, et al. Calcium- and vitamin D-regulated keratinocyte differentiation. Mol Cell Endocrinol. 2001;177:161–71.

Bikle DD, Ng D, Oda Y, et al. The vitamin D response element of the involucrin gene mediates its regulation by 1,25-dihydroxyvitamin D3. J Invest Dermatol. 2002;119:1109–13.

Bikle DD, Oda Y, Xie Z, et al. Calcium and 1,25(OH)2D: interacting drivers of epidermal differentiation. J Steroid Biochem Mol Biol. 2004;89-90:355–60.

Bikle DD, Elalieh H, Chang S, et al. Development and progression of alopecia in the vitamin D receptor null mouse. J Cell Physiol. 2006;207:340–53.

Blanpain C, Fuchs E. Epidermal stem cells of the skin. Annu Rev Cell Dev Biol. 2006;22:339–73.

Cangkrama M, Ting SB, Darido C, et al. Stem cells behind the barrier. Int J Mol Sci. 2013;14:13670–86.

Chapellier B, Mark M, Messaddeq N, et al. Physiological and retinoid-induced proliferations of epidermis basal keratinocytes are differently controlled. EMBO J. 2002;21:3402–13.

Chidgey MA. Desmosomes and disease. Histol Histopathol. 1997;12:1159–68.

Chidgey MA, Yue KK, Gould S, et al. Changing pattern of desmocollin 3 expression accompanies epidermal organisation during skin development. Dev Dyn. 1997;210:315–27.

Chong HC, Tan MJ, Philippe V, et al. Regulation of epithelial-mesenchymal IL-1 signaling by PPARbeta/delta is essential for skin homeostasis and wound healing. J Cell Biol. 2009;184:817–31.

Darwiche N, Celli G, Tennenbaum T, et al. Mouse skin tumor progression results in differential expression of retinoic acid and retinoid X receptors. Cancer Res. 1995;55:2774–82.

Demerjian M, Choi EH, Man MQ, et al. Activators of PPARs and LXR decrease the adverse effects of exogenous glucocorticoids on the epidermis. Exp Dermatol. 2009;18:643–9.

Desvergne B, Wahli W. Peroxisome proliferator-activated receptors: nuclear control of metabolism. Endocr Rev. 1999;20:649–88.

Di-Poi N, Tan NS, Michalik L, et al. Antiapoptotic role of PPARbeta in keratinocytes via transcriptional control of the Akt1 signaling pathway. Mol Cell. 2002;10:721–33.

Di-Poï N, Ng CY, Tan NS, et al. Epithelium-mesenchyme interactions control the activity of peroxisome proliferator-activated receptor beta/delta during hair follicle development. Mol Cell Biol. 2005;25:1696–712.
Donet E, Bayo P, Calvo E, et al. Identification of novel glucocorticoid receptor-regulated genes involved in epidermal homeostasis and hair follicle differentiation. J Steroid Biochem Mol Biol. 2008;108:8–16.
Dusek RL, Getsios S, Chen F, et al. The differentiation-dependent desmosomal cadherin desmoglein 1 is a novel caspase-3 target that regulates apoptosis in keratinocytes. J Biol Chem. 2006;281:3614–24.
Eckert RL, Welter JF. Transcription factor regulation of epidermal keratinocyte gene expression. Mol Biol Rep. 1996;23:59–70.
Elder JT, Fisher GJ, Zhang QY, et al. Retinoic acid receptor gene expression in human skin. J Invest Dermatol. 1991;96:425–33.
Elias PM, Menon GK. Structural and lipid biochemical correlates of the epidermal permeability barrier. Adv Lipid Res. 1991;24:1–26.
Elias PM, Fritsch PO, Lampe M, et al. Retinoid effects on epidermal structure, differentiation, and permeability. Lab Invest. 1981;44:531–40.
Feingold KR. Thematic review series: skin lipids. The role of epidermal lipids in cutaneous permeability barrier homeostasis. J Lipid Res. 2007;48:2531–46.
Feingold KR, Jiang YJ. The mechanisms by which lipids coordinately regulate the formation of the protein and lipid domains of the stratum corneum: role of fatty acids, oxysterols, cholesterol sulfate and ceramides as signaling molecules. Dermatoendocrinol. 2011;3:113–8.
Fisher GJ, Voorhees JJ. Molecular mechanisms of retinoid actions in skin. FASEB J. 1996;10:1002–13.
Fisher GJ, Talwar HS, Xiao JH, et al. Immunological identification and functional quantitation of retinoic acid and retinoid X receptor proteins in human skin. J Biol Chem. 1994;269:20629–35.
Fisher C, Blumenberg M, Tomić-Canić M, et al. Retinoid receptors and keratinocytes. Crit Rev Oral Biol Med. 1995;6:284–301.
Garlick JA, Taichman LB. Fate of human keratinocytes during reepithelialization in an organotypic culture model. Lab Invest. 1994;70:916–24.
Ghyselinck NB, Dupé V, Dierich A, et al. Role of the retinoic acid receptor beta (RARbeta) during mouse development. Int J Dev Biol. 1997;41:425–47.
Godsel LM, Getsios S, Huen AC, et al. The molecular composition and function of desmosomes. Handb Exp Pharmacol. 2004;165:137–93.
Gröne A. Keratinocytes and cytokines. Vet Immunol Immunopathol. 2002;88:1–12.
Hanley K, Jiang Y, Crumrine D, et al. Activators of the nuclear hormone receptors PPARalpha and FXR accelerate the development of the fetal epidermal permeability barrier. J Clin Invest. 1997;100:705–12.
Hanley K, Kömüves LG, Ng DC, et al. Farnesol stimulates differentiation in epidermal keratinocytes via PPARalpha. J Biol Chem. 2000;275:11484–91.
Higashiyama H, Kinoshita M, Asano S, et al. Immunolocalization of farnesoid X receptor (FXR) in mouse tissues using tissue microarray. Acta Histochem. 2008;110:86–93.
Hosomi J, Hosoi J, Abe E, et al. Regulation of terminal differentiation of cultured mouse epidermal cells by 1 alpha,25-dihydroxyvitamin D3. Endocrinol. 1983;113:1950–57.
Icre G, Wahli W, Michalik L, et al. Functions of the peroxisome proliferator-activated receptor (PPAR) alpha and beta in skin homeostasis, epithelial repair, and morphogenesis. J Investig Dermatol Symp Proc. 2006;11:30–5.
Jamora C, Fuchs E. Intercellular adhesion, signalling and the cytoskeleton. Nat Cell Biol. 2002;4:E101–8.
Jiang YJ, Lu B, Kim P, et al. PPAR and LXR activators regulate ABCA12 expression in human keratinocytes. J Invest Dermatol. 2008;128:104–9.
Jiang YJ, Lu B, Tarling EJ, et al. Regulation of ABCG1 expression in human keratinocytes and murine epidermis. J Lipid Res. 2010;51:3185–95.

Jiang YJ, Kim P, Lu YF, et al. PPARgamma activators stimulate aquaporin 3 expression in keratinocytes/epidermis. Exp Dermatol. 2011;20:595–9.

Kerns M, DePianto D, Yamamoto M, et al. Differential modulation of keratin expression by sulforaphane occurs via Nrf2-dependent and -independent pathways in skin epithelia. Mol Biol Cell. 2010;21:4068–75.

Kessner D, Ruettinger A, Kiselev MA, et al. Properties of ceramides and their impact on the stratum corneum structure. Part 2: stratum corneum lipid model systems. Skin Pharmacol Physiol. 2008;21:58–74.

Kömüves LG, Hanley K, Lefebvre AM, et al. Stimulation of PPARalpha promotes epidermal keratinocyte differentiation in vivo. J Invest Dermatol. 2000;115:353–60.

Kömüves LG, Schmuth M, Fowler AJ, et al. Oxysterol stimulation of epidermal differentiation is mediated by liver X receptor-beta in murine epidermis. J Invest Dermatol. 2002;118:25–34.

Koster MI, Roop DR. The role of p63 in development and differentiation of the epidermis. J Dermatol Sci. 2004;34:3–9.

Kypriotou M, Huber M, Hohl D, et al. The human epidermal differentiation complex: cornified envelope precursors, S100 proteins and the 'fused genes' family. Exp Dermatol. 2012;21:643–9.

Lee SS, Pineau T, Drago J, et al. Targeted disruption of the alpha isoform of the peroxisome proliferator-activated receptor gene in mice results in abolishment of the pleiotropic effects of peroxisome proliferators. Mol Cell Biol. 1995;15:3012–22.

Lefort K, Dotto GP. Notch signaling in the integrated control of keratinocyte growth/differentiation and tumor suppression. Semin Cancer Biol. 2004;14:374–86.

Li YC, Pirro AE, Amling M, et al. Targeted ablation of the vitamin D receptor: an animal model of vitamin D-dependent rickets type II with alopecia. Proc Natl Acad Sci U S A. 1997;94:9831–5.

Li M, Indra AK, Warot X, et al. Skin abnormalities generated by temporally controlled RXRalpha mutations in mouse epidermis. Nature. 2000;407:633–6.

Li M, Chiba H, Warot X, et al. RXR-alpha ablation in skin keratinocytes results in alopecia and epidermal alterations. Development. 2001;128:675–88.

Lippens S, Denecker G, Ovaere P, et al. Death penalty for keratinocytes: apoptosis versus cornification. Cell Death Differ. 2005;12(Suppl 2):1497–508.

Livshits G, Kobielak A, Fuchs E, et al. Governing epidermal homeostasis by coupling cell-cell adhesion to integrin and growth factor signaling, proliferation, and apoptosis. Proc Natl Acad Sci U S A. 2012;109:4886–91.

Lohnes D, Kastner P, Dierich A, et al. Function of retinoic acid receptor gamma in the mouse. Cell. 1993;73:643–58.

Lorand L, Graham RM. Transglutaminases: crosslinking enzymes with pleiotropic functions. Nat Rev Mol Cell Biol. 2003;4:140–56.

Lu B, Rothnagel JA, Longley MA, et al. Differentiation-specific expression of human keratin 1 is mediated by a composite AP-1/steroid hormone element. J Biol Chem. 1994;269:7443–49.

Lufkin T, Lohnes D, Mark M, et al. High postnatal lethality and testis degeneration in retinoic acid receptor alpha mutant mice. Proc Natl Acad Sci U S A. 1993;90:7225–9.

Ma S, Rao L, Freedberg IM, et al. Transcriptional control of K5, K6, K14, and K17 keratin genes by AP-1 and NF-kappaB family members. Gene Expr. 1997;6:361–70.

Maas-Szabowski N, Shimotoyodome A, Fusenig NE, et al. Keratinocyte growth regulation in fibroblast cocultures via a double paracrine mechanism. J Cell Sci. 1999;112(12):1843–53.

Madison KC. Barrier function of the skin: "la raison d'être" of the epidermis. J Invest Dermatol. 2003;121:231–41.

Malloy PJ, Pike JW, Feldman D, et al. The vitamin D receptor and the syndrome of hereditary 1,25-dihydroxyvitamin D-resistant rickets. Endocr Rev. 1999;20:156–88.

Man MQ, Choi EH, Schmuth M, et al. Basis for improved permeability barrier homeostasis induced by PPAR and LXR activators: liposensors stimulate lipid synthesis, lamellar body secretion, and post-secretory lipid processing. J Invest Dermatol. 2006;126:386–92.

Manabe M, O'Guin WM. Keratohyalin, trichohyalin and keratohyalin-trichohyalin hybrid granules: an overview. J Dermatol. 1992;19:749–55.

Mao-Qiang M, Fowler AJ, Schmuth M, et al. Peroxisome-proliferator-activated receptor (PPAR)-gamma activation stimulates keratinocyte differentiation. J Invest Dermatol. 2004;123:305–12.
McKenzie RC, Sabin E. Aberrant signalling and transcription factor activation as an explanation for the defective growth control and differentiation of keratinocytes in psoriasis: a hypothesis. Exp Dermatol. 2003;12:337–45.
Michalik L, Wahli W. Involvement of PPAR nuclear receptors in tissue injury and wound repair. J Clin Invest. 2006;116(3):598–606.
Michalik L, Wahli W. Peroxisome proliferator-activated receptors (PPARs) in skin health, repair and disease. Biochim Biophys Acta. 2007;1771:991–8.
Michalik L, Desvergne B, Tan NS, et al. Impaired skin wound healing in peroxisome proliferator-activated receptor (PPAR)alpha and PPARbeta mutant mice. J Cell Biol. 2001;154:799–814.
Nagy L, Schwabe JWR. Mechanism of the nuclear receptor molecular switch. Trends Biochem Sci. 2004;29:317–24.
Nemes Z, Steinert PM. Bricks and mortar of the epidermal barrier. Exp Mol Med. 1999;31:5–19.
Nemes Z, Marekov LN, Steinert PM, et al. Involucrin cross-linking by transglutaminase 1. Binding to membranes directs residue specificity. J Biol Chem. 1999;274:11013–21.
Oda Y, Sihlbom C, Chalkley RJ, et al. Two distinct coactivators, DRIP/mediator and SRC/p160, are differentially involved in VDR transactivation during keratinocyte differentiation. J Steroid Biochem Mol Biol. 2004;89–90:273–6.
Oda Y, Uchida Y, Moradian S, et al. Vitamin D receptor and coactivators SRC2 and 3 regulate epidermis-specific sphingolipid production and permeability barrier formation. J Invest Dermatol. 2009;129:1367–78.
Ohtsuki M, Tomic-Canic M, Freedberg IM, et al. Regulation of epidermal keratin expression by retinoic acid and thyroid hormone. J Dermatol. 1992;19:774–80.
Pal M, Tan MJ, Huang RL, et al. Angiopoietin-like 4 regulates epidermal differentiation. PLoS One. 2011;6:e25377.
Pelletier G, Ren L. Localization of sex steroid receptors in human skin. Histol Histopathol. 2004;19:629–36.
Perez-Moreno M, Fuchs E. Catenins: keeping cells from getting their signals crossed. Dev Cell. 2006;11:601–12.
Peters JM, Lee SS, Li W, et al. Growth, adipose, brain, and skin alterations resulting from targeted disruption of the mouse peroxisome proliferator-activated receptor beta(delta). Mol Cell Biol. 2000;20:5119–28.
Piepkorn M, Pittelkow MR, Cook PW, et al. Autocrine regulation of keratinocytes: the emerging role of heparin-binding, epidermal growth factor-related growth factors. J Invest Dermatol. 1998;111:715–21.
Pillai S, Bikle DD. Role of intracellular-free calcium in the cornified envelope formation of keratinocytes: differences in the mode of action of extracellular calcium and 1,25 dihydroxyvitamin D3. J Cell Physiol. 1991;146:94–100.
Porter RM, Lane EB. Phenotypes, genotypes and their contribution to understanding keratin function. Trends Genet. 2003;19:278–5.
Proksch E, Brandner JM, Jensen JM, et al. The skin: an indispensable barrier. Exp Dermatol. 2008;17:1063–72.
Radoja N, Diaz DV, Minars TJ, et al. Specific organization of the negative response elements for retinoic acid and thyroid hormone receptors in keratin gene family. J Invest Dermatol. 1997;109:566–72.
Radoja N, Gazel A, Banno T, et al. Transcriptional profiling of epidermal differentiation. Physiol Genomics. 2006;27:65–78.
Rees RB. Newer treatment in dermatology. South Med J. 1975;68:1395–400.
Robyr D, Wolffe AP, Wahli W, et al. Nuclear hormone receptor coregulators in action: diversity for shared tasks. Mol Endocrinol. 2000;14:329–47.
Rosen ED, Sarraf P, Troy AE, et al. PPAR gamma is required for the differentiation of adipose tissue in vivo and in vitro. Mol Cell. 1999;4:611–7.

Samarasekera EJ, Sawyer L, Wonderling D, et al. Topical therapies for the treatment of plaque psoriasis: systematic review and network meta-analyses. Br J Dermatol. 2013;168:954–67.

Schmuth M, Jiang YJ, Dubrac S, et al. Thematic review series: skin lipids. Peroxisome proliferator-activated receptors and liver X receptors in epidermal biology. J Lipid Res. 2008;49:499–509.

Schurer NY, Elias PM. The biochemistry and function of stratum corneum lipids. Adv Lipid Res. 1991;24:27–56.

Schürer NY, Plewig G, Elias PM, et al. Stratum corneum lipid function. Dermatologica. 1991;183:77–94.

Segre JA. Epidermal barrier formation and recovery in skin disorders. J Clin Invest. 2006;116: 1150–8.

Smith EL, Walworth NC, Holick MF, et al. Effect of 1 alpha,25-dihydroxyvitamin D3 on the morphologic and biochemical differentiation of cultured human epidermal keratinocytes grown in serum-free conditions. J Invest Dermatol. 1986;86:709–14.

Smola H, Thiekötter G, Fusenig NE, et al. Mutual induction of growth factor gene expression by epidermal-dermal cell interaction. J Cell Biol. 1993;122:417–29.

Tan NS, Michalik L, Noy N, et al. Critical roles of PPAR beta/delta in keratinocyte response to inflammation. Genes Dev. 2001;15:3263–77.

Tan NS, Michalik L, Desvergne B, et al. Peroxisome proliferator-activated receptor (PPAR)-beta as a target for wound healing drugs: what is possible? Am J Clin Dermatol. 2003;4:523–30.

Tan NS, Michalik L, Desvergne B, et al. Peroxisome proliferator-activated receptor-beta as a target for wound healing drugs. Expert Opin Ther Targets. 2004a;8:39–48.

Tan NS, Michalik L, Di-Poi N, et al. Critical roles of the nuclear receptor PPARβ/δ in skin wound healing. Biochem Soc Trans. 2004b;32:97–102.

Tan NS, Michalik L, Desvergne B, et al. Multiple expression control mechanisms of peroxisome proliferator-activated receptors and their target genes. J Steroid Biochem Mol Biol. 2005;93:99–105.

Tan NS, Icre G, Montagner A, et al. The nuclear hormone receptor PPARβ/δ potentiates cell chemotactism, polarization and migration. Mol Cell Biol. 2007;27(20):7161–75.

Thacher SM, Vasudevan J, Chandraratna RA, et al. Therapeutic applications for ligands of retinoid receptors. Curr Pharm Des. 2000;6:25–58.

Tinkle CL, Pasolli HA, Stokes N, et al. New insights into cadherin function in epidermal sheet formation and maintenance of tissue integrity. Proc Natl Acad Sci U S A. 2008;105:15405–10.

Tobin DJ. Biochemistry of human skin-our brain on the outside. Chem Soc Rev. 2006;35:52–67.

Törmä H. Regulation of keratin expression by retinoids. Dermatoendocrinol. 2011;3:136–40.

Tsuruta D, Green KJ, Getsios S, et al. The barrier function of skin: how to keep a tight lid on water loss. Trends Cell Biol. 2002;12:355–7.

Watt FM, Kubler MD, Hotchin NA, et al. Regulation of keratinocyte terminal differentiation by integrin-extracellular matrix interactions. J Cell Sci. 1993;106(1):175–82.

Xie Z, Komuves L, Yu QC, et al. Lack of the vitamin D receptor is associated with reduced epidermal differentiation and hair follicle growth. J Invest Dermatol. 2002;118:11–6.

Yang Q, Yamada A, Kimura S, et al. Alterations in skin and stratified epithelia by constitutively activated PPARalpha. J Invest Dermatol. 2006;126:374–85.

Yoshizawa T, Handa Y, Uematsu Y, et al. Mice lacking the vitamin D receptor exhibit impaired bone formation, uterine hypoplasia and growth retardation after weaning. Nat Genet. 1997;16:391–6.

Part IV
Sebaceous Lipids

Chapter 8
The Brain of the Skin: Sebaceous Gland

Christos C. Zouboulis

Core Messages

- Sebocytes express biologically active cytoplasmic and nuclear receptors for various molecules.
- The hypothalamic–pituitary–adrenal (HPA)-like skin axis is active in the sebaceous glands.
- Sebaceous gland development, cell differentiation, and lipogenesis are mainly regulated by androgens, whereas the sebaceous glands possess the entire enzymatic machinery for steroidogenesis.
- Sebaceous glands are strongly involved in the molecular pathways of intrinsic skin aging.
- There are similarities in the molecular pathways during sebaceous gland/entire skin aging and those of neurodegenerative diseases.

Abstract Among the multiple endocrine functions of the sebaceous gland currently reported, an independent endocrine function and the involvement in a regulatory neuropeptide program are of major importance. Sebocytes express bioactive receptors for peptide hormones, including neurotransmitters, which are mostly arranged on the cell surface (CRHR, MCR, μ-opiate receptors, VPAC receptors, CBR, histamine 1 receptor, IGF-1 receptor, EGFR, and GHR), and for steroid and thyroid class hormones (AR, PR, ER, RAR, RXR, VDR, PPAR, and LXR), which are found in the cytoplasm or nuclear compartment. A hypothalamic–pituitary–adrenal-like axis has been detected in the skin. CRH, the most proximal molecule of the axis, is produced in human sebocytes under stress conditions and is likely to serve as an important autocrine sebocyte hormone with a homeostatic prodifferentiation activity. Sebaceous gland development, cell differentiation, and lipogenesis are mainly

C. C. Zouboulis (✉)
Departments of Dermatology, Venereology, Allergology and Immunology,
Dessau Medical Center, Auenweg 38, 06847 Dessau, Germany
e-mail: christos.zouboulis@klinikum-dessau.de

A. Pappas (ed.), *Lipids and Skin Health*, DOI 10.1007/978-3-319-09943-9_8

regulated by androgens, whereas the sebaceous gland possesses the entire enzymatic machinery for steroidogenesis. During the process of aging, the sebaceous gland undergoes molecular changes, which are representative for the skin and predictive for tissues derived from the same embryologic origin, such as the central nervous system. In the last 40 years, increasing knowledge on sebaceous gland functions has turned the concept of a "living fossil with a past but no future" to that of the "brain of the skin."

Abbreviations

α-MSH	α-Melanocyte stimulating hormone
Δ^{5-4} 3β-HSD	3β-Hyroxysteroid dehydrogenase Δ^{5-4} isomerase
24OHase	1,25-Dihydroxyvitamin D-24-hydroxylase
AR	Androgen receptor
CBR	Cannabinoid receptor
CRH	Corticotropin-releasing hormone
CRHR	Corticotropin-releasing hormone receptor
DHEA	Dehydroepiandrosterone
DHT	5α-Dihydrotestosterone
EGF	Epidermal growth factor
ER	Estrogen receptor
GH	Growth hormone
HPA	Hypothalamic–pituitary–adrenal
IGF-1	Insulin-like growth factor 1
IL	Interleukin
LXR	Liver X receptor
MC1R	Melanocortin 1 receptor
MC5R	Melanocortin 5 receptor
PPAR	Peroxisome proliferator-activated receptor
PR	Progesterone receptor
RA	Retinoic acid
RAR	Retinoic acid receptor
RXR	Retinoid X receptor
VDR	Vitamin D receptor

Introduction: The Birth of Neurodermatology

In 1964, Günter Stüttgen, a famous German dermatologist, proposed that there is an association among the central nervous system, the autonomous nervous system, the endocrine glands, and the skin, especially the sebaceous and sweat glands (Stüttgen 1964; Fig. 8.1). Moreover, he suggested that the central nervous system may

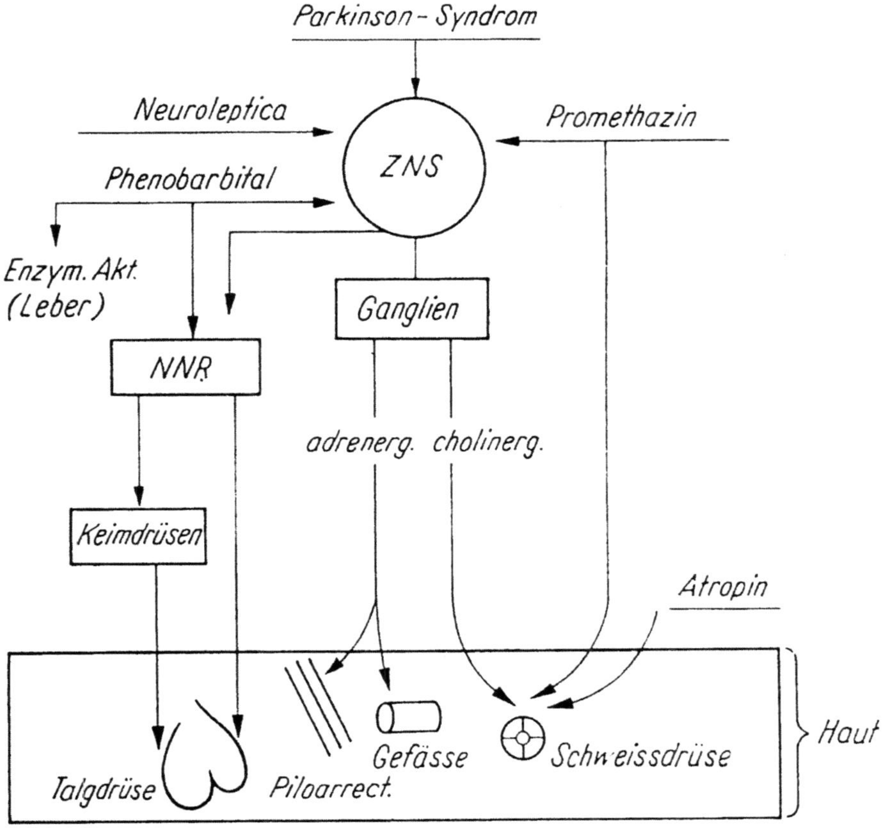

Fig. 8.1 Original figure of Günter Stüttgen's publication (1964). *adrenerg.* adrenergic, *cholinerg.* cholinergic, *Enzym. Akt.* (*Leber*) enzyme activity in the liver, *Ganglien* ganglia, *Gefässe* vessels, *Haut* skin, *Keimdrüsen* sexual glands, *Neuroleptica* neuroleptic drugs, *NNR* adrenals, *Parkinson-Syndrom* Parkinson's disease, *Piloarrect.* musculus arrector pili, *Promethazin* promethazine, *Schweissdrüse* sweat gland, *Talgdrüse* sebaceous gland, ZNS CNS

regulate the secretion of sebum. It took 38 years until a first functional association between the hypothalamic corticotropin-releasing hormone (CRH) and human sebaceous gland cells could be detected (Zouboulis et al. 2002).

Among the several endocrine functions of the sebaceous gland currently reported (Zouboulis 2010; Table 8.1), its independent endocrine function and the involvement in a regulatory neuropeptide program are of major importance.

Table 8.1 Endocrine functions of the sebaceous gland

Synthetic activity
Production of vernix caseosa
Production of sebum
Expression of the histamine-1 receptor and inhibition of squalene through antihistamines
Endocrine properties
Regulation of the independent endocrine function of the skin
Expression of all enzymes that are responsible for steroidogenesis from the circulating lipids
Regulation of local androgen synthesis
Major role in the hormonally induced skin aging process
Modification of lipid synthesis by a combined androgen and peroxisome proliferator-activated receptor ligands, estrogen, and the insulin-like growth factor-1 axis
Expression of vitamin D receptor and the vitamin D metabolizing enzymes
Expression of the retinoid metabolizing cytochrome P450 enzyme system
Selective control of the action of hormones and xenobiotics on the skin
Influenced by a regulatory neuropeptide program

Hormone Receptors in Human Sebaceous Glands and Their Biological Activity

Sebocytes express receptors for peptide hormones, neurotransmitters, which are mostly arranged on the cell surface, and for steroid and thyroid class hormones (Table 8.2), which are found in the cytoplasm or nuclear compartment (Zouboulis 2009a; Fig. 8.2).

Peptide Hormone and Neurotransmitter Receptors

Serpentine or "seven transmembrane domain" receptors, which are expressed and functional in human sebocytes, are:

- CRH receptor (CRHR)1 and 2, whereas CRHR1 is more abundant and seems to regulate CRH activity (Zouboulis et al. 2002; Krause et al. 2007). Through binding to CRHR1, CRH and urocortin reduce sebocyte proliferation. CRH up-regulates Δ^{5-4} 3β-hyroxysteroid dehydrogenase expression, synthesis of neutral lipids and interleukin(IL)6 and IL8 release.
- Melanocortin [α-melanocyte stimulating hormone (α-MSH)]-1 and -5 receptors (MC1R and MC5R), which bind α-MSH and are located at the cellular surface of sebocytes. MC1R regulates inflammation in SZ95 sebocytes (Böhm et al. 2002) and exhibits a stronger expression in acne-involved sebaceous glands (Ganceviciene et al. 2007). The expression of MC5R is weaker than that of MC1R but has been shown to be a marker of human sebocyte differentiation, since it is expressed in differentiated, lipid-containing sebocytes, only (Zhang et al. 2006a).

Table 8.2 Hormone receptors in human sebaceous gland cells

Receptors	Natural ligands	Function on sebocytes
Peptide hormones and neurotransmitter receptors		
Serpentine receptors (seven transmembrane domain)		
CRH receptors 1 and 2 (CRHR1 > CRHR2)	CRH, urocortin	↓ proliferation, ↑ Δ^{5-4} 3β-HSD (CRH), ↑ lipogenesis (CRH), ↑ IL6 and IL8 release (CRH)
Melanocortin-1 and -5 receptors (MC1R and MC5R)	α-MSH	↓ IL1-induced IL8 synthesis (MC5R), differentiation marker (MC1R)
μ-Opiate receptors	β-Endorphin	↓ EGF-induced proliferation, ↑ lipogenesis
VPAC receptors	VIP, neuropeptide Y and CGRP	Neuropeptide Y-stimulated IL6 and IL8 release
Cannabinoid receptors (CR1 and CR2)	Cannabinoids	↑ Lipogenesis
Histamine receptor 1	Histamine	Regulation of squalene synthesis
Single transmembrane domain receptors with endogenous tyrosine kinase activity		
IGF-1 receptor	IGF-1, insulin	↑ Lipogenesis
EGF receptor	EGF	Controls differentiation (↓ adipophilin and MC5R levels), proliferation (↑), lipogenesis and inflammation (↓ IL6, IL8 and TNFα levels)
Single transmembrane domain receptors without endogenous tyrosine kinase activity		
GH receptor	GH	↑ Differentiation, ↑ 5α-DHT activity on lipogenesis
Nuclear receptors		
Steroid receptors		
Androgen receptor	Testosterone, DHT	↑ Proliferation (in association with PPAR ligands: ↑ lipogenesis)
Progesterone receptor	Progesterone	
Thyroid receptors		
Estrogen receptors (ERα and ERβ)	17β-estradiol	↑ Synthesis of polar lipids
Retinoic acid receptors (RARα and RARγ)	all-trans-RA	↓ Proliferation
Retinoid X receptors (RXRα,>RXRβ, RXRγ)	9-cis RA	Regulation of lipogenesis (?)
Vitamin D receptor (VDR)	Vitamin D_3	Regulation of cell proliferation, cell cycle, lipid content, and IL6 and IL8 release
Peroxisome proliferator-activated receptors (PPARα, PPARγ > PPARβ)	LA (RRARβ/d), LTB4 (RRARα) PG-D2, 15-deoxy-Δ12,14-PG-J2 (RRARγ)	↑ Lipogenesis, ↑ PG-E2 release, ↑ IL6 release, ↑ COX-2 synthesis ↑ Lipogenesis, ↑ IL6 release ↑ Eotaxin 3

Table 8.2 (continued)

Receptors	Natural ligands	Function on sebocytes
Liver X receptors (LXRα and LXRβ)	22(R)-Hydroxycholesterol	↓ Proliferation, ↑ lipogenesis, ↓ COX-2-induced nitric oxide synthetase

α-MSH α-melanocyte-stimulating hormone, *CGRP* calcitonin gene-related peptide, *COX* cyclooxygenase, *CRH* corticotropin-releasing hormone, *DHT* 5α-dihydrotestosterone, *EGF* epidermal growth factor, *GH* growth hormone, *HSD* hyroxysteroid dehydrogenase, *IGF* insulin-like growth factor, *IL* interleukin, *LA* linoleic caid, *LTB4* leukotriene B4, *PG* prostaglandine, *PPAR* peroxisome proliferator-activated receptor, *RA* retinoic acid, *TNF* tumor necrosis factor, *VIP* vasoactive intestinal popypeptide

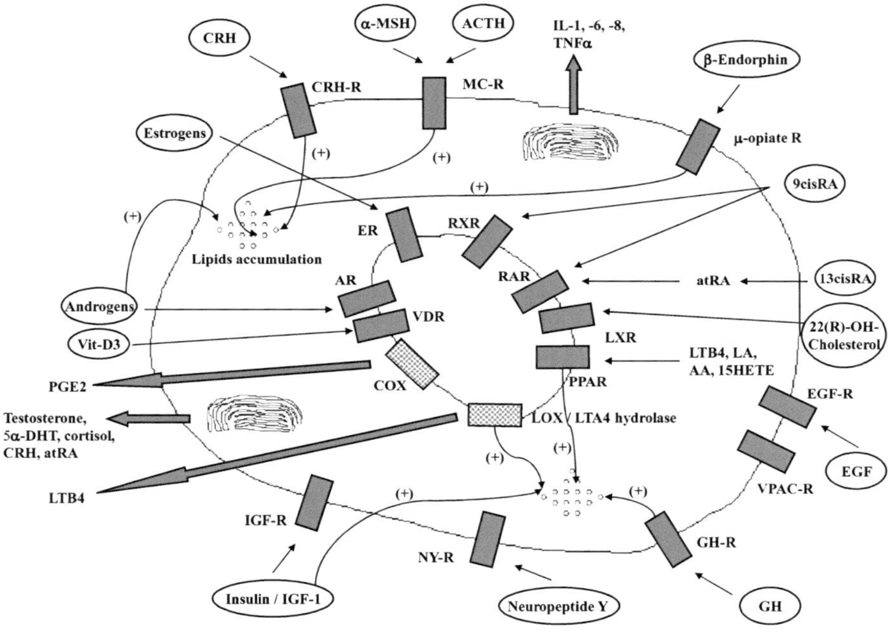

Fig. 8.2 Interaction between enzymes, membrane, and nuclear receptors as well as their ligands in human sebocytes: their influence on lipid accumulation is displayed. The release of various hormones and inflammatory mediators is also seen. *α-MSH* α-melanocyte stimulating hormone, *9-cis-RA* 9-cis-retinoic acid, *AA* arachidonic acid, *ACTH* adrenocotricotropic hormone, *AR* androgen receptor, *at-RA* all-trans-retinoic acid, *COX* cyclooxygenase, *CRH* corticotropin-releasing hormone, *DHT* 5α-dihydrotestosterone, *EGF* epidermal growth factor, *EGR-R* epidermal growth factor receptor, *ER* estrogen receptor, *GH* growth hormone, *IGF-1* insulin like growth factor-1, *IL* interleukin, *LA* linoleic acid, *LOX* lipoxygenase, *LTA4-hydrolase* leukotriene A4 hydrolase, *LTB4* leukotriene B4, *LXR* liver X receptor, *NY* neuropeptide Y, *PG* prostaglandin, *PPAR* peroxisome proliferator-activated receptor, *R* receptor, *RAR* retinoic acid receptor, *TNF-α* tumor necrosis factor-α, *VDR* vitamin D receptor, *Vit-D_3* vitamin D_3, *VPAC-R* vasoactive intestinal peptide receptor. (modified from Zouboulis 2008)

- μ-opiate receptors, which bind β-endorphin. β-endorphin stimulates lipogenesis and specifically increases the amount of C16:0, C16:1, C18:0, C18:1, and C18:2 fatty acids to an extent similar to linoleic acid in sebocytes (Böhm et al. 2004).
- VPAC receptors, which bind vasoactive intestinal polypeptide, receptors for neuropeptide Y, and calcitonin gene-related peptide (Seiffert et al. 2000). Neuropeptide Y activates cytokine synthesis. The calcitonin gene-related peptide is often colocalized with substance P.
- Cannabinoid receptors (CBR) 1 and 2 are expressed in SZ95 sebocytes and sebaceous glands (Ständer et al. 2005; Dobrosi et al. 2008). CBR1 was found in the differentiated sebocytes and CBR2 in the undifferentiated cells, whereas endocannabinoids influence sebocyte differentiation via CBR2.
- Histamine 1 receptor, which binds with histamine and regulates squalene synthesis (Pelle et al. 2008). Antihistamines, ligands of histamine 1 receptor reduced squalene synthesis in SZ95 sebocytes.
  The single-transmembrane domain receptors, insulin-like growth factor (IGF)-1 receptor, and epidermal growth factor (EGF) receptor that harbour intrinsic tyrosine kinase activity are expressed on SZ95 sebocyte cell surface.
- IGF-1 receptor can be activated by IGF-1 and high concentrations of insulin (Makrantonaki et al. 2006). It amplifies lipid accumulation in SZ95 sebocytes in a dose-dependent manner. The activation of the IGF-1 receptor induced lipogenesis in SEB-1 sebocytes by sterol response element-binding protein-dependent and independent pathways (Smith et al. 2006). IGF-1 also stimulates proliferation and differentiation of rat preputial gland cells, which resemble sebocytes, especially in combination with growth hormone (GH; Deplewski and Rosenfield 1999).
- EGF receptor was shown to be expressed in human sebocytes (Nanney et al. 1984; Takata et al. 2012). Its inhibition in SZ95 sebocytes led to upregulation of adipophilin and MC5R expression levels, which are differentiation markers for human sebocytes, and enhanced proinflammatory signaling by induction of IL6, IL8, and tumor necrosis factor-α release. Current data indicate that EGF may play a multimodal role on sebaceous gland cell proliferation, differentiation, lipogenesis, and inflammatory signaling (Zouboulis 2013).
  Activation of GH receptor, which does not possess intrinsic tyrosine kinase activity but appear to function through interaction with soluble transducer molecules, in SZ95 sebocytes, stimulates sebocyte differentiation and augments the effect of 5α-dihydrotestosterone (DHT) on sebum synthesis (Zouboulis et al. 2002; Makrantonaki et al. 2006).

Nuclear Receptors

The steroid receptor family is represented in human sebocytes by the androgen receptor (AR) and the progesterone receptor (PR), which are mostly expressed in basal and early differentiated sebocytes (Zouboulis et al. 2007; Fritsch et al. 2001; Fimmel et al. 2007).

- Human sebocytes exhibit the highest AR density among human skin cells. AR down regulation reduces sebocyte proliferation (Fimmel et al. 2007). Five intracellular enzymes—all of them expressed in sebocytes (Fritsch et al. 2001)—are involved in activation and inactivation of androgens before binding to AR. Dehydroepiandrosterone (DHEA) sulfate is metabolized by the steroid sulfatase to DHEA. DHEA and androstosterone are converted to testosterone and later to DHT by 5α-reductase (Fritsch et al. 2001; Chen et al. 1998). Sebocyte studies of Akamatsu et al. and Zouboulis et al. showed a dose-dependent induction of sebocyte proliferation by testosterone treatment (Akamatsu et al. 1993) and no effect on lipid stimulation (Zouboulis et al. 1999). Investigations by Rosenfield et al. and Makrantonaki et al. proved that the effect of androgens on sebaceous lipids is mediated by peroxisome proliferator-activated receptor (PPAR) ligands (Rosenfield et al. 1998; Makrantonaki and Zouboulis 2007).
- PR was found in nuclei of basal sebocytes of sebaceous glands (Pelletier and Ren 2004). The thyroid receptors, estrogen receptors (ER; α- and β-isotypes; Pelletier and Ren 2004; Thornton et al. 2006; Thornton et al. 2003), retinoic acid receptors (RAR; isotypes α and γ), and retinoid X receptors (RXR; isotypes α, β, γ; Reichrath et al. 1997; Tsukada et al. 2000), vitamin D receptor (VDR; Reichrath et al. 2000), PPAR (Makrantonaki and Zouboulis 2007; Schmuth et al. 2005; Alestas et al. 2006), and liver X receptors (LXR; -α and -β isotypes; Russell et al. 2007; Hong et al. 2008) are expressed in human sebocytes.
- ER-β is expressed in basal and partially differentiated sebocytes. ER-α is expressed in basal and early differentiated sebocytes. One of the natural estrogens, estradiol, is created by oxidative reduction of 4-androstene-3,17-dione. Treatment of sebocytes with 17β-estradiol showed an effect on polar lipid production but no stimulating effect on neutral lipids (Makrantonaki et al. 2008). Other previous in vitro data indicated that estrogens may have an influence on the biological activity of sebaceous glands (Guy et al. 1996).
- RARα and γ and RXRα are the predominant retinoid receptors in human sebocytes. RAR regulates cell proliferation (Tsukada et al. 2000). The natural ligands for RAR and RXR are *all-tans*-retinoic acid (RA) and 9-*cis* RA. 13-*cis* RA inhibits proliferation of SZ95 sebocytes, whereas it was found to be metabolized intracellularly to its isoform and RAR ligand *all-tans*-RA. RXR agonists (rexinoids) are stimulating sebocyte differentiation and proliferation. Rexinoids in combination with specific PPAR agonists, such WY 14643, troglitazone and cabaprostacycline, affected differentiation and growth in cultured primary sebocyte-like rat preputial cells (Kim et al. 2001).
- SZ95 sebocytes were found to express vitamin D-25-hydroxylase, 25-hydroxyvitamin D-1α-hydroxylase, and 1,25-dihydroxyvitamin D-24-hydroxylase (24OHase; Krämer et al. 2009). Vitamin D_3 induces time- and dose-dependent modulation of cell proliferation, cell cycle regulation, lipid content, and IL6 and IL8 secretion by cultured sebocytes. RNA expression of VDR and 24OHase was upregulated along with vitamin D_3 treatment.
- PPARα and γ are the predominant PPAR subtypes in human sebocytes (Alestas et al. 2006). PPAR is present in mitochondria, peroxisomes and microsomes

of sebocytes, and regulate multiple lipid metabolic genes. Sebaceous lipogenesis is maximally induced by androgens and PPAR ligands (Makrantonaki and Zouboulis 2007).

- SZ95 sebocytes express LXRα and β receptors at the mRNA and protein levels. The application of natural 22(R)-hydroxycholesterol or synthetic ligands significantly inhibited sebocyte proliferation and increased lipogenesis. The expression of known LXR targets, such as fatty acid synthase and SREBP1, was induced by the synthetic LXR ligand TO901317, which also decreased the expression of cyclooxygenase 2 and inducible nitric oxide synthase that was induced by lipopolysaccharide treatment (Hong et al. 2008).

Activation of the HPA-Like Skin Axis in Sebaceous Glands

Communication and reciprocal regulation between nervous, endocrine, and immune systems are essential for biological stability and responses to external and internal challenges. In particular, neuropeptides, hormones, and cytokines act as signaling molecules that mediate communication between the three interacting systems. Analogous to central responses to stress, which involve predominantly the hypothalamic–pituitary–adrenal (HPA) axis, it has been proposed that the skin may share similar mediators (Alesci and Bornstein 2000; Slominski and Wortsman 2000; Slominski et al. 2000), whereas the sebaceous gland plays a major role in these procedures (Zouboulis et al. 2002; Zouboulis and Chen 2013; Zouboulis and Böhm 2004; Zouboulis 2009b; Schagen et al. 2011; Elewa et al. 2012).

Indeed, CRH, the most proximal molecule of the cutaneous HPA-like axis, its binding protein and CRHRs, which act as a central regulatory system of the HPA axis (Slominski et al. 2000), have been detected in human sebaceous glands at the mRNA and protein levels in tissue (Elewa et al. 2012; Kono et al. 2001; Ganceviciene et al. 2009a) and in culture (Zouboulis et al. 2002; Krause et al. 2007). CRH is likely to serve as an important autocrine hormone in sebocytes with a homeostatic prodifferentiation activity. It directly inhibits proliferation, induces lipid synthesis, and enhances mRNA expression of 3β-hyroxysteroid dehydrogenase Δ^{5-4} isomerase (Δ^{5-4} 3β-HSD) in human sebocytes in vitro, the enzyme which converts DHEA to testosterone. Testosterone and GH, which also enhance sebaceous lipid synthesis, antagonize CRH in human sebocytes in vitro by down-regulating or modifying CRHR expression, respectively (Zouboulis et al. 2002). The induction of sebaceous lipids by CRH is CRHR1-mediated. CRH also enhances the release of IL-6 and IL-8 in human sebocytes in vitro by an IL-1β-independent pathway (Krause et al. 2007). The CRH analogue urocortin also inhibits sebocyte proliferation in vitro, while urotensin and sauvagine seem to be inactive. These findings implicate a major involvement of CRH in the clinical development of seborrhoea and acne (Ganceviciene et al. 2007, 2009a, b; Fig. 8.3) as well as in other skin disorders and diseases associated with alterations in the formation of sebaceous lipids (Zouboulis 2009b, 2010).

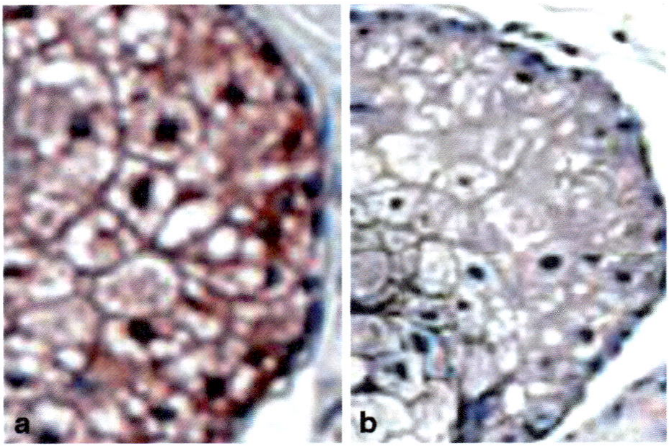

Fig. 8.3 Acne-amplified intracellular CRH levels. Sebaceous glands in **a** acne-involved skin, and **b** healthy skin. Original magnification 400×

Androgens and Sebaceous Gland Development, Cell Differentiation, and Lipogenesis

The development of the sebaceous gland is closely related to the differentiation of the two other epithelial lineages, hair follicles and epidermis (Zouboulis et al. 2003). From the 13th to 15th week the sebaceous gland develops from the hair bulge. Subsequently, the pituitary adrenocorticotropic hormone regulates the steroidogenic activity of the fetal adrenal zone, which is the main source of DHEA-sulfate. The circulating DHEA-sulfate is metabolized by the blood monocytes in DHEA. DHEA can be further metabolized into androstenedione and testosterone, particularly in the sebaceous gland, which expresses—as the single cell of the skin—the necessary enzyme Δ^{5-4} 3β-HSD (Fritsch et al. 2001). The sebaceous gland is probably the most important target organ of fetal androgens in the skin. They reached the peak of their gestational activity in the third trimester of pregnancy (Zouboulis et al. 2003). Its secretion forms a significant part of the vernix caseosa, the variable, adhesive fetal skin surface film. It is composed of lipids, dead keratinocytes, water and other substances in the amniotic fluid, and protects the skin of the developed embryo from the amniotic fluid until birth.

A strong increase in sebum excretion occurs a few hours after birth which peaks during the first week (Agache et al. 1980), whereas maternal and neonatal sebum excretion rates directly correlate (Henderson et al. 2000). This correlation is lost in the following weeks and is independent from breast feeding. At this time, the sebum level per unit skin surface is in the same range as in young adults (Agache et al. 1980) and the sequence of sebaceous transformation seems identical to that in postnatal life. These events suggest an important role of the hormonal environment of the mother on the sebaceous gland of the newborn and indicate that androgenetic stimulus for sebum secretion occurs before birth through the placenta (Zouboulis et al. 2003; Agache et al. 1980). It is interesting that later in life a mother with acne history influences the severity of acne the most (Ghodsi et al. 2009). The sebum excretion then slowly subsides. Females display a different pattern of sebum excretion than males. Directly after birth the levels in females are lower than in the males but a large increase takes place between the 3rd and 6th day, followed by a fall, bringing the levels below that of the males. At 6 months the levels are low in both sexes. Very low levels, approximately 10 $\mu g/cm^2$and often below 0.5 $\mu g/cm^2$, remain constant from 6 months to the prepubertal period (Pochi et al. 1977). A new rise takes place at about 9 years (Zouboulis 2008) with adrenarche and continues up to 17 years when the adult level is reached. It has been suggested that the endocrine environment of the neonate correlates and may influence the sebaceous gland development in puberty (Henderson et al. 2000). Since DHEA also exhibits proinflammatory activity on the sebaceous glands, DHEA-associated acne, such as acne praecox (prepubertal acne) and female acne tarda (Zouboulis and Piquero-Martin 2003; Dréno et al. 2013) present with characteristic inflammatory lesions at the lower part of the face (Fig. 8.4).

The sebaceous gland expressed all the enzymes required for genuine steroidogenesis (Slominski et al. 2013) and is able to produce sexual hormones from circulating cholesterol (Thiboutot et al. 2003). On the other hand, under physiological conditions, it saves energy by using circulating DHEA as favorable substrate for testosterone synthesis (Chen et al. 2010).

Signaling and Metabolic Pathways Operative in Hormonally Aged Sebocytes In Vitro

With age skin undergoes major morphological and physiological changes. In sun-protected nonexposed skin areas, ageing is mainly attributed to intrinsic factors such as genetics and changes in the endocrine environment and reflects degradation processes of the entire organism (Zouboulis and Makrantonaki 2011; Farage et al. 2012). The sebaceous gland undergoes molecular changes, which are representative for the skin (Makrantonaki and Zouboulis 2009) and, moreover, predictive for tissues derived of the same embryologic origin, such as the central nervous system (Makrantonaki et al. 2010). Major molecular pathways associated with neurodegenerative diseases are also modified in intrinsically aging skin (Makrantonaki et al. 2012) and in the hormonally aged SZ95 sebocytes in vitro (Makrantonaki et al. 2006; Table 8.3).

Whereas Wnt signaling seems to be a label of skin aging (Makrantonaki et al. 2012), canonical Wnt signaling activation is induced by prostaglandin E2 (Yoshida et al. 2013), which is produced in the sebaceous glands (Zhang et al. 2006b) and represents one of the most active enzymatic pathways in this organ (Zouboulis 2005). Sebaceous lipogenesis is induced, among others, by a diacylglycerol acyltransferase-triggered increase of triacylglycerol synthesis formation and is accompanied by an increase of prostaglandin levels, including that of prostaglandin E2 (Iwata et al. 2005).

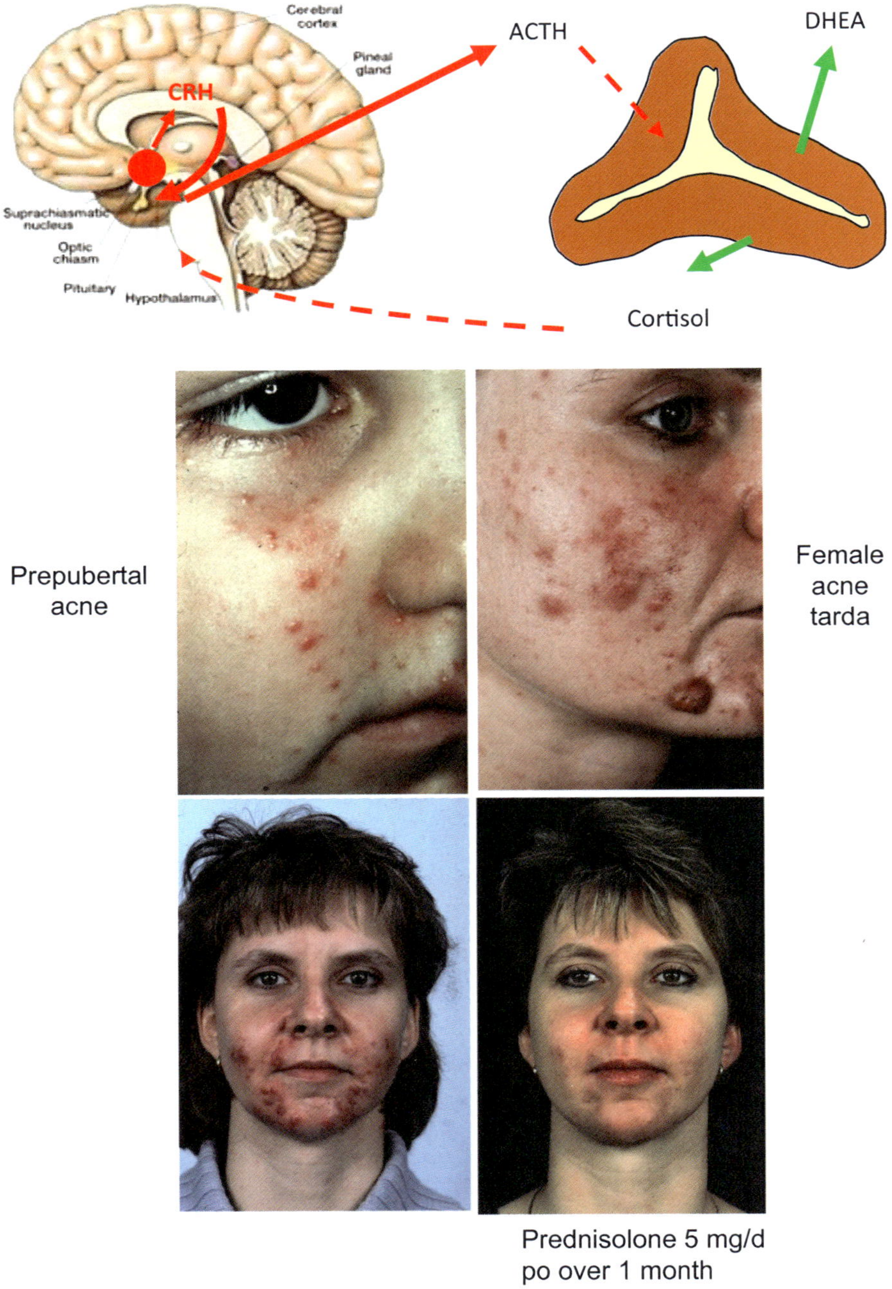

Fig. 8.4 Mechanism of the CRH-induced DHEA-triggered induction of adrenal acne (prepubertal and female late types). The dominant presence of inflammatory lesions at the lower part of the face is a characteristic clinical sign. Lower panel: in premenstrually aggravated female acne tarda with upper normal to increased serum DHEA levels a systemic low-dose prednisolone treatment (2.5–7.5 mg/d) may induce a marked improvement. (lower panel from: Zouboulis and Piquero-Martin 2003)

Table 8.3 Molecular pathways modified in both hormonally aged human SZ95 sebocytes in vitro and in human skin tissue derived from young and elderly donors. The bold letters indicate pathways also modified in neurodegenerative diseases. (from Makrantonaki et al. 2006, 2012)

Integrin signaling pathway
Inflammation mediated by chemokine and cytokine signaling pathway
Wnt signaling pathway
Huntington disease
EGF receptor signaling pathway
P53 pathway
Interleukin signaling pathway
Alzheimer disease-presenilin pathway
Ras pathway
Apoptosis signaling pathway
Cytoskeletal regulation
Alzheimer disease-amyloid secretase pathway
VEGF signaling pathway
TGF-beta signaling pathway
PI3 kinase pathway
P53 pathway feedback loops 2
Oxidative stress response
Insulin/IGF pathway-protein kinase B signaling cascade
Nicotinic acetylcholine receptor signaling pathway
Parkinson disease
Cadherin signaling pathway
Ubiquitin proteasome pathway
Toll receptor signaling pathway
Notch signaling pathway
Muscarinic acetylcholine receptor 2 and 4 signaling pathway
JAK/STAT signaling pathway

Conclusions

Forty years after Albert M. Kligman's declaration that the sebaceous gland is a reminiscence of human development, a "living fossil with a past but no future" (Kligman 1963), we can surely state that the human sebaceous gland has become an increasingly interesting and relevant research subject, which has turned to be recognized as the "brain of the skin." Its multiple physiological functions and its involvement in the most important molecular pathways, which regulate skin health and disease make cutaneous and sebaceous gland research synonymous. Moreover, applications of sebaceous gland models (Zouboulis et al. 1999) have already moved over the borders of skin research (Makrantonaki et al. 2006) and are likely to also be used in the future as representative models for other organs and the entire human body.

References

Agache P, Blanc D, Barrand C, Laurent R. Sebum levels during the first year of life. Br J Dermatol. 1980;103:643–9.

Akamatsu H, Zouboulis CC, Orfanos CE. Spironolactone directly inhibits proliferation of cultured human facial sebocytes and acts antagonistically to testosterone and 5 alpha-dihydrotestosterone in vitro. J Invest Dermatol. 1993;100:660–2.

Alesci S, Bornstein SR. Neuroimmunoregulation of androgens in the adrenal gland and the skin. Horm Res. 2000;54:281–6.

Alestas T, Ganceviciene R, Fimmel S, Müller-Decker K, Zouboulis CC. Enzymes involved in the biosynthesis of leukotriene B_4 and prostaglandin E_2 are active in sebaceous glands. J Mol Med. 2006;84:75–87.

Böhm M, Schiller M, Ständer S, et al. Evidence for expression of melanocortin-1 receptor in human sebocytes in vitro and in situ. J Invest Dermatol. 2002;118:533–9.

Böhm M, Li Z, Ottaviani M, et al. Beta-endorphin modulates lipogenesis in human sebocytes. J Invest Dermatol. 2004;123:A10.

Chen W, Zouboulis CC, Fritsch M, et al. Evidence of heterogeneity and quantitative differences of the type 1 5α-reductase expression in cultured human skin cells. Evidence of its presence in melanocytes. J Invest Dermatol. 1998;110:84–9.

Chen W, Tsai S-J, Sheu H-M, Tsai J-C, Zouboulis CC. Testosterone synthesized in cultured human SZ95 sebocytes mainly derives from dehydroepiandrosterone. Exp Dermatol. 2010;19:470–2.

Deplewski D, Rosenfield R. Growth hormone and insulin like growth factors have different effects on sebaceous cell growth and differentiation. Endocrinology. 1999;140:4089–94.

Dobrosi N, Tóth B, Nagy G, et al. Endocannabinoids enhance lipid synthesis in human sebocytes via cannabinoid receptor-2-mediated signaling. FASEB J. 2008;22:3685–95.

Dréno B, Layton A, Zouboulis CC, et al. Adult female acne: a new paradigm. J Eur Acad Dermatol Venereol. 2013;27:1063–70.

Elewa RM, Abdallah M, Youssef N, Zouboulis CC. Aging-related changes in cutaneous corticotropin-releasing hormone system reflect a defective neuroendocrine-stress response in aging. Rejuvenation Res. 2012;15:366–73.

Farage M, Miller KW, Zouboulis CC, Pierard G, Maibach HM. Sex steroid hormones and gender differences in skin aging. J Steroid Horm Sci. 2012;3:109.

Fimmel S, Saborowski A, Térouanne B, Sultan C, Zouboulis CC. Inhibition of the androgen receptor by antisense oligonucleotides regulates the biological activity of androgens in SZ95 sebocytes. Horm Metab Res. 2007;39:149–56.

Fritsch M, Orfanos CE, Zouboulis CC. Sebocytes are the key regulators of androgen homeostasis in human skin. J Invest Dermatol. 2001;116:793–800.

Ganceviciene R, Graziene V, Böhm M, Zouboulis CC. Increased in situ expression of melanocortin-1 receptor in sebaceous glands of lesional skin of patients with acne vulgaris. Exp Dermatol. 2007;16:547–52.

Ganceviciene R, Böhm M, Fimmel S, Zouboulis CC. The role of neuropeptides in the multifactorial pathogenesis of acne vulgaris. Dermatoendocrinol. 2009a;1:170–6.

Ganceviciene R, Graziene V, Fimmel S, Zouboulis CC. Involvement of the corticotropin-releasing hormone system in the pathogenesis of acne vulgaris. Br J Dermatol. 2009b;160:345–52.

Ghodsi SZ, Orawa H, Zouboulis CC. Prevalence, severity and severity risk factors of acne in high school pupils: a community-based study. J Invest Dermatol. 2009;129:2136–41.

Guy R, Green M, Kealey T. Modeling acne in vitro. J Invest Dermatol. 1996;106:176–82.

Henderson CA, Taylor J, Cunliffe WJ. Sebum excretion rates in mothers and neonates. Br J Dermatol. 2000;142:110–1.

Hong I, Lee M, Na T, Zouboulis CC, Lee M. LXRα enhances lipid synthesis in SZ95 sebocytes. J Invest Dermatol. 2008;128:1266–72.

Iwata C, Akimoto N, Sato T, Morokuma Y, Ito A. Augmentation of lipogenesis by 15-deoxy-$\Delta^{12,14}$-prostaglandin J_2 in hamster sebaceous glands: identification of cytochrome P450-mediated 15-deoxy-$\Delta^{12,14}$-prostaglandin J_2 production. J Invest Dermatol. 2005;125:865–72.
Kim M, Deplewski D, Ciletti N, Michel S, Reichert U, Rosenfield R. Limited cooperation between peroxisome proliferator-activated receptors and retinoid X receptor agonists in sebocyte growth and development. Mol Genet Metab. 2001;74:362–9.
Kligman AM. The uses of sebum? In: Montagna W, Ellis RA, Silver AF, editors. Advances in biology of skin. Vol IV. The sebaceous glands. Oxford:Pergamon; 1963. pp. 110–2.
Kono M, Nagata H, Umemura S, Kawana S, Yoshiyuki R. In situ expression of corticotropin-releasing hormone (CRH) and proopiomelanocortin (POMC) genes in human skin. FASEB J. 2001;15:2297–9.
Krämer C, Seltmann H, Seifert M, Tilgen W, Zouboulis CC, Reichrath J. Characterization of the vitamin D endocrine system in human sebocytes in vitro. J Steroid Biochem Mol Biol. 2009;113:9–16.
Krause K, Schnitger A, Fimmel S, Glass E, Zouboulis CC. Corticotropin-releasing hormone skin signalling is receptor-mediated and is predominant in the sebaceous glands. Horm Metab Res. 2007;39:166–70.
Makrantonaki E, Zouboulis CC. Testosterone metabolism to 5α-dihydrotestosterone and synthesis of sebaceous lipids is regulated by the peroxisome proliferators-activated receptor ligand linoleic acid in human sebocytes. Br J Dermatol. 2007;156:428–32.
Makrantonaki E, Zouboulis CC. The sebaceous gland- a model of hormonal aging. In: Daskalaki A, editor. Handbook of research on systems biology applications in medicine. IGI Global, Hershey, PA, USA, Section V, Chapter XIX, 2009. pp. 328–34.
Makrantonaki E, Adjaye J, Herwig R, et al. Age-specific hormonal decline is accompanied by transcriptional changes in human sebocytes in vitro. Aging Cell. 2006;5:331–44.
Makrantonaki E, Vogel K, Fimmel S, Oeff M, Seltmann H, Zouboulis CC. Interplay of IGF-I and 17ß-estradiol at age-specific levels in human sebocytes and fibroblasts in vitro. Exp Gerontol. 2008;43:939–46.
Makrantonaki E, Schönknecht P, Hossini AM, et al. Skin and brain age together: the role of hormones in the ageing process. Exp Gerontol. 2010;45:801–13.
Makrantonaki E, Brink TC, Zampeli V, et al. Identification of biomarkers of human skin ageing in both genders: Wnt signalling-a label of skin ageing? PLoS One. 2012;7:e50393.
Nanney LB, Magid M, Stoscheck CM, King LE Jr. Comparison of epidermal growth factor binding and receptor distribution in normal human epidermis and epidermal appendages. J Invest Dermatol. 1984;83:385–93.
Pelle E, McCarthy J, Seltmann H, et al. Identification of histamine receptors and reduction of squalene levels by an antihistamine in sebocytes. J Invest Dermatol. 2008;128:1280–5.
Pelletier G, Ren L. Localization of sex steroid receptors in human skin. Histol Histopathol. 2004;19:629–36.
Pochi PE, Strauss JS, Downing DT. Sebum, acne and androgens in children. Clin Res. 1977;25:531A.
Reichrath J, Mittmann M, Kamradt J, Müller S. Expression of retinoid-X receptors (-alpha, -beta, -gamma) and retinoic acid receptors (-alpha, -beta, -gamma) in normal human skin: an immunohistological evaluation. Histochem J. 1997;29:127–33.
Reichrath J, Classen U, Meineke V, et al. Immunoreactivity of six monoclonal antibodies directed against 1,25-dihydroxyvitamin-D3 receptors in human skin. Histochem J. 2000;32:625–9.
Rosenfield R, Deplewski D, Kentsis A, Ciletti N. Mechanisms of androgen induction of sebocyte differentiation. Dermatology. 1998;196:43–6.
Russell L, Harrison W, Bahta A, Zouboulis CC, Burrin J, Philpott M. Characterization of liver X receptor expression and function in human skin and the pilosebaceous unit. Exp Dermatol. 2007;16:844–52.
Schagen SK, Ganceviciene R, Krause K, et al Update on cutaneous stress: the sebocyte own corticotropin-releasing hormone system is an amplifier of inflammation. In: Wollina U, editor.

COSMODERM XVI-European Society for Cosmetic and Aesthetic Dermatology (ESCAD)-International Proceedings. Medimont, Bologna, 2011. pp. 63–6.
Schmuth M, Ortegon A, Mao-Qiang M, Elias P, Feingold K, Stahl A. Differential expression of fatty acid transport proteins in epidermis and skin appendages. J Invest Dermatol. 2005;125:1174–81.
Seiffert K, Zouboulis CC, Seltmann H, Granstein R. Expression of neuropeptide receptors by human sebocytes and stimulatory effect of their agonists on cytokine production. Horm Res. 2000;53:102.
Slominski AT, Wortsman J. Neuroendocrinology of the skin. Endocrin Rev. 2000;21:457–87.
Slominski A, Wortman J, Luger T, Paus R, Solomon S. Corticotropin releasing hormone and proopiomelanocortin involvement in the cutaneous response to stress. Physiol Rev. 2000;80:979–1020.
Slominski A, Zbytek B, Nikolakis G, et al. Steroidogenesis in the skin: implications for local immune functions. J Steroid Biochem Mol Biol. 2013;137:107–23.
Smith T, Cong Z, Gilliland K, Clawson G, Thiboutot D. Insulin-like growth factor-1 induces lipid production in human SEB-1 sebocytes via sterol response element-binding protein-1. J Invest Dermatol. 2006;126:1226–32.
Ständer S, Schmelz M, Metze D, Luger T, Rukwied R. Distribution of cannabinoid receptor 1 (CB1) and 2 (CB2) on sensory nerve fibers and adnexal structures in human skin. J Dermatol Sci. 2005;38:177–88.
Stüttgen G. Zentralnervensystem und Talgsekretion. Arch Klin Exp Dermatol. 1964;219:795–9.
Takata T, Tarutani M, Zouboulis CC, Sano S. Sebaceous glands as the primary target of EGFR-inhibitors in the development of papulopustular eruption. J Dermatol Sci. 2012;66:165–8. (corrigendum: 2013;69:85).
Thiboutot D, Jabara S, McAllister J, et al. Human skin is a steroidogenic tissue: steroidogenic enzymes and cofactors are expressed in epidermis, normal sebocytes, and an immortalized sebocyte cell line (SEB-1). J Invest Dermatol. 2003;120:905–14.
Thornton M, Taylor A, Mulligan K, et al. Oestrogen receptor-β is the predominant oestrogen receptor in human scalp skin. Exp Dermatol. 2003;12:181–90.
Thornton M, Nelson L, Taylor AH, Birch M, Laing I, Messenger A. The modulation of aromatase and estrogen receptor alpha in cultured human dermal papilla cells by dexamethasone: a novel mechanism for selective action of estrogen via estrogen receptor beta? J Invest Dermatol. 2006;126:2010–8.
Tsukada M, Schroder M, Roos T, et al. 13-cis retinoic acid exerts its specific activity on human sebocytes through selective intracellular isomerization to all-trans retinoic acid and binding to retinoid acid receptors. J Invest Dermatol. 2000;115:321–7.
Yoshida GJ, Saya H, Zouboulis CC. Three-dimensional culture of sebaceous gland cells revealing the role of prostaglandin E2-induced activation of canonical Wnt signaling. Biochem Biophys Res Commun. 2013;438:640–6. (corrigendum: 2013;441:271).
Zhang L, Li W, Anthonavage M, Eisinger M. Melanocortin-5 receptor: a marker of human sebocyte differentiation. Peptides. 2006a;27:413–20.
Zhang Q, Seltmann H, Zouboulis CC, Konger RL. Involvement of PPAR-gamma in oxidative stress-mediated prostaglandin E_2 production in SZ95 human sebaceous gland cells. J Invest Dermatol. 2006b;126:42–8.
Zouboulis CC. Sebaceous glands and the prostaglandin pathway-Key stones of an exciting mosaic. J Invest Dermatol. 2005;125:x–xi.
Zouboulis CC. The sebaceous gland in adolescent age. Eur J Ped Dermatol. 2008;18:150–4.
Zouboulis CC. Acne vulgaris and rosacea. In: Granstein RD, Luger T, editors. Neuroimmunology of the skin-basic science to clinical practice. Berlin:Springer; 2009a. pp. 219–32.
Zouboulis CC. Sebaceous gland receptors. Dermato endocrinol. 2009b;1:77–80.
Zouboulis CC. Die Talgdrüse. Hautarzt. 2010;61:467–77.
Zouboulis CC. Epidermal growth factor receptor and the sebaceous gland. Exp Dermatol. 2013;22:695–6.

Zouboulis CC, Böhm M. Neuroendocrine regulation of sebocytes-a pathogenetic link between stress and acne. Exp Dermatol. 2004;13(Suppl 4):31–5.

Zouboulis CC, Chen W. The sebaceous gland and its role as an endocrine organ. World Clin Dermatol. 2013;1(1):37–51.

Zouboulis CC, Makrantonaki E. Clinical aspects and molecular diagnostics of skin aging. Clin Dermatol. 2011;29:3–14.

Zouboulis CC, Piquero-Martin J. Update and future of systemic acne treatment. Dermatology. 2003;206:37–53.

Zouboulis CC, Seltmann H, Neitzel H, Orfanos CE. Establishment and characterization of an immortalized human sebaceous gland cell line (SZ95). J Invest Dermatol. 1999;113:1011–20.

Zouboulis CC, Seltmann H, Hiroi N, et al. Corticotropin releasing hormone: an autocrine hormone that promotes lipogenesis in human sebocytes. Proc Natl Acad Sci U S A. 2002;99:7148–53.

Zouboulis CC, Fimmel S, Ortmann J, Turnbull JR, Boschnakow A. Sebaceous glands. In: Hoath SB, Maibach HI, editors. Neonatal skin-structure and function. 2nd ed. New York: Marcel Dekker; 2003. pp. 59–88.

Zouboulis CC, Chen W, Thornton MJ, Qin K, Rosenfield RL. Sexual hormones in human skin. Horm Metab Res. 2007;39:85–95.

Chapter 9
Sebaceous Lipids

Apostolos Pappas

Core Messages

- Human sebum is a mixture of nonpolar lipids: triglycerides, wax esters, squalene, fatty acids, and small amounts of cholesterol, cholesterol esters, and diglycerides.
- Elevated sebum excretion is a major factor involved in the pathophysiology of acne and seboreeic dermatitis.
- The sebaceous gland produces lipid species that are exclusive to sebum and cannot be found in other areas of the body.
- Complexity and uniqueness characterize sebaceous lipids, since Δ6 monounsaturated chains, wax esters, and squalene accumulation are examples that manifest the uniqueness of sebaceous lipid biology.
- Genetic knockout animal models of sebaceous lipid synthesis demonstrate dramatic changes in skin physiology and pathology. Impairment of sebaceous lipid pathways results to severe skin phenotypes.

Abstract Skin is protected by a layer of lipids, of both sebaceous and keratinocyte origin, which cover the surface of the skin. Different compositions of surface lipids have been reported depending on the method of sampling. Lipids produced by the epidermal cells are usually less per area in regions rich in sebaceous glands. The holocrine nature of the sebaceous gland will eventually result that the sebum will eventually coat the surface of the skin and the fur. The sebaceous lipids are primarily nonpolar lipids as triglycerides, wax esters, and squalene, while the epidermal lipids consist of ceramides, free-fatty acids, and cholesterol in almost equimolar concentrations. The composition of the sebaceous lipids manifests uniqueness and an intriguing biology exclusive to this gland. Elevated sebum excretion is a major factor involved in the pathophysiology of acne, therefore, an attempt to decode

A. Pappas (✉)
The Johnson & Johnson Skin Research Center, CPPW, a division of Johnson & Johnson Consumer Companies, Inc., Skillman, NJ 08558, USA
e-mail: APappas@its.jnj.com

A. Pappas (ed.), *Lipids and Skin Health,* DOI 10.1007/978-3-319-09943-9_9

and elucidate the roles that these unique lipids have on normal skin functions and acne is imperative.

Abbreviations

KO	Knock out
SC	Stratum corneum
VLCFA	Very-long-chain fatty acids

Introduction

The sebaceous gland, although once considered to be a fossilized appendage of the skin, nowadays is considered to be an important endocrine organ; as the previous chapter in this book has already outlined. The holocrine eruption of the sebaceous gland results in the secretion and release of sebum. The sebum that consists of the most diverse and "exotic" lipid mixture found in a specific human gland will eventually coat the surface of the skin and the fur. The majority of the epidermal surface lipids in areas rich in sebaceous glands, are in fact of sebaceous origin, while the lipids produced by the epidermis are an insignificant fraction of the total extractable surface lipid (Greene et al. 1970). In these particular areas, as for example face, the epidermal origin lipids average between 5 and 10 μg/cm^2, compared with average recoveries to an order or two orders of magnitude (50–500 μg of sebum/cm^2) from the forehead. In addition, areas rich in sebaceous glands are usually the areas where acne lesions are manifested.

Human sebum is a unique mixture of nonpolar lipids with a diversity of species that can be found nowhere else in the human body. Although the main component is triglycerides, the major ingredient of fat cells, they also contain wax esters, squalene, fatty acids, and smaller amounts of cholesterol, cholesterol esters, and diglycerides (Downing et al. 1989; Smith and Thiboutot 2008; Stewart 1992; Strauss et al. 1991).

Sebaceous Lipids and Their Uniqueness

The sebaceous lipids are unique, diverse, and intriguing. Nicolaides (1974) has so far offered one of the best statements about them: "two key words characterize the uniqueness of skin lipids: complexity and perversity." The relative composition of sebum though is depending on the method of sampling used to acquire them; as demonstrated by the different reported compositions in various analytical papers. However, it is very well documented that the major components of sebum are the triglycerides. This class of lipids could potentially be sampled before or after their modification by bacteria, which hydrolyze them to free-fatty acids and glycerol (Nicolaides 1974; Downing et al. 1969; Haahti and Horning 1963; James and

Table 9.1 A compilation of the relative sebum composition

Lipid class	Range weight, %	Mean weight, %
Triglycerides	20–60	45
Wax esters	23–29	25
Squalene	10–14	12
Free-fatty acids	5–40	10
Cholesterol and sterol esters	1–5	4
Diglycerides	1–2	2

Wheatley 1956; Knags 2007). The mean weight % that is often cited in the literature is given in Table 9.1.

Remarkably enough, human sebaceous lipids are significantly different in quantity and quality from sebaceous lipids of other species that possess sebaceous glands or excretions (Nicolaides and Ansari 1968; Nikkari 1974; Stewart and Downing 1991). The reason for such a unique sebum composition is not yet understood; however, one also should consider that human skin has a unique texture and physiology. In addition, acne is also unique to humans. These seem to be pieces of the same puzzle, which suggest that the unique sebaceous lipids are associated to this odd and human specific disease. Elevated sebum excretion is clearly a major factor involved in the pathophysiology of acne (Cunliffe 1989; Thiboutot 2004; Zouboulis 2004).

The majority of lipids produced by all other organs of the human body are similar to the sebaceous lipids with the exception of sapienic acid, its metabolites, and the wax esters which are unique to sebaceous species, and there are no validated reports that it can be synthesized by any other organ of the body. In addition to the synthesis of sapienic acid, its metabolites, and wax esters, the accumulation of squalene and the presence of very-long chain hydroxylated fatty acids are certainly uncommon in other organs and unique manifestations in sebum (Nicolaides 1974; Nicolaides and Ansari 1968). In other mammals and rodents, there are also significant levels of unique lipids for example fatty acids with odd numbers of carbon atoms or with a branched chain. It is also possible that some of these molecules are in reality, products of the resident skin microbiome, since they are more common to the bacterial metabolism (Nicolaides 1971). Another possibility for their presence is that they could be synthesized from branched precursors, which stem from the essential branched amino acid catabolism (Stewart 1992).

a) Sapienic Acid

The predominant monounsaturated fatty acid of sebum is the sapienic acid (16:1, Δ6), which has its single double bond at the sixth position from the carboxyl end (Nicolaides 1974; Wertz 2007). In nature, long-chain fatty acids with similar chain length are the most abundant, however, their initial double bond is preferably inserted in the ninth position from the carboxyl end. The 16-carbon isomer with the single double bond at the ninth position is the palmitoleic acid, a naturally found and abundant in many tissues and organisms fatty acid. Sapienic acid in the human body is truly unique to sebum and is not found as previously mentioned in other organs. In addition, most likely humans do not obtain it from the diet since very few plant species have been reported to manufacture this unusual fatty acid (Nicolaides 1971, 1974). Remarkably enough sapienic acid can be elongated by two carbons in human sebaceous cells and further accommodate an additional double bond between the fifth and sixth carbon; process that yields the sebaleic acid (18:2Δ5, 8), which is the most specific reaction and metabolite that has been reported only in human sebaceous cells. Sapienic acid is still the most prevalent and in much higher levels than any of its derivatives, isomers, or other monounsaturated fatty acids found in sebum. However, the potential role of sapienic acid in the etiology of acne is still remained to be decoded and somehow controversial. There are certainly reports with a conflicting message to the presence of sapienic acid in sebum where it could correlate with elevated sebum levels (Smith et al. 2008), while other reports demonstrate its potency against bacteria commonly associated with acne (Drake et al. 2008; Georgel et al. 2005; Wille and Kydonieus 2003).

A more comprehensive summary on the sapienic acid biology and roles is included within this book (Chapter "Sapienic Acid and Sebum Specific Fatty Acid Metabolism of the Human Sebaceous Gland").

b) Wax Esters

Wax esters are also unique molecules to sebaceous cells and are not produced by any other cell type in the human body. They account for about 25–30 % of the total sebaceous gland lipids and their production correlates with sebaceous gland differentiation (Strauss et al. 1991; Wertz 2007). Several animal models have demonstrated a strong correlation between atrophic-sebaceous gland, impaired-wax ester synthesis, and skin and hair abnormalities (Chen et al. 2002; Miyazaki et al. 2001).

Wax ester synthases (Cheng and Russell 2004; Lardizabal et al. 2000) have recently been discovered, however, additional reports (Yen et al. 2005a, b) provided evidence that another family of enzymes can also synthesize waxes. Therefore, there is not a single and unique wax synthase and besides, the wax ester biosynthesis is still not fully explored in humans. DGAT1 and DGAT2 acyl transferases catalyze acylation of alcohols, diols, retinol, etc., as well as a diacylglycerol. The human wax synthases AWAT1 and AWAT2, belonging to the DGAT2 family, can transfer acyl groups to mono or diacylglycerol, fatty alcohol, and retinol (Turkish and Sturley 2009). Most probably, these multifunctional acyl tranferases can account for the majority of the mono- and diester waxes.

Although active wax synthesis correlates with the differentiation of sebaceous cells, it is still uncertain if they are the reason or the outcome of the differentiation process. In vitro, it is one of the distinct pathways that is usually downregulated no matter if explanted sebaceous glands, tissues, cell preps, or transformed cell lines are used, since in vitro experiments that report the synthesis of similar to 25 % of waxes in lipid excretions are missing. Age and sex related differences have been reported in wax ester synthesis, which also correlate with total sebum production

(Downing et al. 1989; Jacobsen et al. 1985; Stewart et al. 1982). In nature, waxes always act as protective layers against excessive dehydration or water permeability and therefore are found on the surface of leaves, fruits, plant stems, algae, or even the skin, feathers, and fur of animals and additionally they also coat the wall of bacteria and fungi (Kolattukudy 1980). Waxes are very stable molecules and are far more resistant to oxidation, hydrolysis, and heat than triglycerides or phospholipids. Their primary role of protection is not the only one as they could also have lubrication properties. They are the most hydrophobic molecules found on the surface of organisms and that is how they help in sealing in the internal moisture of tissues while they are preventing their excessive hydration (Kolattukudy 1980). In certain instances, the packing and physicochemical properties of the wax crystals demonstrate unusual surface self cleaning properties that repel not only moisture, but together with water any kind of physical or biological invader, from dust to pollen, mites, and bacteria. This phenomenon has been termed as the "lotus effect" and is a fundamental property of the way that waxes can pack and form microstructures that could induce water repelling (Koch et al. 2007). For a more comprehensive summary on the wax ester biology and roles the reader should read the included chapter within this book (Chapter "Wax Esters: Chemistry and Biosynthesis").

c) Squalene

It will not be an understatement to claim that there is nothing unique about the synthesis of squalene since it is a precursor of cholesterol, therefore, most mammalian cells that have the capacity to synthesize cholesterol could also synthesize squalene. Cholesterol is an indispensable molecule for cell membranes fluidity and structure. Squalene is its precursor that is a long-unsaturated hydrocarbon. In any other tissue of the human body other than the sebaceous gland, it will readily get converted to the intermediates that would eventually yield the final product, cholesterol (Elias and Feingold 1992; Nicolaides 1974). The uniqueness in human sebum is the presence and accumulation of this cholesterol precursor to unusually high levels (12–15 %); and such levels have never been reported to any other tissue or organ. On the other hand, cholesterol does not account more than 2 % of the total sebaceous gland lipids; although is abundant and to the level of 30 % in the epidermal keratinocyte origin lipids and therefore it could be a contaminating source of cholesterol in sebum. Squalene synthase is the enzyme responsible for the production of squalene and squalene epoxidase or monooxygenase for its further processing to cholesterol. It is speculated that in sebaceous cells the activity of these two enzymes regulates the observed accumulation of squalene.

Squalene is very hydrophobic as a long hydrocarbon, however, highly unsaturated, it attributes fluidity to the molecule, therefore it becomes a natural lubricant with high penetration efficiency; therefore its role could be more diverse than just being a precursor of cholesterol. Past reports demonstrated, but not fully validated, possible roles of squalene oxidation products on UV protection (Ohsawa et al. 1984) and also irritation (Chiba et al. 2000). These oxidation products, together with oxidation products from unsaturated free fatty acids, have been reported to be comedogenic (Kligman et al.1970; Motoyoshi 1983). Perhaps this could be a reason as to why human sebum transports lipophilic antioxidants as vitamin E (Thiele et al. 1999) or humectants as glycerol (Fluhr et al. 2003), which orchestrate important roles in protecting skin from lipid oxidation and proper barrier function, respectively.

A more comprehensive summary on the squalene biology and roles is included within this book (Chapter "Squalene Chemistry and Biology").

Importance of Sebaceous Lipids and Animal Models

Genetic knock out (KO) animal models of lipid synthesis have clearly demonstrated the importance of sebaceous lipids in skin and hair physiology. In most of these studies, skin and fur abnormalities became the common denominator, once a certain sebaceous lipid pathway is disturbed. One of the first models to be reported, the melanocortin-5 receptor (MC5-R) KO, resulted in severe defects in water repulsion and thermoregulation due to decreased production of sebaceous lipids (Chen et al. 1997). The effect of the MC5-R on sebaceous lipid metabolism unveiled an additional path for the melanocortin receptors, besides their anticipated role on pigmentation, obesity, or body weight regulation.

Two years after the MC5-R KO was reported, Zheng et al. (1999) demonstrated by positional cloning that the dramatic alopecia manifested in the asebia mouse is due to the lack of a functional stearoyl-CoA desaturase (Scd1) enzyme activity. The absence of mature sebaceous glands demonstrated the evident significance of the *SCD1* gene and its products (monounsaturated fatty acids) to normal sebaceous gland function and additionally their role in hair development and health. The same findings were further confirmed 2 years later, in 2001, by what we can call as the reverse experiment; where the *SCD1* KO mice were constructed which evidently bared a similar phenotype (Miyazaki et al. 2001). The revelation that in both models the *SCD1* activity is solely responsible for scant to absent hair and hypoplastic to absent sebaceous glands was further supported by the fact that sebaceous glands are also scant in certain forms of alopecias (Sundberg 1994). The skin of the *SCD1* KO mice demonstrated lower than normal levels of triglycerides and wax esters, besides the expected lower than normal levels in the *SCD1* direct products as oleic acid and monounsaturated fatty acids.

An amazingly similar phenotype was demonstrated in the acyl CoA:diacylgylcerol acyltransferase 1 (DGAT1) KO mouse, where sebaceous gland atrophy and hair loss were also evident and the most severe skin manifestation (Chen et al. 2002). DGAT is the primary triglyceride synthase and there are two isoforms, DGAT1 and DGAT2 in mice. However, these two isoforms differ in sequence and localization (Headington 1996). DGAT1 has a distinct role and expression than the DGAT2; since it is also involved in the synthesis of wax esters, unlike DGAT2 (Yen et al. 2005a), and is expressed in most tissues, including the sebaceous gland (Chen et al. 2002; Cases et al. 1998). The demonstrated involvement of DGAT1 in wax ester synthesis is consistent with the observation that there are no or little wax esters in the fur lipids of the DGAT1 KO mouse.

The DGAT2 KO animals (Stone et al. 2004), in a similar fashion to SCD2 KO mice (Miyazaki et al. 2005), do not survive due to an impaired skin barrier function. The animals deficient in SCD-2 demonstrated abnormal skin barrier function due to abnormal lamellar bodies and epidermal maturation. This further proved that the presence of monounsaturated fatty acids is vital also for skin's barrier component and besides the formation of the sebaceous glands.

Additional animal models that demonstrated the importance of sebaceous and also epidermal lipids to skin functions are the KO of the elongases 3 and 4 that are responsible for the synthesis of very-long-chain fatty acids (VLCFA) (Westerberg et al. 2004; Vasireddy et al. 2007). The *Elovl3* gene product is involved in the formation of VLCFA and has a distinct expression in the skin that is restricted to sebaceous and epithelial cells of hair follicles. Disruption of that gene generated a KO model with impaired formation of neutral lipids necessary for skin functions but in addition resulted in disturbed water barrier and increased transepidermal water loss. This was caused to a certain extent from a disruption in normal lamellar body formation that the deficiency of *Elovl3* has caused but most profoundly the *Elovl3*-ablated mice demonstrated sparse hair coats and hyperplastic sebaceous glands with unusual lipid content mainly in monounsaturated fatty acid with 20 carbons.

Furthermore, the importance of the elongated fatty acids was recently further confirmed in a similar animal model where deficiency of the elongase of very-long chain fatty acid -like 4 (ELOVL4) displayed a scaly and wrinkled skin (Vasireddy et al. 2007). In addition, this model demonstrated a severely compromised epidermal permeability barrier function, which resulted in death within a few hours after birth (and similar to SCD2 and DGAT2 KO models). However, in this particular model the skin histology showed an abnormally compacted outer epidermis (SC), and electron microscopy revealed deficient epidermal lamellar body contents, decreased levels in VLCFA (>C28) in ceramide, glucosylceramide, and the free-fatty acid fractions, demonstrating the necessity of VLCFA for the synthesis of skin ceramides rather than a sebaceous gland deficiency.

In addition to the above preclinical models, few other models have additionally demonstrated the importance of sebaceous and epidermal lipid pathways for skin's integrity and physiology. Fatty acid transport proteins are fundamental to nonpolar lipid synthesis and out of the six mammalian FATPs, FATP4 seems to be the most important one involved in skin lipid biosynthesis (Schmuth et al. 2005). Deletion of FATP4 in mice resulted in perturbations in the biosynthesis of skin lipids mainly by keratinocytes that caused barrier dysfunction and neonatal fatality (Lin et al. 2013). One of the pathways that was drastically reduced by the absence of FATP4 was the wax-ester synthesis.

In addition, mammalian sebaceous glands also produce some wax diesters that require 2-hydroxylation of fatty acids which is catalyzed by the mammalian fatty acid 2-hydroxylase (FA2H). When *FA2H* gene, that was originally thought to be important for the synthesis of hydroxylated fatty acid containing sphingolipids of the murine skin, was knocked out, the *FA2H* deficient mice did not show any effect on the level of such sphingolipids (Maier et al. 2011). *FA2H* expression was found to be restricted to sebaceous glands and *FA2H* deficiency caused a drastic reduction not only in 2-hydroxylated glucosylceramides but in addition to diester waxes. The subsequent change in surface lipids caused a blockage in hair canals and thus severely interfered with hair growth and health.

Besides the lipid processing enzymes, we also have evidence that the most important lipid transcriptional factor, the peroxisome proliferating activated receptor gamma (PPARγ) is also required for the formation of sebaceous glands (Fu et al. 2010), but in addition regulates sebocyte functions such as cellular responses to oxidative stress (Zhang et al. 2006) and eotaxin production (Nakahigashi et al. 2012).

The Gsdma3(Dfl)/+ mice are defolliculated mice, characterized by defective sebaceous glands associated with a hair loss phenotype. This is a similar phenotype to the one observed in scarring alopecia, since aberrant hair cycle, chronic inflammation, reduced lipogenesis in sebaceaous glands, loss of hair follicles, and reduced expression of PPARγ were all features present in defolliculated mice. Consequently, the use of these mice could serve as a good model to test the effects on PPARγ activation (Ruge et al. 2011) and their effect of lipid synthesis in the pilosebaceous follicle.

Prostaglandins besides being proinflammatory mediators involved in acne lesion development (Niemann and Horsley 2012) are also PPAR agonists. Mice with an increased cyclooxygenase-2 (COX-2) expression and prostaglandins E2 levels demonstrated sebaceous gland hyperplasia and subsequently enhanced sebum production (Zhang et al. 2006). This directly suggests an important role for COX-2 signaling pathway in sebocyte biology as expression and activation of COX-2 has been shown in in vitro models to be PPARγ-mediated as well. It has been reported that additional lipid oxidizing agents that activate PPARγ *could further* induce lipogenesis in sebocytes (Nakahigashi et al. 2012; Ruge et al. 2011; Niemann and Horsley 2012). All these findings together allow the hypothesis that sebocyte differentiation and lipogenesis could be regulated by PPARγ-mediated pathways.

Nutrition and Essential Fatty Acids

Linoleic (18:2, Δ9,12) and α-linolenic acid (18:2, Δ9,12,15) are the two most important polyunsaturated and essential fatty acids that come strictly form the diet. They are the precursors of all eicosanoids and bioactive lipid mediators that stem from the arachidonic acid and prostaglandin metabolism, as captured in Chapter "Polyunsaturated Fatty Acid Oxygenated Metabolites in Skin." Like vitamins and minerals they also survive the digestive tract and are delivered via the systemic circulation to various organs. Sebum analysis demonstrates that these dietary essential fatty acids and their derivatives constitute small amounts in surface lipid samplings (Nicolaides 1974). Not long ago, two intriguing studies that took place in the beginning of this millennium (Fu and Sinclair 2000; Fu et al. 2001), revealed a tight association of these essential fatty acids with the skin. When guinea pigs were dosed with radioactively labeled linoleic and linolenic acids, their skin and fur became the most heavily labeled tissue, via the action of the sebaceous glands, suggesting an essential role for essential fatty acids in sebaceous and skin biology.

The essential fatty acid metabolism in sebaceous cells has not been fully elucidated. One study has claimed that acne patients had a linoleic acid deficiency (Downing et al. 1986). On the other hand, it is well-known that linoleic acid is an essential structural component of skin ceramides. An additional study (Pappas et al. 2002) revealed that linoleic acid undergoes a rapid oxidation and selective degradation in sebaceous cells. Consequently, this selective and specific to the sebaceous cells activity will allow palmitic acid to be available as the sole substrate to the delta 6 desaturase of the sebaceous cells and allow them to synthesize the sebaceous specific fatty acid, the sapienic acid. The delta 6 desaturase is the predominant desaturase of the human sebaceous cells (Ge et al. 2003) that otherwise would have catalyzed the synthesis of more omega-6 derivatives from linoleic acid, since it is the enzyme's preferred substrate. The observed selective β-oxidation of linoleic acid correlated with the ability of the sebaceous glands to synthesize wax esters, which are differentiation markers for the sebaceous cells. Thus, oxidation of linoleic acid is specific to sebaceous cells and correlates with the two most important differentiation properties, wax ester and sapienic acid synthesis.

Epilogue

Sebaceous lipids are important for healthy skin and fur. The idea that the unusual lipid species found on sebaceous gland could make the microenvironment unfriendly to fungi and bacteria has gained more attention. Even if bacteria are able to consume the major component of sebum, the triglycerides, these upon hydrolysis will yield fatty acids that are unusual enough; in combination with other perverse lipids they could orchestrate unique mechanisms that will select which organism is an enemy and which is desirable on our skin. Additional studies of course are in need to decode the complete roles of these unique mixtures of lipids and their relationship to acne.

Better analytical techniques and lipidomics would certainly increase our understanding on their role and clarify their complexity.

References

Cases S, Smith SJ, Zheng YW, Myers HM, Lear SR, Sande E, Novak S, Collins C, Welch CB, Lusis AJ, Erickson SK, Farese RV Jr. Identification of a gene encoding an acyl CoA: diacylglycerol acyltransferase, a key enzyme in triacylglycerol synthesis. Proc Natl Acad Sci U S A. 1998;95:13018–23.

Chen W, Kelly MA, Opitz-Araya X, Thomas RE, Low MJ, Cone RD. Exocrine gland dysfunction in MC5-R-deficient mice: evidence for coordinated regulation of exocrine gland function by melanocortin peptides. Cell. 1997;91(6):789–98.

Chen HC, Smith SJ, Tow B, Elias PM, Farese RV Jr. Leptin modulates the effects of acyl CoA: diacylglycerol acyltransferase deficiency on murine fur and sebaceous glands. J Clin Invest. 2002;109:175–81.

Cheng JB, Russell DW. Mammalian wax biosynthesis. II. Expression cloning of wax synthase cDNAs encoding a member of the acyltransferase enzyme family. J Biol Chem. 2004;279(36):37798–807.

Chiba K, Yoshizawa K, Makino I, Kawakami K, Onoue M. Comedogenicity of squalene monohydroperoxide in the skin after topical application. J Toxicol Sci. 2000;25(2):77–83.

Cunliffe WJ. Acne. London: Martin Dunitz; 1989.

Downing DT, Strauss JS, Pochi PE. Variability in the chemical composition of human skin surface lipids. J Invest Dermatol. 1969;53(5):322–7.

Downing DT, Stewart ME, Wertz PW, Strauss JS. Essential fatty acids and acne. J Am Acad Dermatol. 1986;14(2 Pt 1):221–5.

Downing DT, Stewart ME, Strauss JS. Changes in sebum secretion and the sebaceous gland. Clin Geriatr Med. 1989;5:109–14.

Drake DR, Brogden KA, Dawson DV, Wertz PW. Thematic review series: skin lipids. Antimicrobial lipids at the skin surface. J Lipid Res. 2008;49(1):4–11.

Elias PM, Feingold KR. Lipids and the epidermal water barrier: metabolism, regulation, and pathophysiology. Semin Dermatol. 1992;11(2):176–82.

Fluhr JW, Mao-Qiang M, Brown BE, Wertz PW, Crumrine D, Sundberg JP, Feingold KR, Elias PM. Glycerol regulates stratum corneum hydration in sebaceous gland deficient (asebia) mice. J Invest Dermatol. 2003;120(5):728–37.

Fu Z, Sinclair AJ. Increased alpha-linolenic acid intake increases tissue alpha-linolenic acid content and apparent oxidation with little effect on tissue docosahexaenoic acid in the guinea pig. Lipids. 2000;35(4):395–400.

Fu Z, Attar-Bashi NM, Sinclair AJ. 1-14C-linoleic acid distribution in various tissue lipids of guinea pigs following an oral dose. Lipids. 2001;36(3):255–60.

Fu G, et al. Committed differentiation of hair follicle bulge cells into sebocytes: an in vitro study. Int J Dermatol. 2010;49(2):135–40.

Ge L, Gordon JS, Hsuan C, Stenn K, Prouty SM. Identification of the delta-6 desaturase of human sebaceous glands: expression and enzyme activity. J Invest Dermatol. 2003;120(5):707–14.

Georgel P, Crozat K, Lauth X, Makrantonaki E, Seltmann H, Sovath S, Hoebe K, Du X, Rutschmann S, Jiang Z, Bigby T, Nizet V, Zouboulis CC, Beutler B. A toll-like receptor 2-responsive lipid effector pathway protects mammals against skin infections with gram-positive bacteria. Infect Immun. 2005;73(8):4512–21.

Greene RS, Downing DT, Pochi PE, Strauss JS. Anatomical variation in the amount and composition of human skin surface lipid. J Invest Dermatol. 1970;54(3):240–7.

Haahti E, Horning EC. Isolation and characterization of saturated and unsaturated fatty acids and alcohols of human skin surface lipids. Scand J Clin Lab Invest. 1963;15:73–8.

Headington JT. Cicatricial alopecia. Dermatol Clin. 1996;14:773–82.

Jacobsen E, Billings JK, Frantz RA, Kinney CK, Stewart ME, Downing DT. Age-related changes in sebaceous wax ester secretion rates in men and women. J Invest Dermatol. 1985;85:483–5.

James AT, Wheatley VR. Studies of sebum. 6. The determination of the component fatty acids of human forearm sebum by gas-liquid chromatography. Biochem J. 1956;63(2):269–73.

Kligman AM, Wheatley VR, Mills OH. Comedogenicity of human sebum. Arch Dermatol. 1970;102(3):267–75.

Knags H. Cell biology of the pilosebaceous unit. In: Webster GF, Rawlings AV, editors. Acne and its therapy. New York: Informa Healthcare; 2007. pp. 9–36.

Koch K, Dommisse A, Barthlott W, Gorb SN. The use of plant waxes as templates for micro- and nanopatterning of surfaces. Acta Biomater. 2007;3(6):905–9.

Kolattukudy PE. Cutn, suberin and waxes. In: Stumpf PK, editor. Comprehensive biochemistry of plants. Vol. IV. London: Academic; 1980. pp. 600–45.

Lardizabal KD, Metz JG, Sakamoto T, Hutton WC, Pollard MR, Lassner MW. Purification of a jojoba embryo wax synthase, cloning of its cDNA, and production of high levels of wax in seeds of transgenic arabidopsis. Plant Physiol. 2000;122(3):645–55.

Lin MH, Hsu FF, Miner JH. Requirement of fatty acid transport protein 4 for development, maturation, and function of sebaceous glands in a mouse model of ichthyosis prematurity syndrome. J Biol Chem. 2013;288(6):3964–76.

Maier H, Meixner M, Hartmann D, Sandhoff R, Wang-Eckhardt L, Zöller I, Gieselmann V, Eckhardt M. Normal fur development and sebum production depends on fatty acid 2-hydroxylase expression in sebaceous glands. J Biol Chem. 2011;286(29):25922–34.

Miyazaki M, Man WC, Ntambi JM. Targeted disruption of stearoyl-CoA desaturase1 gene in mice causes atrophy of sebaceous and meibomian glands and depletion of wax esters in the eyelid. J Nutr. 2001;131(9):2260–8.

Miyazaki M, Dobrzyn A, Elias PM, Ntambi JM. Stearoyl-CoA desaturase-2 gene expression is required for lipid synthesis during early skin and liver development. Proc Natl Acad Sci U S A. 2005;102(35):12501–6.

Motoyoshi K. Enhanced comedo formation in rabbit ear skin by squalene and oleic acid peroxides. Br J Dermatol. 1983;109(2):191–8.

Nakahigashi K, et al. PGD2 induces eotaxin-3 via PPARgamma from sebocytes: a possible pathogenesis of eosinophilic pustular folliculitis. J Allergy Clin Immunol. 2012;129(2):536–43.

Nicolaides N. The structures of the branched fatty acids in the wax esters of vernix caseosa. Lipids. 1971;6(12):901–5.

Nicolaides N. Skin lipids: their biochemical uniqueness. Science. 1974;186(4158):19–26. (Review).

Nicolaides N, Ansari MN. Fatty acids of unusual double-bond positions and chain lengths found in rat skin surface lipids. Lipids. 1968;3(5):403–10.

Niemann C, Horsley V. Development and homeostasis of the sebaceous gland. Semin Cell Dev Biol. 2012;23(8):928–36.

Nikkari T. Comparative chemistry of sebum. J Invest Dermatol. 1974;62:257–67.

Ohsawa K, Watanabe T, Matsukawa R, Yoshimura Y, Imaeda K. The possible role of squalene and its peroxide of the sebum in the occurrence of sunburn and protection from the damage caused by U.V. irradiation. J Toxicol Sci. 1984;9(2):151–9.

Pappas A, Anthonavage M, Gordon JS. Metabolic fate and selective utilization of major fatty acids in human sebaceous gland. J Invest Dermatol. 2002;118(1):164–71.

Ruge F, et al. Delineating immune-mediated mechanisms underlying hair follicle destruction in the mouse mutant defolliculated. J Invest Dermatol. 2011;131(3):572–9.

Schmuth M, Ortegon AM, Mao-Qiang M, Elias PM, Feingold KR, Stahl A. Differential expression of fatty acid transport proteins in epidermis and skin appendages. J Invest Dermatol. 2005;125(6):1174–81.

Smith KR, Thiboutot DM. Sebaceous gland lipids: friend or foe? J Lipid Res. 2008;49(2):271–81.

Smith RN, Braue A, Varigos GA, Mann NJ. The effect of a low glycemic load diet on acne vulgaris and the fatty acid composition of skin surface triglycerides. J Dermatol Sci. 2008;50(1):41–52.

Stewart ME. Sebaceous glands lipids. Semin Dermatol. 1992;11:100–5.

Stewart ME, Downing DT. Chemistry and function of mammalian sebaceous lipids. Adv Lipid Res. 1991;24:263–301.

Stewart ME, Quinn MA, Downing DT. Variability in the fatty acid composition of wax esters from vernix caseosa and its possible relation to sebaceous gland activity. J Invest Dermatol. 1982;78(4):291–5.

Stone SJ, Myers HM, Watkins SM, Brown BE, Feingold KR, Elias PM, Farese RV Jr. Lipopenia and skin barrier abnormalities in DGAT2-deficient mice. J Biol Chem. 2004;279(12):11767–76.

Strauss JS, Downing DT, Ebling JF, Stewart ME. Sebaceous glands. In: Goldsmith LA, editor. Physiology, biochemistry and molecular biology of the skin. New York: Oxford University Press; 1991. pp. 712–40.
Sundberg JP. The asebia (ab, ab1) mutations, chromosome 19. In: Sundberg JP, editor. Handbook of mouse mutations with skin and hair abnormlities. Bar Harbor: CRC Press; 1994. pp. 171–8.
Thiboutot D. Regulation of human sebaceous glands. J Invest Dermatol. 2004;123:1–12.
Thiele JJ, Weber SU, Packer L. Sebaceous gland secretion is a major physiologic route of vitamin E delivery to skin. J Invest Dermatol. 1999;113(6):1006–10.
Turkish AR, Sturley SL. The genetics of neutral lipid biosynthesis: an evolutionary perspective. Am J Physiol Endocrinol Metab. 2009;297(1):E19–27.
Vasireddy V, Uchida Y, Salem N Jr, Kim SY, Mandal MN, Reddy GB, Bodepudi R, Alderson NL, Brown JC, Hama H, Dlugosz A, Elias PM, Holleran WM, Ayyagari R. Loss of functional ELOVL4 depletes very long-chain fatty acids (> or = C28) and the unique omega-O-acylceramides in skin leading to neonatal death. Hum Mol Genet. 2007;16(5):471–82.
Wertz PW. Sebum secretions and acne. In: Webster GF, Rawlings AV, editors. Acne and its therapy. New York: Informa Healthcare; 2007. pp. 37–43.
Westerberg R, Tvrdik P, Undén AB, Månsson JE, Norlén L, Jakobsson A, Holleran WH, Elias PM, Asadi A, Flodby P, Toftgård R, Capecchi MR, Jacobsson A. Role for ELOVL3 and fatty acid chain length in development of hair and skin function. J Biol Chem. 2004;279(7):5621–9.
Wille JJ, Kydonieus A. Palmitoleic acid isomer (C16:1delta6) in human skin sebum is effective against gram-positive bacteria. Skin Pharmacol Appl Skin Physiol. 2003;16(3):176–87.
Yen CL, Monetti M, Burri BJ, Farese RV Jr. The triacylglycerol synthesis enzyme DGAT1 also catalyzes the synthesis of diacylglycerols, waxes, and retinyl esters. J Lipid Res. 2005a;46(7):1502–11.
Yen CL, Brown CH 4th, Monetti M, Farese RV Jr. A human skin multifunctional O-acyltransferase that catalyzes the synthesis of acylglycerols, waxes, and retinyl esters. J Lipid Res. 2005b;46(11):2388–97.
Zhang Q, et al. Involvement of PPARgamma in oxidative stress-mediated prostaglandin E(2) production in SZ95 human sebaceous gland cells. J Invest Dermatol. 2006;126(1):42–8.
Zheng Y, Eilertsen KJ, Ge L, Zhang L, Sundberg JP, Prouty SM, Stenn KS, Parimoo S. Scd1 is expressed in sebaceous glands and is disrupted in the asebia mouse. Nat Genet. 1999;23:268–70.
Zouboulis CC. Acne and sebaceous gland function. Clin Dermatol. 2004;22:360–6.

Chapter 10
Sapienic Acid: Species-Specific Fatty Acid Metabolism of the Human Sebaceous Gland

Stephen M. Prouty and Apostolos Pappas

Core Messages

- Sapienic acid is the most prevalent unsaturated fatty acid in human sebum.
- A high concentration of naturally occurring sapienic acid is documented in only one place in the animal kingdom: human sebum.
- Fatty Acid Desaturase 2 (FADS2) is the rate limiting enzyme in the formation of polyunsaturated fatty acids (termed here as the "PUFA-type" reaction) and only in the human sebaceous gland does FADS2 catalyze a "sebaceous-type" reaction to produce high levels of sapienic acid.
- Multiple tissue-specific mechanisms are required for production of sapienic acid, one of which is reduction of competing desaturase activity in human sebaceous glands.
- Sapienic acid is a "first-line" component of the innate immune system at the cutaneous surface, where it functions as an antimicrobial agent.
- Elucidation of sapienic acid metabolism in human sebaceous cells can provide valuable insight into the etiology and pathophysiology of human skin diseases such as acne and atopic dermatitis.

Abstract Hair follicle-associated sebaceous glands secrete sebum, a highly complex lipid mixture that covers the skin surface and hair shafts. The functional versatility of lipids, combined with the wide array of sebaceous lipid classes and aliphatic moieties, provide mammals with a substrate that facilitates adaptation to their diverse environments, including interaction with animals and microbes. Unique among the complexity of sebaceous lipids is sapienic acid, a 16 carbon monounsaturated fatty

S. M. Prouty (✉)
1075 BRB II/III, Department of Dermatology, Perelman School of Medicine,
University of Pennsylvania, 421 Curie Blvd., Philadelphia, PA 19104, USA
e-mail: proutysm@mail.med.upenn.edu

A. Pappas
The Johnson & Johnson Skin Research Center, CPPW, a division of Johnson & Johnson Consumer Companies, Inc., Skillman, NJ 08558, USA
e-mail: APappas@its.jnj.com

A. Pappas (ed.), *Lipids and Skin Health,* DOI 10.1007/978-3-319-09943-9_10

acid with an extremely rare position of the double bond, located between carbons 6 and 7 from the carboxyl terminal. Human sebum is the only documented location in the animal kingdom where sapienic acid is abundant and naturally occurring. It is produced by fatty acid desaturase 2 (*FADS2*), the same enzyme that is rate-limiting in the formation of polyunsaturated fatty acids. Multiple tissue-specific mechanisms are utilized in the human sebaceous gland in order to "repurpose" *FADS2* for the production of sapienic acid, chief among which is the reduction of competing desaturase activity. Among mammals, human sebum has the highest amount of free fatty acids, of which sapienic acid is the most abundant monounsaturated fatty acid. Consistent with the role of fatty acids in modulating host-microbe interactions, sapienic acid has the highest antimicrobial activity among free fatty acids in human sebum, while also demonstrating selectivity for *Staphylococcus aureus,* an opportunistic pathogen. Increased infection by *Staphylococcus aureus* is associated with a reduction in sapienic acid in sebum of patients with atopic dermatitis, and topical application of sapienic acid is correlated with decreased bacterial load and amelioration of symptoms. Taken together, this strongly suggests that sapienic acid functions as a "first-line" component of the innate immune system at the cutaneous surface. The species-specific nature of sapienic acid in human sebum is related to the unique architecture of human skin and its microbial environment. Insight into pathogenesis of human skin disease will benefit from further investigation into the biochemistry of sapienic acid production in human sebaceous glands.

Abbreviations

AA	arachidonic acid, 20:4n-6, 20:4Δ5,8,11,14
ALA	α-linolenic acid, 18:3n-3, 18:3Δ9,12,15
AWAT1	acyl-CoA wax alcohol acyltransferase 1
AWAT2	acyl-CoA wax alcohol acyltransferase 2
DGAT1	diacylglycerol O-acyltransferase 1
DGAT2	diacylglycerol O-acyltransferase 2
DHA	docosahexaenoic acid, 22:6n-3, 22:6Δ4,7,10,13,16,19
EFA	essential fatty acid
EPA	eicosapentaenoic acid, 20:5n-3, 20:5Δ5,8,11,14,17
EST	expressed sequence tag
FADS1	fatty acid desaturase 1
FADS2	fatty acid desaturase 2
FFA	free fatty acid
6-HA	*cis*-6-hexadecenoic acid, 6*Z*-hexadecenoic acid, 16:1n-10, 16:1Δ6
15-HETE	15-hydroxyeicosatetraenoic acid
5-HODE,	5-hydroxy-(6*E*,8*Z*)-octadecadienoic acid
13-HODE	13-hydroxyoctadecadienoic acid
LA	linoleic acid, 18:2n-6, 18:2Δ9,12
15-LOX-2	15-lipoxygenase-2
OA	oieic acid, 18:1n-9, 18:1Δ9
5-oxo-ODE	5-oxo-(6*E*,8*Z*)-octadecadienoic acid

PA	palmitic acid, 16:0
POA	palmitoleic acid, 16:1n-7, 16:1Δ9
PPAR-γ	peroxisome proliferator-activated receptor-γ
MUFA	monounsaturated fatty acid
PUFA	polyunsaturated fatty acid
SA	sapienic acid, 16:1n-10, 16:1Δ 6
SCD	stearoyl-CoA desaturase
SFA	saturated fatty acid
TG	triacylglycerol
UFA	unsaturated fatty acid

Introduction

A striking feature of mammalian skin is the profusion of glands (both in type and quantity), a development that occurred after the separation of the two major lineages of amniotes, one leading to mammals and the other to reptiles/birds (Dhouailly 2009). The skin of birds contains only one major lipid-producing gland (uropygial gland; for conditioning of feathers) and reptiles have only a few localized skin glands that produce pheromones (Jared et al. 1999; Dhouailly 2009; Schempp et al. 2009). In contrast, mammalian skin glands include those that produce (1) a lipid mixture known as sebum (sebaceous glands), (2) sweat (apocrine and eccrine glands), and (3) milk (mammary glands). Whereas the roles of milk (neonatal nutrition) and eccrine sweat (thermoregulation) are well defined, the role(s) of sebum is less clear. This is partly due to the greater anatomical and chemical diversity of sebaceous glands and sebum, respectively (both between and within species). The lipids of sebum are (1) highly diverse in lipid class and aliphatic moieties, (2) are not typically found in internal tissues, and (3) are species-specific. This heterogeneity parallels the diversity of mammalian speciation and suggests that sebum plays a multifunctional role in adaptation of the animal to its environment (terrestrial and/or aquatic) and for interactions with other organisms, including microbes. The chemical properties of lipids are well suited for this as they are hydrophobic (waterproofing of skin and hair; Lupi 2008; Lin et al. 2013), are more volatile than carbohydrates or proteins (pheromonal communication; Hadley 1985), and their amphiphilic properties can emulsify and disrupt cell membranes (antimicrobial activity; Desbois and Smith 2010; Parsons et al. 2012).

There are two main types of sebaceous glands in mammals: (1) those that are associated with hair follicles and (2) free sebaceous glands. As most mammals have abundant hair follicles, follicle-associated sebaceous glands are much more numerous; thus this sebum accounts for the vast majority of skin surface lipids. Follicle-associated sebaceous glands secrete sebum into the sebaceous duct, which then continues via the follicular infundibulum to the surface of the skin and hair shaft. In contrast to the generalized anatomical distribution of hair follicle-associated sebaceous glands, free sebaceous glands are more localized and, usually lacking a hair follicle, secrete sebum directly to the surface (Thody and Shuster 1989). Two main categories of free sebaceous glands can be distinguished. The first category are those that are relatively small, are devoid of hair follicles (with rare exceptions), and are located at or near mucocutaneous junctions such as the anogenital region, oral region (vermillion border of lip and in the oral mucosa), eyelid, and areolae (Miles 1958; Hyman and Guiducci 1963). These locations (areas of ingress into internal body regions), as well as the nature of the lipid secretion of these sebaceous glands suggest, in part, a protective function (Brasser et al. 2011). The second category of free sebaceous glands are large aggregates of sebaceous glandular tissue that variably associate with few hair follicles, and are located in specific body regions such as the ventral gland of gerbils, the costovertebral gland (flank organ) of hamsters, the preputial gland of mice, the tarsal gland of black-tailed deer, and the brachial gland of lemurs (Montagna and Yun 1962; Brownlee et al. 1969; Thody and Shuster 1989). In many cases, these large, free sebaceous glands are sexually dimorphic, show changes associated with hormonally controlled reproductive activity, and their lipid secretion functions as pheromones to signal a broad range of information such as an individual's reproductive and social status, species, age, and gender (Quay 1977; Knapp et al. 2006). Whereas the pheromonal function of this type of free sebaceous gland is well established, and an antimicrobial role for the smaller free sebaceous glands in the oral mucosa has been proposed (Brasser et al. 2011), the role of the more ubiquitous follicle-associated sebaceous gland in skin homeostasis is less clear. The cellular function of follicular sebaceous glands is production of sebum, the lipid constituents of which have been analyzed in many species. The fundamental observation from these studies (Nicolaides 1974; Stewart and Downing 1991) is that the lipids of sebum are highly complex and specific not only to a given species, but to individuals within a species. Taking into account this heterogeneity, as well as some basic commonalities (such as hydrophobicity), major roles of sebum are (1) waterproofing of surface of skin and hair shafts, (2) skin hydration, and (3) modulation of microbes. Additional functions may reflect adaptations specific to the organism, its skin architecture, and its environment.

Beyond this level of lipid complexity and heterogeneity, human sebum is unique among animals in possessing a 16-carbon monounsaturated fatty acid (MUFA) with a *cis* double bond located at the sixth carbon from the carboxyl end (16:1Δ6), or the tenth carbon from the methyl end (16:1n-10; Fig. 10.1). As there are several types of nomenclature for lipids, this fatty acid is also referred to as *cis*-6-hexadecenoic acid (6-HA) or 6Z-hexadecenoic acid[1] (systematic names), as well as sapienic acid[2] (SA), the common name (Stewart 1986) that reflects a highly restricted distribution in the animal kingdom, namely the sebum of the species *Homo sapiens*. In contrast, palmitoleic acid (POA) is a positional isomer of SA, with the *cis* double bond

[1] The name *cis*-6-hexadecenoic acid or 6Z-hexadecenoic acid is often used when referring to 16:1Δ6 (or 16:1n-10) originating from nonhuman sources. In this chapter, reference to this monounsaturated fatty acid will use the abbreviation 6-HA.

[2] The name sapienic acid is often used when referring to 16:1Δ6 (or 16:1n-10) originating in human sebum. In this chapter, reference to this monounsaturated fatty acid will use the abbreviation SA.

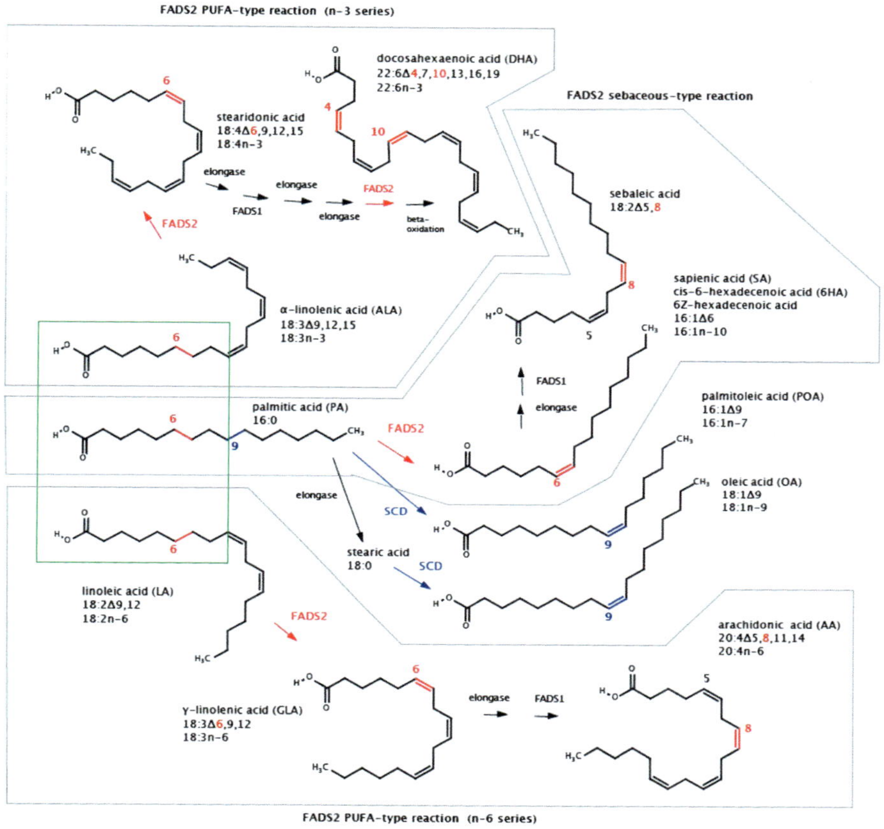

Fig. 10.1 Enzymatic pathway for sapienic acid and related fatty acids. Fatty acid desaturase 2 (*FADS2*) catalyzes two distinctly different types of desaturation reactions, depending on the tissue type. In most tissues, *FADS2* is the rate-limiting enzyme that catalyzes the polyunsaturated (PUFA)-type reaction, of which there are two series of noninterconvertible PUFAs. The essential fatty acid (diet-derived) α-linolenic acid (with a double bond located three carbons from the methyl end) is the parent PUFA of the n-3 series (*upper grey box*), with the metabolite docosahexaenoic acid having a critical role in nervous system function. The essential fatty acid (diet-derived) linoleic acid (with a double bond located six carbons from the methyl end) is the parent PUFA of the n-6 series (*lower grey box*), with the metabolite arachidonic acid having a critical role in eicosanoid signaling. Under normal conditions, the sebaceous-type reaction (*middle grey box*) of *FADS2* is only found in the human sebaceous gland, where palmitic acid, the parent saturated fatty acid, is converted to the monounsaturated fatty acid, sapienic acid. Sebaleic acid is the predominant PUFA in the human sebaceous gland, being formed by elongation of sapienic acid followed by *FADS1*-mediated desaturation. Note that the only difference between sebaleic acid and linoleic acid are the positions of the double bonds. Competing reactions for palmitic acid as substrate (outside of the *grey boxes*) are: (1) desaturation by stearoyl-CoA desaturase (*SCD*) to form the monounsaturated fatty acid, palmitoleic acid, and (2) elongation to form the saturated fatty acid, stearic acid, which can then be desaturated by *SCD* to produce oleic acid. Note that the only difference between sapienic acid and palmitoleic acid is the position of the double bond. The *green box* highlights the similarity of the first 9 carbons (from the carboxyl terminal) of the *FADS2* substrates with chain length of 16 or 18 carbons: the PUFAs linoleic and α-linolenic acids and the saturated fatty acid,

located at the ninth carbon from the carboxyl end (16:1Δ9), or the seventh carbon from the methyl end (16:1n-7; Fig. 10.1). POA is ubiquitous among most living organisms (Gurr et al. 2002a) and the sebum of many mammals (including the human fetus) contains POA and/or its metabolites (Stewart and Downing 1991).

Given the rare distribution of SA/6-HA in nature, and the biochemical complexity of its production (discussed below), a central question arises: "With regard to human sebum, what is the advantage of SA versus its positional isomer, POA, given that fatty acids with a double bond located at the ninth carbon from the carboxyl end (Δ9-desaturated) are present in the sebum of most other species, including the vernix caseosa (embryonic human sebum)?" This is a difficult question to answer at this time, since the precise function(s) of human sebum as a whole are unclear. However, clues may be gained by the examination of related questions such as: (1) where is SA produced, (2) how is SA produced, (3) are there human disorders that correlate with altered levels of SA, and (4) what are the functions of similar MUFAs in other systems?

Occurrence of Sapienic Acid

High concentrations of naturally occurring SA/6-HA are documented in only two places in the biosphere: human sebum (Nicolaides 1974) and the seed oil of the black-eyed Susan vine, *Thunbergia alata* Bojer ex Sims (Cahoon et al. 1994). Up to 47% of human sebum is free fatty acid (FFA; Kohler et al. 2009), of which SA is dominant, second only to palmitic acid (PA), the substrate for its production (Nicolaides 1974; Ge et al. 2003; Fig. 10.1). In *Thunbergia alata*, 6-HA comprises over 80% of seed oil by weight (Spencer et al. 1971; Cahoon et al. 1994). That abundant SA is found in human skin and 6-HA in a plant seed suggests that other natural sources may be found, especially with development of increasingly sensitive analytical methods. Indeed, 6-HA has been detected at low levels in oceanic particulate matter (Nichols et al. 1989), mushrooms (Marekov et al. 2012), neonatal rat brain (Spence 1970), and human blood plasma (Sansone et al. 2013). In addition, the geometric isomer, *trans*-6-hexadecenoic acid is present, again at low levels, in

palmitic acid. Despite the large structural differences towards the methyl end of these fatty acids (which plays a role in the higher affinity of PUFAs for *FADS2*), the molecular similarity toward the carboxyl end (*green box*) allows these fatty acids to be desaturated by a single *FADS2* enzyme, in which the catalytic site (between carbon 6 and 7) is determined relative to the carboxyl terminal. All *FADS2*-relevant reactions and structures are indicated in *red*, with relevant carbons greater than 6 resulting from elongation and/or beta-oxidation. The various naming conventions for the fatty acids are indicated: common name (including the abbreviation used in this article) and/or systematic name, shorthand notation indicating chain length-/number of double bonds-/position of double bonds, and shorthand notation indicating chain length-/number of double bonds-/fatty acid series based on double bond closest to methyl end (n-3 or n-6 PUFA families; n-7, n-9, n-10 for MUFA families). All *SCD*-relevant reactions and structures are indicated in *blue*. Not all metabolites are shown for the PUFA (n-3, n-6) and MUFA (n-7, n-9, n-10) fatty acid families some marine animals (Ackman et al. 1971; Hooper and Ackman 1971) and in human blood plasma (Sansone et al. 2013). (It is noted that the vast majority of biologically active unsaturated fatty acids (UFAs) have the *cis* (or Z) conformation). In contrast to the abundant sources of SA in human sebum and 6-HA in the seed oil of *Thunbergia alata* (where the synthetic pathways are known), the formation, and in some cases the source, of the low abundance 6-HA isomers is not defined.

Biosynthesis of Sapienic Acid

Despite the evolutionary distance between animals and plants, there are similarities with regard to the synthesis and high accumulation of SA/6-HA: (1) PA (16:0) is directly desaturated between the sixth and seventh carbons (from the carboxyl end) by a Δ6-desaturase in both human (*FADS2*; *F*atty *A*cid *D*esaturase *2*; also known as delta(Δ)6-desaturase; Fig. 10.1) and plant (Δ6 16:0-*a*cyl *c*arrier *p*rotein (ACP) desaturase, (2) triacylglycerol is the most prevalent lipid class in human sebum and *T. alata* seed oil and is the main lipid to which SA/6-HA is esterified, and (3) specialized biochemical pathways are required for the generation and utilization of this highly unusual fatty acid.

FADS2 and the "PUFA-Type" Reaction

FADS2 is a member of the fatty acid desaturase gene family, which in mammals includes *SCD* (*S*tearoyl-*C*oA *D*esaturase; also known as *FADS5* and Δ9-desaturase), *FADS1* (also known as Δ5-desaturase), and *FADS3* (putative Δ13-desaturase; Guillou et al. 2010; Rioux et al. 2013). The combined activity of the human gene family is responsible for the production of (1) the endogenous MUFAs POA (16:1Δ9) and OA (18:1Δ9) by *SCD*; SA (16:1Δ6) by *FADS2*; (2) the polyunsaturated fatty acids (PUFAs) such as arachidonic (AA, 20:4n-6), eicosapentaenoic (EPA, 20:5n-3), and docosahexaenoic (DHA, 22:6n-3) acids, via a multi-enzyme pathway that includes *FADS1, FADS2,* and elongases (Guillou et al. 2010; Fig. 10.1). With the exception of SA (which is mainly secreted to the skin surface), these MUFAs and PUFAs are used internally for multiple functions that include cellular structure, energy storage, and lipid signaling, as exemplified by membrane phospholipids, triacylglycerols, and eicosanoids, respectively (Nakamura and Nara 2004).

FADS2 is the rate-limiting enzyme in the formation of PUFAs, in which the essential fatty acids (EFA; obtained from diet) linoleic acid (LA; 18:2Δ9,12) and α-linolenic acid (ALA; 18:3Δ9,12,15) are desaturated between the sixth and seventh carbons from the carboxyl terminal to give γ-linolenic acid (18:3Δ6,9,12) and stearidonic acid (18:3Δ6,9,12,15), respectively (Pereira et al. 2003; Fig. 10.1). γ-linolenic acid undergoes further rounds of elongation, desaturation (by *FADS1* and *FADS2*), and beta-oxidation to form the biologically important lipids AA (20:4n-6, 20:4Δ5,8,11,14) and docosapentaenoic acid (22:5n-6, 22:5Δ4,7,10,13,16). Since these PUFAs are derived from LA (in which the first double bond counting from the methyl end is at carbon 6), they are part of the n-6 series of EFAs (Pereira et al. 2003; Fig. 10.1). Stearidonic acid undergoes similar rounds of elongation, desaturation, and beta oxidation to form EPA (20:5n-3, 20:5Δ5,8,11,14,17) and DHA (22:6n-3, 22:6Δ4,7,10,13,16,19). These PUFAs are derived from ALA (in which the first double bond counting from the methyl end is at carbon 3) and are part of the n-3 series of EFAs (Fig. 10.1). Severe multiorgan dysfunction occurs when the diet is deficient in the essential fatty acids LA and ALA, a condition known as essential fatty acid deficiency (Gurr et al. 2002b). For example, a deficit of EFAs during critical periods of brain development can result in permanent deficits in a wide spectrum of neurological function (Richardson 2004). Similarly, the epidermal barrier requires LA, without which it becomes excessively permeable to water, resulting in life-threatening dehydration (Gurr et al. 2002b). Given the fundamental roles of the n-6 and n-3 PUFAs for membrane homeostasis and for lipid signaling, and the position of *FADS2* at the beginning of these metabolic pathways, expression of *FADS2* and the concomitant Δ6-desaturase activity is widespread among mammalian organs (Cho et al. 1999; Ge et al. 2003). Consistent with substrates that are PUFAs themselves, *FADS2* is a member of the "front-end" subgroup of desaturases that introduce a double bond into a fatty acid already containing two or more double bonds, and does so at the "front-end" of the fatty acid, or between the carboxyl end and the first double bond closest to this end, usually located between the ninth and tenth carbons (Meesapyodsuk and Qiu 2012; Fig. 10.1). This "PUFA-type" catalytic activity is present (albeit at variable levels) in all organs, with one notable exception, the sebaceous glands of human skin. In this singular human tissue, *FADS2* catalyzes a "sebaceous-type" reaction to produce high levels of naturally occurring SA, which is secreted to the skin surface as the major MUFA of sebum (Ge et al. 2003; Fig. 10.1).

FADS2 and the "Sebaceous-Type" Reaction

The discovery of *FADS2* as the major desaturase of human sebaceous glands and the demonstration of SA production by cells overexpressing human *FADS2* conclusively demonstrated that the same *FADS2* gene that carries out the PUFA-type reaction also carries out the sebaceous-type reaction (Ge et al. 2003). However, prior to this discovery, the existence of the sebaceous-type reaction catalyzed by a Δ6-desaturase was postulated by several independent lines of investigation. Researchers characterizing sebaceous lipids were the first to identify (and name) SA, and based on the family of sebum fatty acids with the highly unusual double bond position between the sixth and seventh carbons (from the carboxyl terminal), they correctly postulated that a Δ6-desaturase activity directly desaturates PA (16:0) to form SA (16:1Δ6; Nicolaides 1974; Stewart and Downing 1991). Similarly, early work characterizing the catalytic mechanism of PUFA formation in rat liver microsomes demonstrated the formation of low levels of 6-HA (16:1Δ6) when *SCD* activity was inhibited in the same preparation by an inhibitor specific to Δ9-desaturase (Pollard et al. 1979). Without the genomic information of the *FADS2* gene and transcripts, it was not possible to conclusively determine if these Δ6-desaturase activities were carried out by one or multiple *FADS2* isoforms. Indeed, studies of substrate preference (18:3Δ9,12,15 > 18:2Δ9,12 > 18:1Δ9) and competitive inhibition (18:3Δ9,12, 15 > 18:2Δ9,12 > 18:1Δ9; Brenner and Peluffo 1966) would be more consistent with different Δ6-desaturases, one acting on PUFA substrates and the other acting on saturated fatty acid (SFA) substrates. Finally, as more Δ9-desaturases were being cloned from various organisms including nematodes (Watts and Browse 2000) and mice (Zheng et al. 2001; Miyazaki et al. 2006), it was demonstrated that different Δ9-desaturated MUFAs (POA and OA) were generated by different *SCD* isoforms, each with a different substrate specificity. In combination, the above data was insufficient to link a single *FADS2* gene to both the PUFA-type and sebaceous-type reactions, a conclusion that required cloning and characterization of the *FADS2* locus (Cho et al. 1999; Marquardt et al. 2000), tissue localization of *FADS2* mRNA in skin (Ge at al. 2003), and an *in vitro* enzyme assay using overexpressed human *FADS2* enzyme and PA as a substrate (Ge et al. 2003).

Mechanism of Sapienic Acid Biosynthesis in Human Sebaceous Glands

Given the above information, how does the human sebaceous gland produce such high levels of this unusual MUFA using a desaturase whose preferred substrates are PUFAs? In this case, the biochemical objective is to enable the *FADS2* sebaceous-type reaction (Fig. 10.1), which due to its inherent inefficiency, requires multiple mechanisms to modulate substrate, product, and enzyme, such that the equilibrium of the reaction is shifted to production of SA. Several factors contribute to the need for this complex tissue-specific regulation: (1) *SCD* and *FADS2* are ubiquitously expressed due to the cellular requirement of *de novo* synthesis of MUFAs and biosynthesis of PUFAs, respectively, for membrane homeostasis, energy, and signaling, (2) PA (16:0) is one of two natural SFA substrates for *SCD* (the other being stearic acid, 18:0; Fig. 10.1), and as such is much more efficiently desaturated by *SCD* than by *FADS2* (Guillou et al. 2003), (3) *FADS2* has evolved to recognize fatty acid substrates with two or more *cis* double bonds (Sperling et al. 2003; Gostincar et al. 2010), thus SFAs have very low affinity, as indicated by the low Δ6-desaturation of SFAs (even in the presence of an *SCD* inhibitor) as compared to the high Δ6-desaturation of ALA to stearidonic acid (PUFAs) in the same microsomal preparation (Pollard et al. 1979; Fig. 10.1). The adaptations to overcome these factors include (1) removal of competing desaturase activity for PA, (2) removal of PUFA substrates that compete for binding of PA to *FADS2*, (3) accumulation of high levels of the substrate PA, (4) efficient sequestration of the product SA (esterification into triacylglycerols, wax esters, and cholesterol esters), and (5) high levels of *FADS2* enzyme activity.

Of these various mechanisms, removal of competing *SCD* activity for PA (Guillou et al. 2004) is likely the most critical. Early evidence linking SA to desaturase competition between *FADS2* and *SCD* for PA was (1) the observation that the vernix caseosa (human embryonic sebum) contains Δ9-desaturated POA (16:1Δ9) which decreases in the perinatal period, and with onset of puberty, is replaced with Δ6-desaturated SA (16:1Δ6) in sebum (Nicolaides et al. 1972; Nazzaro-Porro et al. 1979), and (2) using rat liver microsomes, 6-HA could be produced from PA only when an inhibitor of *SCD* was included in the reaction (Pollard et al. 1979). More recently, when primary sebocytes (pediatric source) undergo differentiation and lipogenesis, lipid analysis shows a high SA to POA ratio, similar to adult sebum (McNairn et al. 2013). Inhibition of *SCD* mRNA expression/accumulation is likely the primary means of reducing *SCD* competition for PA, suggested by weak/undetectable levels of *SCD* mRNA in sebaceous glands of human scalp hair follicles (Ge at al.2003; Fig. 10.2), and is further supported by evidence indicating that *SCD* activity is primarily controlled by transcription (Mauvoisin and Mounier 2011) and mRNA turnover (Ntambi 1999). Findings in other species reinforce this concept of desaturase competition. 6-HA is only detected in the preputial gland of mice with a targeted deletion of the *Scd1* gene (Miyazaki et al. 2002), and transfection of recombinant rat *Fads2* in COS-7 cells produced more 6-HA from PA when *Fads2* was expressed alone versus when *Fads2* and *Scd1* were co-expressed (Guillou et al. 2003). With the lack of *SCD* activity in the human sebaceous gland, *FADS2* is now in the unique situation of having to desaturate PA (a SFA) in the presence of its natural substrate LA (a PUFA). This is likely accomplished by removal of LA via expression of 15-lipoxygenase-2 (*15-LOX-2*) in human sebaceous glands. This mechanism is strongly supported by two lines of evidence: (1) localization studies of human sebaceous glands *in vivo* show that *FADS2* (Ge et al. 2003; Fig. 10.2) and *15-LOX-2* (Shappell et al. 2001) mRNA and protein are both expressed in differentiated sebocytes and not in the undifferentiated stem cells adjacent to the basement membrane (Fig. 10.2), and (2) explant cultures of human sebaceous glands specifically oxidize and degrade LA but not SFAs or MUFAs (Pappas et al. 2002)–an activity which correlates with the ability of the explant to produce sebaceous-specific lipids such as wax esters and squalene (Nicolaides 1974). LA and AA (metabolite of LA) are the natural substrates for *15-LOX-2,* which results in their oxidation and multistep conversion to their respective hydroxy fatty acids, 13-hydroxyoctadecadienoic acid (13-HODE) and 15-hydroxyeicosatetraenoic acid (15-HETE; Brash et al. 1997). Oxidized PUFAs (via oxygenases or non-enzymatic peroxidation) are termed oxylipins and are characterized by potent bioactivity, short half-lives, and tight regulation (Tourdot et al. 2014). Once formed, these bioactive lipids have several roles: signaling, induction of structural changes by provoking secondary oxygenations, and entering beta-oxidation for energy production (Feussner et al. 1997; Brash 1999). This multifunctionality could impart a dual function for *15-LOX-2* in human sebaceous glands: (1) shielding *FADS2* from LA, its preferred PUFA substrate by metabolizing LA to oxylipins that are not substrates for *FADS2* and/or promoting catabolism of LA by beta-oxidation, and (2) promoting sebocyte differentiation by activating peroxisome proliferator-activated receptor gamma (*PPARG*)-dependent

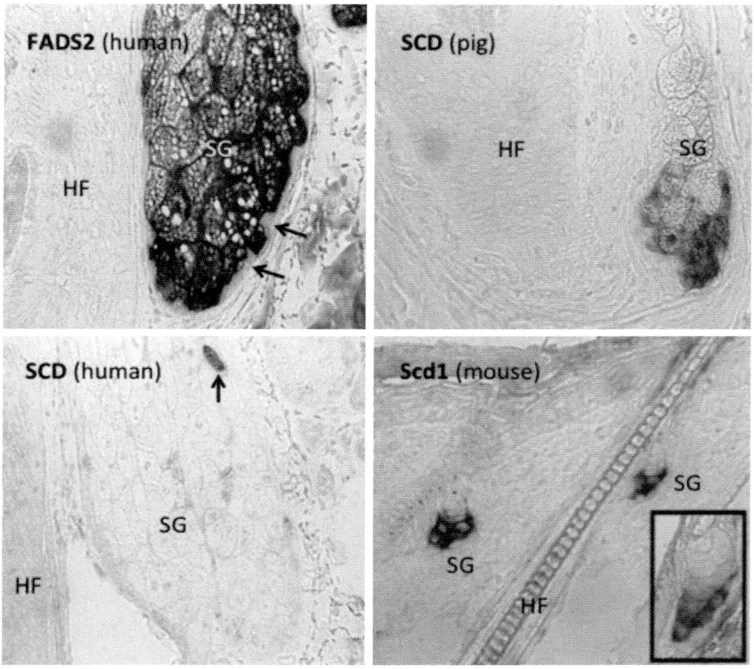

Fig. 10.2 *In situ* hybridization reveals expression of *FADS2* and *SCD1* mRNA in sebaceous glands. *Upper left*: Fatty acid desaturase 2 (*FADS2*) is abundantly expressed in human sebaceous glands (SG), with highest level in the early differentiating sebocytes immediately centripetal to the basal stem cells (*arrows*), which do not express *FADS2* mRNA. HF, hair follicle. *Lower left*: *SCD* mRNA is not expressed in human sebaceous glands (SG), with the exception of rare sebocytes (*arrow*). HF, hair follicle. *Upper right*: *SCD* mRNA is expressed in pig sebaceous glands (SG). HF, hair follicle. *Lower right*: *Scd1* mRNA is expressed mouse sebaceous glands (SG). Inset demonstrates *Scd1* expression is restricted to the *lower half* of the mouse sebaceous gland. The lack of expression of *SCD* in the majority of sebocytes in the human sebaceous gland, and the strong expression of *SCD* in pig and *Scd1* in mouse sebaceous glands, suggest that *SCD* gene expression is repressed in human sebaceous glands.

transcription of lipid metabolic genes. Several independent lines of research, when considered in aggregate, suggest a link between (1) the PUFAs LA, AA, and their oxylipin metabolites generated by *15-LOX-2,* (2) a differential effect of *PPARγ* on *FADS2* and *SCD*, and (3) the role of *PPARγ* in sebaceous gland homeostasis. 13-HODE is an endogenous activator of *PPARγ* in macrophages following uptake of oxidized LDL (Nagy et al. 1998) and 15-HETE activates *PPARγ* and inhibits proliferation (promotes differentiation) in prostate cancer cells (Tang et al. 2002). Treatment of HepG2 cells with pioglitazone (*PPAR-γ* agonist) increases *FADS2* mRNA and enzyme activity but not that of *SCD* (Saliani et al. 2013). Using primary sebocytes generated from explanted sebaceous glands from pediatric donors, treatment with LA induces lipogenesis that correlates with increased mRNA for *PPARγ* and *FADS2*, and lipid analysis indicates that SA (*FADS2*-derived) predominates over

POA (*SCD*-derived), demonstrating a differential desaturase response, as well as remarkable lipid similarity to *in vivo* human sebum (McNairn et al. 2013). Cultured rat preputial sebocytes undergo differentiation and lipogenesis when treated with LA or with a *PPARγ* agonist (Rosenfield et al. 1999).

Driving the sebaceous-type reaction toward product formation (SA; 16:1Δ6) requires efficient removal of product as well as efficient accumulation of substrate (PA; 16:0). Early work on characterization of sebaceous lipids demonstrated that sebaceous glands preferentially accumulate fatty acids with a chain length of sixteen carbons. Triacylglycerols and wax esters together account for ~50% of total sebum lipids in humans (Nicolaides 1974), with PA (16:0) and SA (16:1Δ6) comprising more than 80% of the fatty acids that are esterified in these two lipid classes (Green et al. 1984; Marzouki et al. 1988). Additionally, human sebaceous gland organ cultures preferentially incorporate PA over other fatty acids during synthesis of wax esters (Pappas et al. 2002), indicating selective utilization of 16 carbon fatty acids in sebum production. It is also possible that human sebaceous glands are less efficient in elongating PA (16:0) to stearic acid (18:0), thus leading to accumulation of PA. While several *Elovl* (*e*longation *o*f *v*ery *l*ong chain fatty acids) genes are expressed and/or function in mammalian skin [(*Elovl1* (Tvrdik et al. 2000; Wang et al. 2005; Sassa et al. 2013), *Elovl3* (Wang et al. 2005), *Elovl4* (Umeda et al. 2003)], including the sebaceous gland [(*Elovl3* (Westerberg et al. 2004) and *Elovl4* (McMahon et al. 2014)], no such evidence is available for *Elovl6*, which is the most efficient elongase for converting PA (16:0) to stearic acid (18:0; Matsuzaka et al. 2007; Green et al. 2010; Kihara 2012). Reduced *ELOVL6* in human sebaceous glands may be related to lack of *SCD* expression; these two genes are coordinately regulated *in vivo* (Matsuzaka et al. 2007; Shimano 2012). Targeted deletion of *Elovl6* in mice results in decreased expression of *Scd1* and a shift of from 18-carbon predominant FA to 16-carbon predominant FA (Matsuzaka et al. 2007).

In addition to substrate accumulation, maintenance of such high levels of MUFA production requires efficient sequestration of SA. Esterification into the various classes of sebaceous lipids provides the "sink" for SA, which at the same time prevents accumulation of free fatty acids which are toxic to cells (Park et al. 2014). Once PA is desaturated by *FADS2* to produce SA, there are three main pathways for further metabolism: (1) esterification into triacylglycerol (TG), for which *DGAT1* and *DGAT2* are the rate limiting enzymes (Wakimoto et al. 2003), (2) esterification into wax esters (WE), which is catalyzed by *AWAT1* and *AWAT2* (Turkish et al. 2005), and (3) conversion into sebaleic acid, a PUFA that is specific to human sebaceous glands (discussed below).

Finally, since the sebaceous-type *FADS2* reaction is inefficient compared to the PUFA-type reaction, increased *FADS2* mRNA and protein may provide compensation. *FADS2* mRNA is the most abundant transcript in a normal human skin cDNA library (NCI_CGAP_Skn3, NCBI Unigene library 8848), comprising 150 out of 7979 (1.9%) expressed sequence tags (ESTs). For comparative purposes, α-actin and type I myosin heavy chain comprise 1.9 and 1.5%, respectively of mRNAs from a normal skeletal muscle cDNA library (Welle et al. 1999). Myofibril transcripts are highly abundant in skeletal muscle and they represent a class of mRNAs that are "most often observed when a cell is producing an enormous quantity of a particular protein or is highly specialized or differentiated to perform a unique function" (Farrell 2005). *FADS2* expression is highly restricted in human skin to a subset of sebocytes (Ge et al. 2003). Thus the abundance of *FADS2* mRNA in sebocytes is even greater than the 1.9% detected in normal human skin. This suggests that a major function of human sebaceous glands is to produce and secrete SA to the skin surface, and that extremely high levels of *FADS2* mRNA and protein are required to support this "industrial scale" production.

Possible Roles of Sapienic Acid

Sapienic acid is involved in the synthesis of the other human-specific sebaceous fatty acid, sebaleic acid (18:2Δ5,8; Nicolaides 1974; Stewart et al. 1986; Fig. 10.1). This is the product of SA elongation by 2 carbons, followed by insertion of a *FADS1*-mediated double bond between the fifth and sixth carbon from the carboxyl end, a reaction and metabolite that occurs only in human sebaceous cells and has not been reported elsewhere. It has been demonstrated that neutrophil 5-lipoxygenase can convert sebaleic acid to 5-HODE, which can be further metabolized to 5-oxo-ODE by 5-hydroxyeicosanoid dehydrogenase, not only in neutrophils but also in keratinocytes (Cossette et al. 2008). Because of its chemoattractant properties, sebum-derived 5-oxo-ODE could be involved in neutrophil infiltration, which is relevant to inflammatory skin diseases such as acne, and atopic dermatitis, though this has not been elucidated as yet. Regarding acne, as sebum secretion increases with the onset of puberty, LA in epidermal acylceramides (required for barrier function) can be replaced with SA (Stewart et al. 1986), which could result in a compromised barrier and hyperkeratinization of the infrainfundibulum, an early event in comedogenesis (Knutson 1974; Downing et al. 1986). Additional suggestive evidence for a pathogenic role of sebum in acne, and in particular *FADS2*-derived SA and/or sebaleic acid, is the dramatic suppression of sebum and improvement in subjects treated with eicosatetraynoic acid (Strauss et al. 1967), an alkyne isomer of AA and an inhibitor of *FADS2* (Nakamura et al. 2000).

A prominent role of fatty acids is modulation of host interaction with microorganisms; the fundamental nature of this role is exemplified by its presence across different taxonomic kingdoms (Desbois and Smith 2010). Indeed FFAs are an important component of the innate immune system and play a prominent role in first-line defense of the cutaneous and mucosal surfaces (Drake et al. 2008). Of all mammals, human skin surface lipids contain the highest amount of FFA, which are generated, in large part, by microbial lipase-mediated hydrolysis of sebum TG (Reisner et al. 1968; Shalita 1974). Of these FFA, SA is the most abundant MUFA and also has the highest antimicrobial activity, with selectivity for *Staphylococcus aureus* (Wille et al 1997; Wille and Kydonieus 2003), a resident bacteria and dangerous opportunistic pathogen due to increasing prevalence of methicillin-resistant strains (Arsic et al. 2012). Selectivity for *Staphylococcus aureus* is related to teichoic acid of the cell wall, and the "stringent discrimination against 16:1 utilization for phospholipid synthesis" in this bacterium (Parsons et al. 2012). Lack of metabolism of 16:1 leads to its accumulation in the cell membrane, disruption of the lipid bilayer, loss of essential metabolites and low-molecular-weight proteins through channels in the cell wall, resulting in abrupt cessation of most major metabolic pathways (Parsons et al. 2012). In this same study, a striking difference in antimicrobial activity of various FFAs was seen, such that 16:1Δ6 (SA) and 16:1Δ9 (POA) were most (and equally) active (discussed below), yet the related FFAs 16:0 (PA), 18:0 (stearic acid), 18:1Δ6, 18:1Δ9 (OA) were completely inactive (Parsons et al. 2012; Fig. 10.1). Increased infection by *Staphylococcus aureus* is associated with atopic dermatitis, in which levels of SA are decreased, and when supplied topically, results in decreased bacterial load and amelioration of symptoms (Takigawa et al. 2005).

Further evidence of an antimicrobial role for SA is its *in vitro* bactericidal activity against the oral pathogens *Streptococcus salivarius* (Wille and Kydonieus 2003), *Fusobacterium nucleatum* (Fischer et al. 2012), and *Porphyromonas gingivalis* (Fischer et al. 2013). As discussed in an earlier section, free SG are found in the oral mucosa and are thought to contribute to the neutral lipids of saliva, the lipid species of which are strikingly similar to sebum (Brasser et al. 2011). Although squalene, wax esters, and TG were identified in saliva (Brasser et al. 2011), the fatty acid moieties were not characterized, leaving open the possibility that POA (16:1Δ9) or lauric acid (12:0, the second most active antimicrobial FFA in human sebum) is the main antimicrobial oral FFA. The lack of documentation of abundant naturally occurring SA in any mammalian tissue other than human sebum lends further support to this tentative conclusion.

When tested head-to-head in antimicrobial assays *in vitro*, SA (16:1Δ6, *FADS2*-derived) and POA (16:1Δ9, *SCD*-derived) are equally effective against *Staphylococcus aureus* (Wille and Kydonieus 2003; Parsons et al. 2012). The specificity of the structure–function relationship indicates the main fatty acid determinants are chain length and the presence of a single double bond, but not the position of the double bond (Parsons et al. 2012). This startling result brings us back to a variation of the central question presented earlier in this chapter: "If the main function of SA is antimicrobial, and POA is an equally effective antimicrobial that is present in the sebum of most other mammals, what is the selective advantage to humans in having *FADS2* produce SA, an extremely rare enzymatic reaction requiring a complex biochemical infrastructure?" Clearly, the presence of SA at the skin surface is important, given the sheer amount of SA in sebum and the very high expression level of *FADS2* in the sebaceous gland. Indeed, one could argue that this is the single most important function of human sebaceous glands and well worth the effort of repurposing *FADS2*. Although the mechanism of SA formation and its likely role as an antimicrobial are becoming more clear, the full answer to the question stated above requires further investigation into the specific features of human skin, sebaceous gland lipid biology, and the microbial environment that sets *Homo sapiens* apart from all other mammals.

As discussed at the outset, the extreme heterogeneity of sebaceous fatty acids is not only species-specific, but it can distinguish among individuals, including humans (Green et al. 1984). The function of human sebaceous glands exhibit changes with age (Nazzaro-Porro et al. 1979; Yamamoto et al. 1987) and gender (Nazzaro-Porro et al. 1979), and is also variable between racial groups (Pappas et al. 2013), as suggested by differential prevalence and sequelae of acne vulgaris between races (Perkins et al. 2011). Given the central role of sapienic acid and *FADS2* in human sebaceous gland biology, further understanding of this unique lipid metabolic pathway may shed light on the heterogeneity of skin disorders among individuals of different age groups, gender, and ethnicities (Pappas et al. 2013), and possibly lead to new treatments.

Acknowledgements The authors wish to thank the members of the The Skin Research Center of Johnson and Johnson who participated in this research, and especially Kurt Stenn, M.D. who had the vision to pursue sebaceous gland gene discovery and lipid biology.

References

Ackman RG, Hooper SN, Frair W. Comparison of the the fatty acid compositions of depot fats from fresh-water and marine turtles. Comp Biochem Physiol B. 1971;40:931–44.

Arsic B, Zhu Y, Heinrichs DE, McGavin MJ. Induction of the staphylococcal proteolytic cascade by antimicrobial fatty acids in community acquired methicillin resistant Staphylococcus aureus. PLoS ONE. 2012;7:e45952.

Brash AR. Lipoxygenases: occurrence, functions, catalysis, and acquisition of substrate. J Biol Chem 1999;274:23679–82.

Brash AR, Boeglin WE, Chang MS. Discovery of a second 15S-lipoxygenase in humans. Proc Natl Acad Sci U S A. 1997;94:6148–52.

Brasser AJ, Barwacz CA, Dawson DV, Brogden KA, Drake DR, Wertz PW. Presence of wax esters and squalene in human saliva. Arch Oral Biol. 2011;56:588–91.

Brenner RR, Peluffo RO. Effect of saturated and unsaturated fatty acids on the desaturation in vitro of palmitic, stearic, oleic, linoleic, and linolenic acids. J Biol Chem. 1966;241:5213–19.

Brownlee RG, Silverstein RM, Muller-Schwarze D, Singer AG. Isolation, identification, and function of the chief component of the male tarsal scent in black-tailed deer. Nature. 1969;221:284–85.

Cahoon EB, Cranmer AM, Shanklin J, Ohlrogge JB. Δ6 hexadecenoic acid is synthesized by the activity of a soluble Δ6 palmitoyl-acyl carrier protein desaturase in Thunbergia alata endosperm. J Biol Chem. 1994;269:27519–26.

Cho HP, Nakamura MT, Clarke SD. Cloning, expression, and nutritional regulation of the mammalian Delta-6 desaturase. J Biol Chem. 1999;274:471–7.

Cossette C, Patel P, Anumolu JR, Sivendran S, Lee GJ, Gravel S, Graham FD, Lesimple A, Mamer OA, Rokach J, Powell WS. Human neutrophils convert the sebum-derived polyunsaturated fatty acid sebaleic acid to a potent granulocyte chemoattractant. J Biol Chem. 2008;283:11234–43

Desbois AP, Smith VJ. Antibacterial free fatty acids: activities, mechanisms of action and biotechnological potential. Appl Microbiol Biotechnol. 2010;85:1629–42.

Dhouailly D. A new scenario for the evolutionary origin of hair, feather, and avian scales. J Anat. 2009;214:587–606.

Downing DT, Stewart ME, Wertz PW, Strauss JS. Essential fatty acids and acne. J Am Acad Dermatol. 1986;14:221–5.

Drake DR, Brogden KA, Dawson DV, Wertz PW. Thematic review series: skin lipids. Antimicrobial lipids at the skin surface. J Lipid Res. 2008;49:4–11.

Farrell RE Jr. RNA methodologies: a laboratory guide for isolation and characterization. 3rd ed. Amsterdam: Academic Press; 2005. p. 33.

Feussner I, Kühn H, Wasternack C. Do specific linoleate 13-lipoxygenases initiate beta-oxidation? FEBS Lett. 1997;406:1–5.

Fischer CL, Drake DR, Dawson DV, Blanchette DR, Brogden KA, Wertz PW. Antibacterial activity of sphingoid bases and fatty acids against Gram-positive and Gram-negative bacteria. Antimicrob Agents Chemother. 2012;56:1157–61.

Fischer CL, Walters KS, Drake DR, Dawson DV, Blanchette DR, Brogden KA, Wertz PW. Oral mucosal lipids are antibacterial against Porphyromonas gingivalis, induce ultrastructural damage, and alter bacterial lipid and protein compositions. Int J Oral Sci 2013;5:130–40.

Ge L, Gordon JS, Hsuan C, Stenn K, Prouty SM. Identification of the Δ-6 desaturase of human sebaceous glands: expression and enzyme activity. J Invest Dermatol. 2003;120:707–14.

Gostincar C, Turk M, Gunde-Cimerman N. The evolution of fatty acid desaturases and cytochrome b5 in eukaryotes. J Membr Biol. 2010;233:63–72.

Green SC, Stewart ME, Downing DT. Variation in sebum fatty acid composition among adult humans. J Invest Dermatol. 1984;83:114–7.

Green CD, Ozguden-Akkoc CG, Wang Y, Jump DB, Olson LK. Role of fatty acid elongases in determination of de novo synthesized monounsaturated fatty acid species. J. Lipid Res. 2010;51:1871–7.

Guillou H, D'andrea S, Rioux V, Jan S, Legrand P. The surprising diversity of Δ6-desaturase substrates. Biochem Soc Trans. 2004;32:86–7.

Guillou H, Rioux V, Catheline D, Thibault J-N, Bouriel M, Jan S, D'Andrea S, Legrand P. Conversion of hexadecanoic acid to hexadecenoic acid by rat Δ6-desaturase. J Lipid Res. 2003;44:450–4.

Guillou H, Zadravec D, Martin PGP, Jacobsson A. The key roles of elongases and desaturases in mammalian fatty acid metabolism: insights from transgenic mice. Prog Lipid Res. 2010;49:186–99.

Gurr MI, Harwood JL, Frayn KN. Fatty acid structure and metabolism. In: Lipid biochemistry. Oxford: Blackwell; 2002a. pp. 13–92.

Gurr MI, Harwood JL, Frayn KN. Dietary lipids. In: Lipid Biochemistry. Oxford: Blackwell; 2002b. pp. 140–5.

Hadley NF. Communication. The adaptive role of lipids in biological systems. New York: Wiley; 1985. p. 253.

Hooper SN, Ackman RG. Trans-6-hexadecenoic acid and the corresponding alcohol in lipids of the sea anemone Metridium dianthus. Lipids. 1971;6:341–6.

Hyman AB, Guiducci AV. Ectopic sebaceous glands. In: Montagna W, Ellis RA, Silver AF, editors. The sebaceous glands. New York: Macmillan; 1963. pp. 78–93.

Jared C, Antoniazzi MM, Silva JR, Freymüller E. Epidermal glands in squamata: microscopical examination of precloacal glands in Amphisbaena alba (Amphisbaenia, Amphisbaenidae). J Morphol. 1999;241:197–206.

Kihara A. Very long-chain fatty acids: elongation, physiology and related disorders. J Biochem. 2012;152:387–95.

Knapp LA, Robson J, Waterhouse JS. Olfactory signals and the MHC: a review and a case study in Lemur catta. Am J Primatol. 2006;68:568–84.

Knutson DD. Ultrastructural observations in acne vulgaris: the normal sebaceous follicle and acne lesions. J Invest Dermatol. 1974;62:288–307.

Kohler T, Weidenmaier C, Peschel A. Wall teichoic acid protects Staphylococcus aureus against antimicrobial fatty acids from human skin. J Bacteriol. 2009;191:4482–4.

Lin M-H, Hsu F-F, Miner JH. Requirement of fatty acid transport protein 4 for development, maturation, and function of sebaceous glands in a mouse model of ichthyosis prematurity syndrome. J Biol Chem. 2013;288:3964–76.

Lupi O. Ancient adaptations of human skin: why do we retain sebaceous and apocrine glands? Int J Dermatol. 2008;47:651–4.

Marekov I, Momchilova S, Grung B, Nikolova-Damyanova B. Fatty acid composition of wild mushroom species of order agaricales-examination by gas chromatography-mass spectrometry and chemometrics. J Chromatogr B Analyt Technol Biomed Life Sci. 2012;910:54–60.

Marquardt A, Stöhr H, White K, Weber BH. cDNA cloning, genomic structure, and chromosomal localization of three members of the human fatty acid desaturase family. Genomics. 2000;66:175–83.

Marzouki ZM, Taha AM, Gomaa KS. Fatty acid profiles of sebaceous triglycerides by capillary gas chromatography with mass-selective detection. J Chromatogr. 1988;425:11–24.

Matsuzaka T, Shimano H, Yahagi N, Kato T, Atsumi A, Yamamoto T, Inoue N, Ishikawa M, Okada S, Ishigaki N, Iwasaki H, Iwasaki Y, Karasawa T, Kumadaki S, Matsui T, Sekiya M, Ohashi K, Hasty AH, Nakagawa Y, Takahashi A, Suzuki H, Yatoh S, Sone H, Toyoshima H, Osuga J, Yamada N. Crucial role of a long-chain fatty acid elongase, Elovl6, in obesity-induced insulin resistance. Nat Med. 2007;13:1193–202.

Mauvoisin D, Mounier C. Hormonal and nutritional regulation of SCD1 gene expression. Biochimie. 2011;93:78–86.

McMahon A, Lu H, Butovich IA. A role for ELOVL4 in the mouse meibomian gland and sebocyte cell biology. Invest Ophthalmol Vis Sci. 2014;55:2832–40.

McNairn AJ, Doucet Y, Demaude J, Brusadelli M, Gordon CB, Uribe-Rivera A, Lambert PF, Bouez C, Breton L, Guasch G. TGFβ signaling regulates lipogenesis in human sebaceous glands cells. BMC Dermatol. 2013;13:2.

Meesapyodsuk D, Qiu X. The front-end desaturase: structure, function, evolution and biotechnological use. Lipids. 2012;47:227–37.

Miles AEW. Sebaceous glands in the lip and cheek mucosa of man. Br Dent J. 1958;105:235–48.

Miyazaki M, Gomez FE, Ntambi JM. Lack of stearoyl-CoA desaturase-1 function induces a palmitoyl-CoA Delta6 desaturase and represses the stearoyl-CoA desaturase-3 gene in the preputial glands of the mouse. J Lipid Res. 2002;43:2146–54.

Miyazaki M, Bruggink SM, Ntambi JM. Identification of mouse palmitoyl-coenzyme A Δ9-desaturase. J Lipid Res. 2006;47:700–4.

Montagna W, Yun JS. The skin of primates. X. The skin of the ring-tailed lemur (Lemur catta). Am J Phys Anthropol. 1962;20:95–117.

Nagy L, Tontonoz P, Alvarez JG, Chen H, Evans RM. Oxidized LDL regulates macrophage gene expression through ligand activation of PPARgamma. Cell. 1998;93:229–40.

Nakamura MT, Cho HP, Clarke SD. Regulation of hepatic Δ-6 desaturase expression and its role in the polyunsaturated fatty acid inhibition of fatty acid synthase gene expression in mice. J Nutr. 2000;130:1561–5.

Nakamura MT, Nara TY. Structure, function, and dietary regulation of Δ6, Δ5, and Δ9 desaturases. Annu Rev Nutr. 2004;24:345–76.

Nazzaro-Porro M, Passi S, Boniforti L, Belsito F. Effects of aging on fatty acids in skin surface lipids. J Invest Dermatol. 1979;73:112–7.

Nichols PD, Volkman JK, Everitt DA. Occurence of cis-6-hexadecenoic acid and other unusual monounsaturated fatty acids in the lipids of oceanic particulate matter. Oceanol Acta. 1989;12:393–403.

Nicolaides, N. Skin lipids: their biochemical uniqueness. Science. 1974;186:19–26.

Nicolaides N, Fu HC, Ansari MN, Rice GR. The fatty acids of wax esters and sterol esters from vernix caseosa and from human skin surface lipid. Lipids. 1972;7:506–17.

Ntambi JM. Regulation of stearoyl-CoA desaturase by polyunsaturated fatty acids and cholesterol. J Lipid Res. 1999;40:1549–58.

Pappas A, Anthonavage M, Gordon JS. Metabolic fate and selective utilization of major fatty acids in human sebaceous gland. J Invest Dermatol. 2002;118:164–71.

Pappas A, Fantasia J, Chen T. Age and ethnic variations in sebaceous lipids. Dermatoendocrinology. 2013;5:319–24.

Park E-J, Lee AY, Park S, Kim J-H, Cho M-H. Multiple pathways are involved in palmitic acid-induced toxicity. Food Chem Toxicol. 2014;67:26–34.

Parsons JB, Yao J, Frank MW, Jackson P, Rock CO. Membrane disruption by antimicrobial fatty acids releases low-molecular-weight proteins from staphylococcus aureus. J Bacteriol. 2012;194:5294–304.

Pereira SL, Leonard AE, Mukerji P. Recent advances in the study of fatty acid desaturases from animals and lower eukaryotes. Prostaglandins Leukot Essent Fatty Acids. 2003;68:97–106.

Perkins AC, Cheng CE, Hillebrand GG, Miyamoto K, Kimball AB. Comparison of the epidemiology of acne vulgaris among Caucasian, Asian, Continental Indian and African American women. J Eur Acad Dermatol Venereol. 2011;25:1054–60.

Pollard MR, Gunstone FD, James AT, Morris LJ. Desaturation of positional and geometric isomers of monoenoic fatty acids by microsomal preparations of rat liver. Lipids. 1979;15:306–14.
Quay WB. Structure and function of skin glands. In: Muller-Schwarze D, Mozell MM, editors. Chemical signals in vertebrates. New York: Plenum Press; 1977. pp. 1–12.
Reisner RM, Silver DZ, Puhvel M, Sternberg TH. Lipolytic activity of Corynebacterium acnes. J Invest Dermatol. 1968;51:190–6.
Richardson AJ. Long-chain polyunsaturated fatty acids in childhood developmental and psychiatric disorders. Lipids. 2004;39:1215–22.
Rioux V, Pédrono F, Blanchard H, Duby C, Boulier-Monthéan N, Bernard L, Beauchamp E, Catheline D, Legrand P. Trans-vaccenate is Δ13-desaturated by FADS3 in rodents. J Lipid Res. 2013;54:3438–52.
Rosenfield RL, Kentsis A, Deplewski D, Ciletti N. Rat preputial sebocyte differentiation involves peroxisome proliferator-activated receptors. J Invest Dermatol. 1999;112:226–32.
Saliani N, Darabi M, Yousefi B, Baradaran B, Khaniani MS, Darabi M, Shaaker M, Mehdizadeh A, Naji T, Hashemi M. PPARγ agonist-induced alterations in Δ6-desaturase and stearoyl-CoA desaturase 1: role of MEK/ERK1/2 pathway. World J Hepatol. 2013;5:220–5.
Sansone A, Melchiorre M, Chatgilialoglu C, Ferreri C. Hexadecenoic fatty acid isomers: a chemical biology approach for human plasma biomarker development. Chem Res Toxicol. 2013;26:1703–9.
Sassa T, Ohno Y, Suzuki S, Nomura T, Nishioka C, Kashiwagi T, Hirayama T, Akiyama M, Taguchi R, Shimizu H, Itohara S, Kihara A. Impaired epidermal permeability barrier in mice lacking elovl1, the gene responsible for very-long-chain fatty acid production. Mol Cell Biol. 2013;33:2787–96.
Schempp C, Emde M, Wölfle U. Dermatology in the Darwin anniversary. Part 1: Evolution of the integument. J Dtsch Dermatol Ges. 2009;7:750–7.
Shalita AR. Genesis of free fatty acids. J Invest Dermatol. 1974;62:332–5.
Shappell SB, Keeney DS, Zhang J, Page R, Olson SJ, Brash AR. 15-Lipoxygenase-2 expression in benign and neoplastic sebaceous glands and other cutaneous adnexa. J Invest Dermatol. 2001;117:36–43.
Shimano H. Novel qualitative aspects of tissue fatty acids related to metabolic regulation: lessons from Elovl6 knockout. Prog Lipid Res. 2012;51:267–71.
Spence MW. Monoenoic fatty-acid isomers of brain in adult and newborn rats. Biochim Biophys Acta. 1970;218:347–56.
Spencer GF, Kleiman R, Miller RW, Earle FR. Occurence of cis-6-hexadecenoic acid as the major component of Thunbergia alata seed oil. Lipids. 1971;6:712–4.
Sperling P, Ternes P, Zank TK, Heinz E. The evolution of desaturases. Prostaglandins Leukot Essent Fatty Acids. 2003;68:73–95.
Stewart ME. Sebaceous gland lipids. In: Bereiter-Hahn J, Matoltsy AG, Richards KS, editors. Biology of the Integument. 2. Vertebrates. Berlin: Springer-Verlag; 1986. pp. 824–32.
Stewart ME, Downing DT. Chemistry and function of mammalian sebaceous lipids. Adv Lipid Res. 1991;24:263–301.
Stewart ME, Grahek MO, Cambier LS, Wertz PW, Downing DT. Dilutional effect of increased sebaceous gland activity on the proportion of linoleic acid in sebaceous wax esters and in epidermal acylceramides. J Invest Dermatol. 1986;87:733–6.
Strauss JS, Pochi PE, Whitman EN. Suppression of sebaceous gland activity with eicosa-5:8:11:14-tetraynoic acid. J Invest Dermatol. 1967;48:492–3.
Takigawa H, Nakagawa H, Kuzukawa M, Mori H, Imokawa G. Deficient production of hexadecenoic acid in the skin is associated in part with the vulnerability of atopic dermatitis patients to colonization by Staphylococcus aureus. Dermatology. 2005;211:240–8.
Tang S, Bhatia B, Maldonado CJ, Yang P, Newman RA, Liu J, Chandra D, Traag J, Klein RD, Fischer SM, Chopra D, Shen J, Zhau HE, Chung LWK, Tang DG. Evidence that arachidonate 15-lipoxygenase 2 is a negative cell cycle regulator in normal prostate epithelial cells. J Biol Chem. 2002;277:16189–201.
Thody AJ, Shuster S. Control and function of sebaceous glands. Physiol Rev. 1989;69:383–416.

Tourdot BE, Ahmed I, Holinstat M. The emerging role of oxylipins in thrombosis and diabetes. Front Pharmacol. 2014;4(176):1–9.

Turkish AR, Henneberry AL, Cromley D, Padamsee M, Oelkers P, Bazzi H, Christiano AM, Billheimer JT, Sturley SL. Identification of two novel human acyl-CoA wax alcohol acyltransferases: members of the diacylglycerol acyltransferase 2 (DGAT2) gene superfamily. J Biol Chem. 2005;280:14755–64.

Tvrdik P, Westerberg R, Silve S, Asadi A, Jakobsson A, Cannon B, Loison G, Jacobsson A. Role of a new mammalian gene family in the biosynthesis of very long chain fatty acids and sphingolipids. J Cell Biol. 2000;149:707–18.

Umeda S, Ayyagari R, Suzuki MT, Ono F, Iwata F, Fujiki K, Kanai A, Takada Y, Yoshikawa Y, Tanaka Y, Iwata T. Molecular cloning of ELOVL4 gene from cynomolgus monkey (Macaca fascicularis). Exp Anim. 2003;52:129–35.

Wakimoto K, Chiba H, Michibata H, Seishima M, Kawasaki S, Okubo K, Mitsui H, Torii H, Imai Y. A novel diacylglycerol acyltransferase (DGAT2) is decreased in human psoriatic skin and increased in diabetic mice. Biochem. Biophys. Res. Commun. 2003;310:296–302.

Wang Y, Botolin D, Christian B, Busik J, Xu J, Jump D.B. Tissue-specific, nutritional, and developmental regulation of rat fatty acid elongases. J. Lipid Res. 2005;46:706–15.

Watts JL, Browse J. A palmitoyl-CoA-Specific Δ9 fatty acid desaturase from Caenorhabditis elegans. Biochem Bioph Res Co. 2000;272:263–9.

Welle S, Bhatt K, Thornton.CA. Inventory of high-abundance mRNAs in skeletal muscle of normal men. Genome Res. 1999;9:506–13.

Wille JJ, Kydonieus A. Palmitoleic acid isomer (C16:1Δ6) in human skin sebum is effective against gram-positive bacteria. Skin Pharmacol Appl Skin Physiol. 2003;16:176–87.

Wille JJ, Drake D, Wertz PW. Identification of cis-palmitoleic acid as the active antimicrobial in human skin sebum. J Invest Dermatol. 1997;108:677.

Westerberg R, Tvrdik P, Undén A-B, Månsson J-E, Norlén L, Jakobsson A, Holleran WH, Elias PM, Asadi A, Flodby P, Toftgård R, Capecchi MR, Jacobsson A. Role for ELOVL3 and fatty acid chain length in development of hair and skin function. J Biol Chem. 2004;279:5621–9.

Yamamoto A, Serizawa S, Ito M, Sato Y. Effect of aging on sebaceous gland activity and on the fatty acid composition of wax esters. J Invest Dermatol. 1987;89:507–12.

Zheng Y, Prouty SM, Harmon A, Sundberg JP, Stenn KS, Parimoo S. Scd3-a novel gene of the stearoyl-CoA desaturase family with restricted expression in skin. Genomics. 2001;71:182–91.

Chapter 11
Wax Esters: Chemistry and Biosynthesis

P. E. Kolattukudy

Core Messages

- Wax esters are unique surface lipids found on the surface of terrestrial organisms.
- These unusual lipids that coat the surface of animals offer unique biochemical challenges that are associated with skin health issues.
- Availability of modern analytical technologies, such as GC/LC/MS, allows the elucidation of their complex surface chemistry.
- The great advances in the genomic technology and gene knockout technology allowed identification of the genes involved and their function in the biosynthesis of some of the surface lipids.
- Elucidating the signaling molecules and mechanisms involved in maintaining the function of sebaceous glands and advances in the regulation of the sebum biosynthetic process may lead to novel ways to manipulate the process to yield skin health benefits.

Abstract Wax esters are unique surface lipids found on the surface of terrestrial organisms. After a brief review of the composition, the mechanisms used in their biosynthesis by sebaceous glands are summarized. The molecular biology of biosynthesis of methyl-branched, short chain and very long chain acids, and the biochemical mechanism of the loss of the carboxyl carbon of the elongated very long chain acids to generate alkanes are reviewed. The molecular biology of biosynthesis of the major types of molecules contains the hydroxyl group(s) to which the acids are esterified and the molecular biology of the esterification process are reviewed. The molecular bases of seasonal and hormonal regulation of biosynthesis of wax esters that implicate some functions of the wax esters are reviewed.

P. E. Kolattukudy (✉)
Burnett School of Biomedical Science, College of Medicine,
University of Central Florida, Orlando, USA
e-mail: PK@ucf.edu

A. Pappas (ed.), *Lipids and Skin Health,* DOI 10.1007/978-3-319-09943-9_11

Introduction

Animals and plants that live on land use lipids on their body surface to prevent desiccation and to influence interaction with other organisms. These surface lipids are quite different in composition form the internal lipids. Different types of wax esters are significant components of surface lipids. Spreadability and stability, both metabolic and chemical, required of the surface lipids are achieved via their chemical composition. To achieve these objectives shorter chain lengths and methyl branches are used. In humans, triacylglycerols, wax esters and squalene are key components of the surface lipids (Pappas 2009). In water birds that encounter colder environment multiple methyl branched fatty acid esters are used (Jacob 1976). For example geese use wax esters composed of 2,4,6,8-tetramethyldecanoic and undecanoic acid while ducks use esters of short-chain fatty acids. Specialized sebaceous glands that line the eyelids produce lipids to coat the tear film to prevent evaporation. The most common cause of certain dry eye conditions is meibomian gland dysfunction (Lee and Tong 2012).

Surface lipids are produced by specialized glands. Human sebaceous glands, distributed throughout the body surface, except the sole and palm, produce sebum (Downing 1976). Avian uropygial gland produces lipids that the birds use to preen their feathers (Jacob 1976). Uropygial gland is a large-bilobed structure located on the base of the tail. Each lobe contains numerous holocrine alveoli that open with a central cavity and the secretion passes through ducts opening at the tip of the papilla. As the cells that line the basement membrane of the gland divide and differentiate into lipid producing cells, these large lipid-loaded cells move towards the lumen. The lipid bodies coalesce and the cells lyse emptying the content into the secretory lumen. All sebaceous glands including mammalian meibomian glands have these three type of cells, the peripheral undifferentiated cells, the large lipid body producing cells and the lipid loaded lysing cells that empty their contents into the lumen (Knop et al. 2011). In addition to the secretion of the sebaceous glands including uropygial glands, the entire avian skin and animal skin serves as a lipogenic sebaceous secretion organ. In the mammalian epidermis, the outer nucleated keratinocyte cell layer generates lipids that constitute epidermal lamellar bodies which increases in the cytoplasm as the epidermal differentiation proceeds and are secreted into the intracellular space where they are processed to form the extracellular lamellar bilayers that form the permeability barrier (Pappas 2009; Proksch and Jensen 2012). Avian epidermal cells, also called sebokeratinocytes, perform functions akin to those performed by the mammalian holocrine sebaceous glands. The membrane bound lipid-enriched organelles, called multigranular bodies, appear as multiple stacks of discs resembling the lamellar bodies of mammalian epidermis (Menon and Menon 2000; Salibian and Montalti 2009).

Chemical Composition

Wax esters found in the surface lipids are composed of fatty acids esterified to the hydroxyl groups of fatty alcohols, hydroxy fatty acids, alkane-1,2-diols, alkane-2,3-diols and alkane-α,ω-diols. The fatty acids found in ester linkages include n- and branched acids usually with very low amounts of unsaturated acids. Detailed descriptions of surface waxes are available (Smith and Thiboutot 2008; Nicolaides 1974; Camera et al. 2010).

Methyl branched fatty acids are widely distributed in glands. Early gas chromatographic analysis of human surface lipids revealed the presence of methyl branched fatty acids in free fatty acids, wax esters, and triacylglycerols. Such acids constitute almost 80 % of the saturated acids in the surface lipids in newborn human (vernix caseosa) but much smaller amounts (12 %) in adult human surface lipids (Strauss et al. 1975). The methyl branches found in mammalian and avian surface lipids often contain monomethyl at iso or anteiso or at 2-,4-,6-, and/or 8-positions. Dimethyl, trimethyl or tetra methyl branched acids, in which the methyl branches are at even numbered positions in the chain, are also found.

In specialized sebaceous glands, such as mammalian meibomian glands, wax esters constitute about one third of the total lipids (Knop et al. 2011; Butovich 2011). The fatty acids range from C12 to C29 saturated and monounsaturated acids with C18:1 ω-9 as the most abundant acid with lesser amounts of C18:1 ω-7, C16:1 ω-7, anteiso C17:0, and iso-C16:0. The major alcohols are iso-C26:0, iso-24:0, anteiso-C25:0, and anteiso-C27:0. O-Acyl ω-hydroxyfatty acids composed of very long chain (C28:1–C34:1) hydroxyacids acylated with C16 and C18 fatty acids with a dominance of C18:1ω9. C30:1, C32:1, and C34:1 ω-hydroxy acids esterified to sterol on one end and with oleate at the other end constitute a minor component. Another related minor component is C32:1-α, ω-diol esterified at both ends with C18:1 and C16:1 fatty acids.

Biosynthesis of Wax Esters in Sebaceous Gland Lipids

Tissues Suitable for Biosynthetic Studies

The small size of mammalian sebaceous glands make their use for biosynthetic studies difficult. Mammalian Meibomian glands that line the eyelids and produce lipids can be used for biosynthetic studies (Anderson and Kolattukudy 1985; Kolattukudy et al. 1985a; Schirra et al. 2005). In recent years sebocytes-derived immortalized cell lines that are capable of differentiation into sebum producing cells have been developed. For example, human facial sebocytes were transfected with simian virus-40 large T antigen and the resulting immortalized cell lines called SZ95 and SEB-1 have been shown to retain major characteristics of human sebocytes including differentiation into lipid loaded cells, responsiveness to androgens and retinoids and gene expression changes (Zouboilis et al. 1999; Thiboutot et al. 2003; Wróbel et al. 2003). However, it is unclear why these cell lines do not produce large amounts of waxes in culture.

Since the avian uropygial glands are specialized in lipid biosynthesis they provide a convenient source to elucidate the enzymology of lipid synthesis. For example, the uropygial glands of water fowl such as ducks and geese yield glands that weigh 2–6 g each and extracts from such glands yield large amounts of enzymes that are dedicated to lipid biosynthesis. Thus, purification and characterization of such enzymes is feasible (Kolattukudy 1981). Hormonal regulation of avian uropygial glands bear resemblance to what is observed with mammalian sebaceous glands and thus the uropygial gland has been suggested as a suitable model to study hormonal action on mammalian sebaceous gland (Abalain et al. 1986).

Biosynthesis of Fatty Acids

The iso and anteiso acids are produced by using appropriately branched chain starters derived from branched chain amino acids instead of the usual acetyl starter. Thus, valine and isoleucine provide the isobranched C4 and anteisobranched C5 starters, respectively to generate isobranched even chain acids and anteisobranched odd chain acids, respectively. The methyl branches at 2, 4, 6, and 8 positions arise by the use of methylmalonyl-CoA instead of malonyl CoA in chain building by the fatty acids synthase (FAS). Evidence indicates that the availability of methylmalonyl-CoA is the key that controls such branched acid synthesis (Buckner and Kolattukudy 1976b).

Evolutionary Mechanisms Used to Generate Multiple Methyl Branched Fatty Acids

Multiple methylbranched acids are constituents of wax esters generated by avian sebaceous glands (Jacob 1976). For example, the major fatty acid component of the major secretion product of the goose uropygial glands are 2,4,6,8-tetramethyldecanoic and undecanoic acids. Obviously, these fatty acids are produced from four sequential condensation reactions with methylmalonyl-CoA molecules with acetyl-CoA or propionyl-CoA as the starter, respectively. Gland extract was demonstrated to catalyze such a synthesis (Buckner and Kolattukudy 1975a). Purification of the protein that catalyzed this synthesis yielded a 500 kDa protein composed of two identical subunits. In fact this protein turned out to be the normal multifunctional FAS that catalyzes the synthesis of n-C16 fatty acids like the other vertebrate FASs (Buckner and Kolattukudy 1976a).

Since the uropygial gland specializes in fatty acid synthesis the major protein in the gland is FAS that can be purified by a single step gel filtration (Buckner and Kolattukudy 1976a). This avian fatty acid synthase was the first reported case

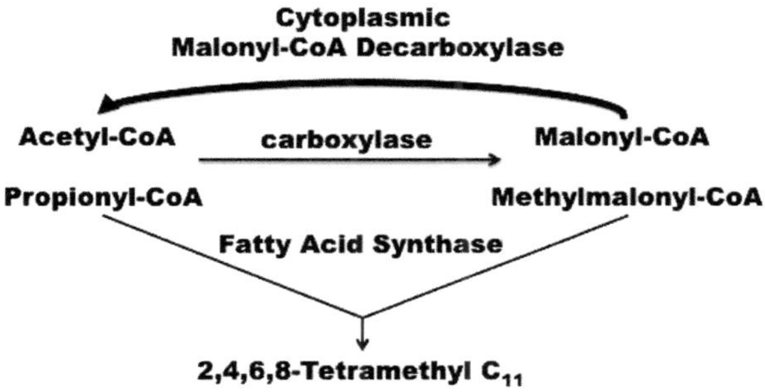

Fig. 11.1 Cytoplasmic malonyl-CoA decarboxylase assures that acetyl-CoA and methylmalonyl-CoA are the only substrate available for FAS

that demonstrated that each subunit of a vertebrate fatty acid synthase contained a covalently attached phosphopantothein providing the first evidence for the identity of the two subunits of vertebrate fatty acid synthase. Subsequently each subunit was found to have all the component activities, acyl transferase, ketoacyl synthase, enoyl reductase, dehydratase and enoyl reductase, acid carrier protein, and thioesterase. Fatty acid synthase from the uropygial gland, that produces multiple methylbranched acids, was found to be identical to that from the liver that produces only n-fatty acids (Buckner and Kolattukudy 1976b). The synthase purified from gland and liver can generate n-fatty acids from acetyl-CoA and malonyl CoA as well as 2,4,6,8-tetramethyl decanoic acid from acetyl-CoA and methylmalonyl-CoA as the only chain elongating substrate, although the rate of branched acid synthesis is low when compared to the rate of n-fatty acid synthesis.

Why does the synthase generate n-fatty acids in the liver while the same enzyme generates multiple methyl branched acids in the uropygial gland? The availability of malonyl-CoA or methylmalonyl-CoA would determine whether a tissue generates n-fatty acids or branched fatty acids. The crude extract of the gland cannot produce malonyl-CoA from acetyl-CoA but generates methylmalonyl-CoA from propionyl-CoA (Buckner and Kolattukudy 1975b). However, purified carboxylase from the gland carboxylates both acetyl-CoA and propionyl-CoA. What prevents the production of malonyl-CoA in the crude extract is an active cytoplasmic malonyl-CoA decarboxylase that is present only in the uropygial gland. Thus, in the gland, the abundant fatty acid synthase would have acetyl-CoA and methylmalonyl-CoA as the substrates from which the synthase produces multiple methyl branched fatty acids (Fig. 11.1).

The cytoplasmic malonyl-CoA decarboxylase has been purified, its cDNA has been cloned and the gene that encodes it has been sequenced (Jang et al. 1989). The results showed that the gene is transcribed to produce an mRNA that would be translated into a protein with a mitochondrial targeting leader sequence in the liver (Courchesne-Smith et al. 1992). In the gland, a different transcription start site is used to generate an mRNA that would encode a decarboxylase protein that lacks the leader sequence and thus remains in the cytoplasm (Fig. 11.2). Thus, a change in the transcription initiation site in the gland causes accumulation of malonyl-CoA decarboxylase in the cytoplasm and causes the production of multiple methylbranched fatty acids using the same acyl-CoA carboxylase and fatty acid synthase present

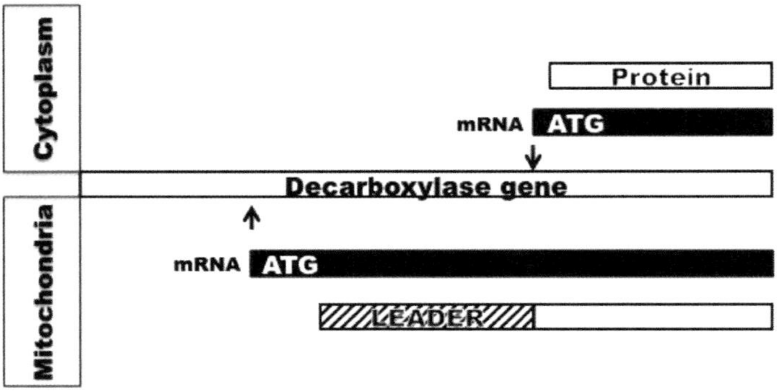

Fig. 11.2 Use of alternate transcription initiation site in the goose uropygial gland leads to an mRNA that is translated into malonyl-CoA decarboxylase protein lacking leader sequence causing accumulation in the cytoplasm. *Arrows* indicate transcription initiation site

in other tissues. This was an unexpected discovery of a minimal change in gene expression that leads to a major change in the composition of the lipids produced by the specialized sebaceous gland. It is interesting to note that the malonyl-CoA decarboxylase gene sequence, first revealed by the work on the uropygial gland, was used to clone the homologous mammalian gene that turned out to be an important player in cardiovascular health by regulating fatty acid oxidation (Lopaschuk and Stanley 2006; Ussher and Lopaschuk 2008) and thus, is currently under extensive investigation (Froese et al. 2013).

It is interesting to note that mycobacteria, that produce 2,4,6,8 tetramethyl branched long-chain acids as one of the main components of its cell wall-associated lipids, has evolved a unique multifunctional synthase that uses n-C_{20}CoA as the starter, in place of acetyl or propionyl CoA, and generate 2,4,6,8 tetramethyl tetracosanoic acid called mycocerosic acid (Rainwater and Kolattukudy 1985). In this case the acyl transferase and condensing domains of the multifunctional peptide have evolved to enable the synthase to use n-C_{20}CoA as the starter, and methylmalonyl-CoA as the chain extending substrate to build the multiple methylbranched mycocerosic acid (Fernandes and Kolattukudy 1997). Thus, when a small specialized tissue within an animal needs to make multiple methyl branched acids, a tissue-specific switch in transcription initiation site allows the gland to use the highly evolved acyl-CoA carboxylase and FAS to produce the unusual lipids. When a whole organism, like the Mycobacterium, needs to make an unusual multiple methyl branched acid, it has evolved a new multifunctional fatty acid synthase uniquely for the synthesis of such an unusual acid.

Very Long Chain Acids

Presence of compounds with very long alkyl chains is a characteristic feature of surface lipids. The very long chain fatty acids are generated by chain elongation enzymes that are located in the membranes, primarily in the endoplasmic reticulum (ER). These enzymes use n-C16 acid that is generated by the cytoplasmic FAS as the starter that is activated by the ER-located enzyme, and then elongate the thioester using malonyl-CoA and NADPH as the other substrates using reactions analogous to those catalyzed by FAS: condensation, ketoreduction, dehydration, and enoyl reduction. These enzymes probably function as a complex attached to the membrane. Membrane preparations from sebaceous glands such as meibomian glands have been demonstrated to catalyze such elongation reactions generating very long chain fatty acids (Anderson and Kolattukudy 1985). The condensing enzyme determines the specificity of the elongating enzyme and a family of elongases, ELOVL1-7, have been identified (Kihara 2012). They show selectivity for the starting acyl-CoA they accept for elongation. ELOVL3 and ELOVL4 have been shown to play important roles in skin lipid biosynthesis (Westerberg et al. 2004; Vasireddy et al. 2007). Disruption of elovl3 gene, that shows restricted expression in the sebaceous gland and hair follicular epithelial cells, displayed a sparse hair coat with disturbed hair lipid content (Westerberg et al. 2004). Fatty acids longer than 20 carbons were virtually undetectable. Consequently, elovl3 deficient mice showed a severe defect in water repulsion and high levels of trans-epidermal water loss. Mutation in elovl4 resulted in the absence of very long chain >C28 acids and ω-O-acyl ceramide that are important components of the extracellular lamellar membrane that constitutes a major permeability barrier (Vasireddy et al. 2007; Uchida 2011). Consequently, these mice displayed scaly, wrinkled skin with severely compromised epidermal permeability barrier causing death within a few hours after birth.

Fatty acids used for the biosynthesis of skin lipids are not only generated within the skin lipid-synthesizing cells, but also transported into such cells. A family of fatty acid transport proteins (FATP) is involved in this process (Khnykin et al. 2011). Of the six such mammalian FATPs, FATP4 seems to be the most important one involved in skin lipid biosynthesis (Schmuth et al. 2005). Thus, deletion of FATP4 in mice results in perturbations in the biosynthesis of skin lipids by keratinocytes that causes barrier dysfunction and consequently neonatal fatality (Lin et al. 2013a). Keratinocyte targeted expression of FATP4 reverses these abnormalities. Wax diester synthesis is drastically reduced by the absence of FATP4. Biosynthesis of the very long chain acids is preferentially inhibited by the absence of FATP4, as the membrane localized elongating enzyme system responsible for the biosynthesis of such acids probably receives the acyl chains for elongation in the activated form from the membrane localized FATP4.

Short-Chain Fatty Acid Synthesis

Chain length of fatty acids generated by multifunctional synthases is determined by the chain length selectivity of the chain-terminating thioesterase that is a segment of FAS. The thioesterase segment of vertebrate FAS, including avian FAS, releases the acyl chains when they reach about 16 carbon length (Bedord et al. 1978). Certain avian uropygial glands, such a mallard duck gland, produce esters of shorter chain fatty acids such as C8, C10, and C12 (Jacob 1976). In such glands a small (30 kDa) thioesterase, uniquely present in the glands, interacts with the FAS and releases the shorter acyl chains from the synthase (deRenobales et al. 1980). Such a thoiesterase was purified to homogeneity from mallard uropygial glands and its cDNA cloned (Rogers et al. 1982; Poulose et al. 1985). This thioesterase reacted with pyrenelbutylmethane phosphono fluoridate with covalent attachment of the pyrene derivative to the active serine resulting in the inactivation of the enzyme (Foster et al. 1985a). Addition of avian fatty acid synthase to pyrene-tagged thiosterase caused a dramatic increase in fluorescence anisotropy of the pyrene, demonstrating the physical interaction between the thiosterase and FAS. The association constant for binding of the two proteins was calculated to be 1 μM with a one to one stoichiometry. This association could be used to conveniently purify the thioesterase from gland extracts using fatty acid synthase as an affinity ligand (Rogers and Kolattukudy 1984).

The small S-acyl-FAS thioesterase was functionally compatible with vertebrate FAS from other sources (Rogers et al. 1982). FAS from murine source and goose were inactivated by PMSF treatment that covalently modified the active serine in the chain-terminating thioesterase segment of FAS. These inactive FAS preparations were reactivated by the addition of the duck small S-acyl FAS thioesterase that released the acyl chains from FAS. Thus, it would appear that the terminal thioesterase domain of multifunctional FAS polypeptide can be functionally replaced by the small s-acyl FAS thioesterase. The interaction between the resident thioesterase domain of FAS with the fatty acid chain built on the thiol at the condensing domain was examined by determining the distance between the active sites of the thioesterase and the pantetheine thiol using Florence Resonance Energy Transfer (Foster et al. 1985b). The active serine was tagged by treatment of the synthase with pyrenebutylmethane phosphono fluoridate. The pantetheine thiol was tagged by treatment with 3- (4 meleinidylphenyl)-7- diethylamino -4- methyl coumarin. When FAS was thus tagged at the thioesterase active site with pyrene and ACP domain active site with coumarin, fluorescence energy transfer occur between the pyrene and the coumarin. The distance between the pyrene and coumarin was estimated to be 37 A from the efficiency of energy transfer. Similar results were observed when the small S-acyl-FAS thioesterase tagged with pyrene was used with coumarin-tagged FAS, supporting the conclusion that the small thioesterase physically interacts with FAS and functionally displaces the resident thioesterase of FAS.

Biosynthesis of Fatty Alcohols

Fatty acyl-CoA reductase that catalyzes fatty alcohol production was first demonstrated in *Euglena gracilis* in which a particulate enzyme catalyzes direct conversions of fatty acyl-CoA to fatty alcohol with NADPH as the reductant without releasing fatty aldehyde intermediate (Kolattukudy 1970). Since then fatty acid acyl-CoA reductases have been found and purified from bacteria, plants, and animals, including sebaceous glands (Hellenbrand et al. 2011; Cheng and Russell 2004a; Moto et al. 2003; Honsho et al. 2010). Microsomes from bovine meibomian gland and avian uropygial glands also catalyze reduction of fatty acyl-CoA to alcohol without release of any aldehyde intermediates (Kolattukudy and Rogers 1986). A microsomal enzyme solubilized and purified form the duck uropygial gland was found to catalyze conversion of fatty acyl-CoA to fatty alcohol without detectable free fatty aldehyde intermediate (Wang and Kolattukudy 1995a). Fatty acyl-CoA reductase that generates aldehyde as the final product was separated from an aldehyde reductase from a plant that generates both fatty alcohols and alkanes from aldehyde (Kolattukudy 1971). A fatty acyl-CoA reductase that generates aldehydes and alcohols was found in Mycobacteria that makes alcohols for wax ester synthesis (Sirakova et al. 2012). Thus, reductases that reduce fatty acyl-CoA to alcohol via free fatty aldehyde intermediate are known.

In more recent years, bioinformatics approaches have been used to identify and clone acyl-CoA reductases. Two fatty acyl-CoA reductases have been identified in human and mouse with an insilico approach (Cheng and Russell 2004a). When the cDNAs encoding FAR1 and FAR2 were expressed in cells, these proteins catalyzed acyl-CoA reduction with NADPH as the reductant. FAR1 showed a preference for C_{16} and C_{18} saturated and unsaturated acyl-CoA whereas FAR2 preferred saturated C_{16} and C_{18} acyl-CoA. FAR1 was found in a variety of murine tissue but FAR2 showed a more restricted occurrence with highest expression in the eyelid that has the wax ester-producing meibomium glands. Both FAR1 and FAR2 that use alcohols for the synthesis of ether lipids were found in the brain. cDNA encoding two fatty acyl-CoA reductases have been cloned from the uropygial glands of several bird species (Hellenbrand et al. 2011). The products of these reductases, FAR1 and FAR2, revealed some differences in substrate specificity. Avian FAR1 showed highest expression in the uropygial gland whereas the avian FAR2 was highly expressed in the brain that is rich in ether lipids.

Biosynthesis of Diesters

Five types of diesters are found in sebaceous secretions. The two most common diesters produced by animal sebaceous glands are 2-hydroxy acids esterified with a fatty alcohol and a fatty acid, and alkane-1,2-diol with both hydroxyl groups in ester linkage with fatty acids (Downing 1976). Diesters of alkane α, ω-diols found in the mammalian meibomian gland constitute a third type of diesters (Nicolaides and Santos 1985). A fourth type of diesters, 3-hydroxyfatty acids with both the hydroxyl group and carboxyl group in ester linkages with a fatty acid and fatty alcohol, respectively, are produced as the major secretion product of mallard duck uropygial gland during the mating season (Kolattukudy et al. 1987a). A fifth type of diester alkane-2,3-diol diesters are the major secretion products of the uropygial glands of galliformis, such as chicken, quail, pheasant, etc. (Jacob 1976).

α-Hydroxylation of fatty acids to generate α-hydroxy fatty acid was found to be catalyzed by a particulate fraction from the uropygial glands of a white-crowned sparrow, Zonotrichia leucophrys (Kolattukudy 1972). The same particulate preparation also catalyzed α-hydroxyfatty acyl-CoA reduction to generate alkane-1,2-diol

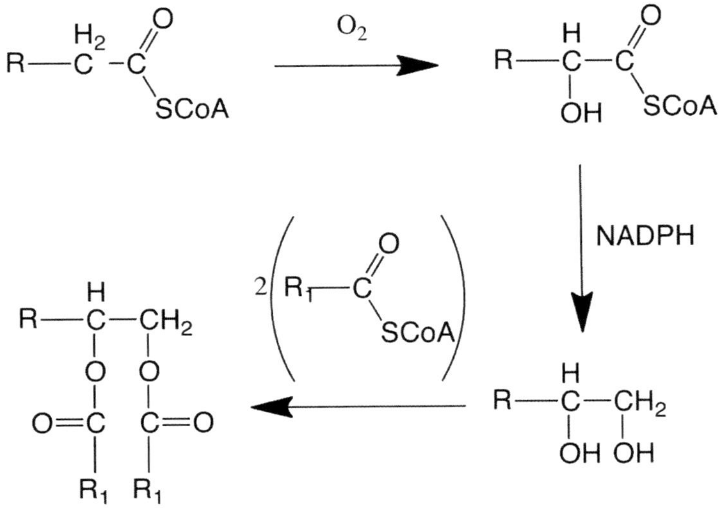

Fig. 11.3 Biosynthesis of Alkane-1,2-diol diesters

with NADPH as the reductant with hydride transfer from the B-side of the nicotinamide ring. Mammalian sebaceous glands also produce diesters that require 2-hydroxylation of fatty acids followed by reduction of the activated carboxyl group to generate alkane-1,2-diols (Fig. 11.3). 2-Hydroxylated fatty acid containing sphingolipids are also abundant in mammalian skin. Mammalian fatty acid 2-hydroxylase (FA2H) is an NADPH dependent enzyme that catalyzes stereospecfic hydroxylation to yield R-enantiomer (Maier et al. 2011; Guo et al. 2012). When FA2H gene, that was thought to be important for the synthesis of hydroxylated fatty acid containing sphingolipids of the murine skin, was knocked out, the FA2H deficient mice did not show any effect on the level of such sphingolipids (Maier et al. 2011). FA2H expression was found to be restricted to sebaceous glands and FA2H deficiency caused a drastic reduction in 2-hydroxylated glucosylceramides and in diester waxes. The consequent change in composition of the surface lipids caused a blockage in hair canals and thus severely interfered with hair growth.

Alkane-α, ω-diols probably arise by ω-hydroxylation of a very long fatty acid followed by a reduction of the activated carboxyl group by a reductase. Biosynthesis of the esters containing very long chain ω-hydroxyfatty acid present in meibomian gland (Kolattukudy et al. 1985a) and the very long chain ω-hydroxyceramides that are highly significant constituents of the extracellular multilamellar bodies in the outer layer of the skin involve ω-hydroxylation by cytochrome P450 (CYP) (Hardwick 2008). Among the super family of CYPs, CYP4F8, and CYP4F22 have been implicated in skin lipid biosynthesis (Kelly et al. 2011). CYP4F8 is expressed in the mammalian epidermis and its expression level is elevated in psoriasis (Stark et al. 2006). Mutations in CYP4F22 gene were found in patients with ichtyosis with abnormality in permeability barrier due to deficiency in very long chain acids and

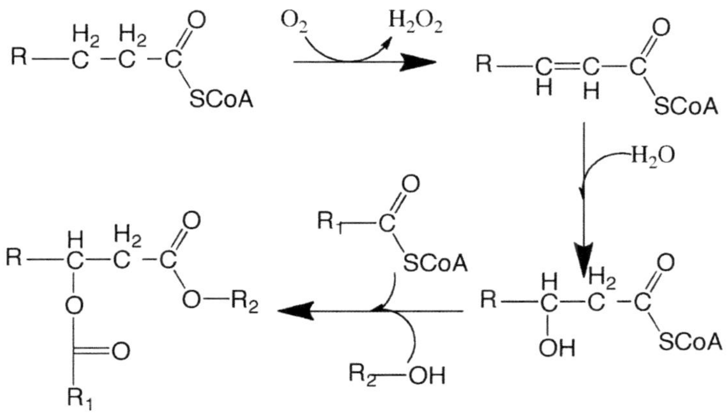

Fig. 11.4 Biosynthesis of 3 hydroxyfatty acid diesters

absence of ω-O-acylceramides (Lefèvre et al. 2006). Deficiency in cytochrome b5 that is an electron transport mediator involved in ω-hydroxylation also results in compromised permeability barrier, probably by a combination of changes in unsaturated fatty acid synthesis and a reduction in ω-hydroxylation required for the biosynthesis of the constituents of the barrier (Finn et al. 2011).

Diesters of 3-hydroxy C8, C10, and C12 acids are produced by female mallard ducks during the breeding period (Kolattukudy and Rogers 1987). A particulate fraction from the gland extract was shown to catalyze 3-hydroxy-C12 fatty acyl-CoA synthesis. The 3-hydroxy fatty acids are produced by hydration of enoyl-CoA generated by peroxisomal fatty acyl-CoA oxidase as shown by biochemical experimental evidence (Bohnet et al. 1991; Fig. 11.4). As indicated later, these processes occur in peroxisomes that proliferate in the female duck uropygial gland during the mating season.

Biosynthesis of alkane-2,3-diol was postulated to involve condensation between fatty aldehyde and the thiamine pyrophosphate derivative of acetaldehyde to generate an acyloin that could undergo reduction to alkane-2,3-diol (Fig. 11.5; Sawaya and Kolattukudy 1972). Results obtained with chicken uropygial glands were consistent with this hypothesis (Tang and Hansen 1976). The uropygial gland of ring-necked pheasant (*Phasianus colchicus*), produces octadecane-2,3-diol as the major diol (85 %). Location of the label in the diol derived from [1-^{14}C] palmitic acid in the gland at C-3 of octadecane-2,3-diol was consistent with the proposed biosynthetic pathway (Sawaya and Kolattukudy 1972). Furthermore, synthetic 3-hydroxy-[3-^{14}C]-octadecane-2-one, the predicted acyloin intermediate, was converted directly into octadecane-2,3-diol when injected into the uropygial gland of the pheasant (Buckner and Kolattukudy 1976b). A cell-free preparation from the gland catalyzed the transfer of hydride from the B-side of the nicotinamide ring of NADPH to the carbonyl carbon of synthetic R,S mixture of the acyloin to generate a mixture of threo- and erythro-octadane-2,3-diol.

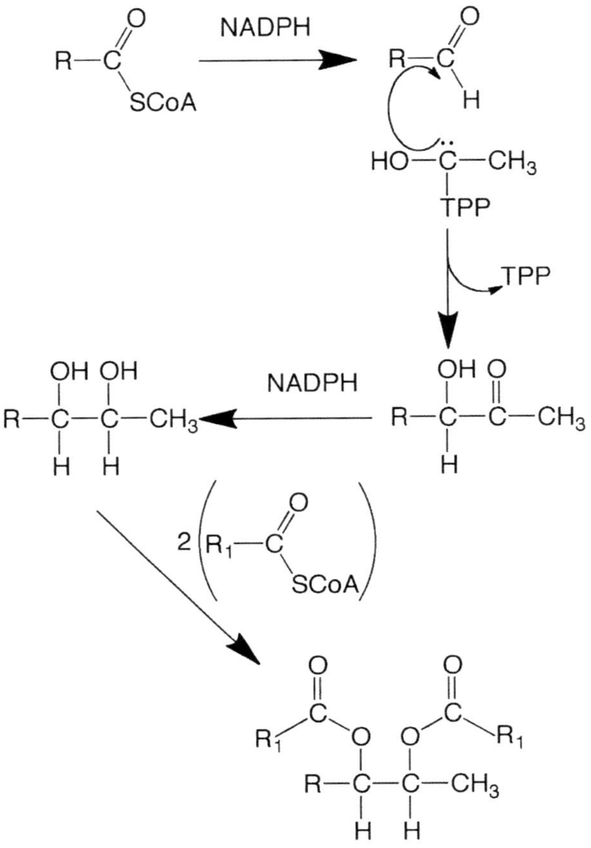

Fig. 11.5 Biosynthesis of Alkane-2,3-diol diesters

Esterification

We have come a long way from the days when direct esterification of a free fatty acid with fatty alcohol was thought to be the mechanism of wax ester synthesis (Friedberg and Greene 1967). With plant enzymes it was demonstrated for the first time that fatty acyl-CoA is used to esterify fatty alcohol (Kolattukudy 1967) (Fig. 11.6). It has been shown that particulate preparations from the avian uropygial glands and mammalian meibomian glands catalyze esterification of fatty alcohols, alkane-1,2-diols, alkane-2,3-diols, 2-hydroxyfatty acids, and 3-hydroxyfatty acids using acyl-CoA (Sawaya and Kolattukudy 1973, Cheng and Russell 2004b; Turkish et al. 2005; Biester et al. 2012). cDNA for wax ester synthases from human, mouse, and avian sources have been cloned and their expression products examined (Cheng and Russell 2004; Miklaszewska et al. 2013). Wax synthase was first cloned from jojoba embryo and expressed in arabidopsin with the demonstration of abundant wax ester synthesis in the seeds of transgenic plants (Lardizabal et al. 2000). Wax synthase-encoding sequences in animals were first identified by expression cloning

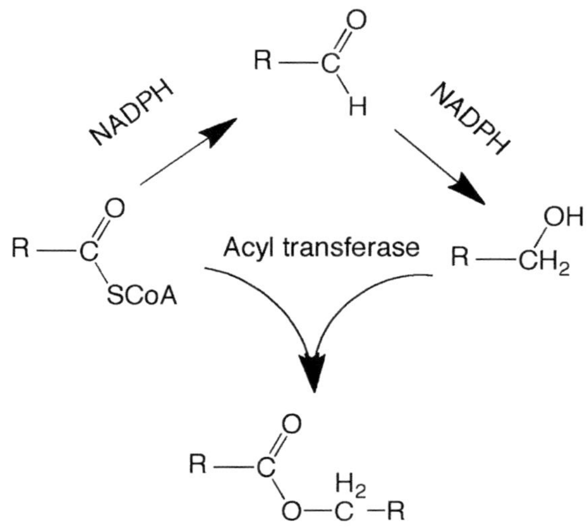

Fig. 11.6 Biosynthesis of monoester wax

of cDNA from mouse preputial glands (Cheng and Russell 2004b). This enzyme showed esterifying activity with a variety of alcohols, including polyisoprenol, and acyl-CoA, but did not show ability to acylate cholesterol or monoacylated or diacylated glycerol. Members of DGAT1 and DGAT2 families of acyl transferases from mammalian and avian sources catalyze acylation of alcohols, diols, retinol, etc. as well as a diacylglycerol. For example, human DGAT1 can catalyze formation of monoesters, diesters, and retinyl esters (Yen et al. 2005a; Yen et al. 2005b). Human wax synthases AWAT1 and AWAT2, belonging to DGAT2 family, can transfer acyl groups to mono or diacylglycerol, fatty alcohol and retinol (Turkish and Sturley 2009). Most probably, the multifunctional acyl tranferases can account for the synthesis of mono- and diester waxes. The presence of transmembrane domains in the wax synthase explains the membrane localization of the acylating enzyme activity in the cell. In cell homogenates however, the nature of the esterifying enzymes catalyzing these reactions have not been elucidated in many cases. Whether specific acyltransferases are involved in the esterification to produce each type of ester remains to be established. However, the possibility of transacylation to diacyl glycerol, fatty alcohol and alkane-1,2-diol by the same enzyme has been strongly suggested. Two acyl transferases encoded by two separate genes catalyze acylation of diacylglycerol to yield triacylglycerol. Of these, DGAT2 seems to be specializing on triacylglycerol synthesis whereas DGAT1 can catalyze acylation of not only diacylglycerol but also alkanol, alkane-1,2-diol and retinol using fatty acyl-CoA as a substrate (Yen et al. 2005a). Deletion of DGAT1 gene in mice shows abnormalities in skin lipid composition with complete absence of diol diesters reflecting its in vivo role in acylation reaction involved in skin lipid biosynthesis. Whether such multifunctional acyl transferases are involved in the synthesis of esters in sebaceous glands in other species remains to be established.

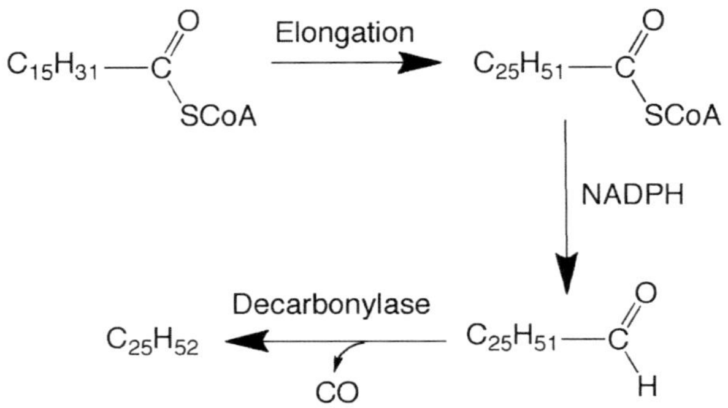

Fig. 11.7 Elongation decarbonylation mechanism for the biosynthesis of alkanes

Biosynthesis of Alkanes

Presence of alkanes as a major component of sebaceous gland lipids is not common, and the alkanes reported to be present in animal surface lipids are thought to be of exogenous origin. However, uropygial gland lipids of a water fowl, eared grebe (*Podiceps nigricollis*), contain large amounts (35–40 %) C21, C23, C25 and C27 alkanes (Cheesbrough and Kolattukudy 1988).

Mechanism of biosynthesis of alkanes, the simplest organic compounds, remained unknown for a long time. In plants, very long chain alkanes are present in significant amounts, together with ketones and secondary alcohols with the functional group in the middle of the carbon chain. This prompted the proposal that head to head condensation between two fatty acids would lead to the formation of the ketone that would give rise to secondary alcohols and ultimately to alkanes. Experiments with ^{14}C-labeled precursors disproved this hypothesis and showed that chain elongation of fatty acids followed by the loss of the carboxyl carbon would give rise to long chain alkanes (Kolattukudy 1987). The mechanism by which the carboxyl carbon of a fatty acid is lost remained a mystery for a long period, as direct decarboxylation of an alkanoic acid seemed mechanistically unlikely. The first clue about a possible mechanism came from the finding that inhibition of alkane synthesis by dithioerythritol resulted in the accumulation of aldehydes with one carbon more than the alkanes. The aldehydes were then shown to be the immediate precursors of alkanes. In alkane producing tissues an acyl-CoA reductase that generates aldehydes has been found (Kolattukudy 1971; Vioque and Kolattukudy 1997; Wang and Kolattukudy 1995b; Lin et al. 2013b). Decarbonylation, a novel biochemical reaction, first discovered in plants (Kolattukudy 1987; Cheesbrough and Kolattukudy 1984; Dennis and Kolattukudy 1992), was shown to be catalyzed by particulate preparations from the uropygial glands of eared grebe that converted an aldehyde to an alkane releasing CO as the other product (Fig. 11.7) (Cheesbrough and Kolattukudy 1988). Even though loss of the carbonyl carbon of an aldehyde (decarbonylation) to yield an alkane has been demonstrated in animals, insects and plants including alga (Kolattukudy 1987; Yoder et al. 1992; Schirmer et al. 2010;

Qiu et al. 2012; Das et al. 2011), the chemical form of the lost carbonyl carbon (CO, CO_2 or formate) and the mechanistic details of how the carbonyl carbon is removed remain to be firmly established.

Regulation of Wax Ester Biosynthesis in Sebaceous Glands

Developmental Changes

There are compositional changes in sebaceous gland lipids that suggest developmental changes. The differences in chemical composition of the sebaceous gland secretion found in newborn babies and those of adult humans suggest developmental changes in the synthesis of sebaceous gland lipids. As humans age the squalene and diester wax content of the sebum decreases (Strauss et al. 1975). Analysis of sebrum production in humans of different ages showed a range of levels with a general pattern of decrease with age with some differences in the level and composition between males and females (Jacobsen et al. 1985; Nazzaro-Porro et al. 1979). The chain lengths and diastereoisomer composition of alkane-2,3 diol diesters of the chicken uropygial glands change significantly as the birds become physiologically mature (Kolattukudy and Sawaya 1974). In developing embryonic goose uropygial glands malonyl-CoA decarboxylase transcripts, a major player in the synthesis of the goose uropygial gland lipids, appears several days prior to hatching and reaches maximal levels by hatching (Kolattukudy et al. 1987b).

Duck uropygial gland provides a system that illustrates developmental changes and hormonal influence on the composition of sebaceous gland wax esters. In the developing duck embryonic uropygial glands, malic enzyme and fatty acid synthase transcripts increase dramatically several days before hatching (Goodridge et al. 1984). The uropygial gland lipids in 2- to 21-day-old ducklings show a composition very different from that of the adult (Kolattukudy et al. 1991). The major components of the duckling glands are long-chain wax esters. As the ducklings approach adulthood, shorter chain esters increase. At 23 days of age shorter chain (<C12) constitute only about 5 %of the acids in the wax ester, whereas the content of such short-chain acids approach 90 % of the acids in 50-day-old ducks. Until juvenile feathers begin to appear at about 3 weeks of age the waxes are composed of extremely complex mixture of n-, mono- and dimethyl branched acids with no obviously dominant components. The complexity decreases with the appearance of adult feather patterns. Monomethyl-branched acids become the dominant acids constituting 94 % of the acids in the wax. These compositional changes, that occur during the period when the down of hatchlings is being replaced with adult feathers, is reflected in the changes in the level of the key enzyme uniquely involved in short-chain acid synthesis. S-Acyl FAS thioesterase that releases the shorter acid from FAS increases dramatically, both at the transcript level and protein level (Kolattukudy et al. 1985b) whereas the FAS level remains relatively constant.

Seasonal Changes in the Adults

Male Mallard

Immediately after the discovery of acyl-FAS thioesterase, we purified this enzyme. During a period in this effort we encountered repeated failures to get enzyme from duck gland extracts for reasons unknown to us at that time. This period turned out to be the postnuptial molt-period called eclipse. During this period the male mallards lose their colorful plumage that becomes replaced with a dull looking plumage giving the males a drab female-like appearance. Thus, we suspected that the difficulty in purifying the enzyme might be related to the physiological changes accompanying eclipse. A systematic examination of the uropygial gland lipid composition of the male mallard ducks over an entire year revealed dramatic seasonal changes in the uropygial gland lipid composition (Kolattukudy et al. 1985c). From February until April, 50–90 % of the wax acids have short chains. In May this percentage began to drop and reached less than 5 % in June and July and then recovered to > 50 % by August. These changes in short-chain acid content was found to be reflected in the S-acyl FAS thioesterase level that dropped to about 10 % during the eclipse period (July) (Kolattukudy et al. 1985c). The S-acyl-FAS thioesterase is a key regulator of the lipid composition in the duck uropygial glands. During eclipse thioesterase level drops, at transcript level and protein level. Transcription is regulated at least in part by steroid hormones. Estradiol and corticosteroid receptor binding consensus sites are present in the first intron of the thioesterase gene (Sasaki et al. 1988). During eclipse estradiol is at a maximal level in the duck. Administration of estradiol or dexamethasone, a synthetic steroid that is more potent than cortisol, reduced the thioesterase transcript and protein levels in the gland (Sasaki and Kolattukudy, unpublished results). Nuclear run off experiments showed that the rate of transcription of thioesterase gene is lower (~ 10 %) in nuclei from glands from animals treated with either estradiol or dexamethasone. This suppression of thioesterase gene transcription results in a drastic decrease in short chain esters in the gland. In fact, the wax ester composition of estradiol or dexamethasone-treated animal glands is very similar to the composition found during eclipse (Fig. 11.8).

The negative regulation of the thioesterase gene transcription by steroids is most probably mediated via DNA methylation. This conclusion is based on several observations. Administration of 5-azacytidine, an inhibitor of DNA methylation, during eclipse, elevated thioesterase gene transcription rates as revealed by nuclear run off experiments and elevated the thioesterase transcript level as measured by RNA blot analysis. The thioesterase transcript and protein levels were reduced to only at about 10 % of that in normal during the eclipse and by dexamethasone injection during non-eclipse period. Azacytidine injection during eclipse recovered the thioesterase transcription rate as measured by nuclear runoff, transcript level measured by RNA blot, and immunologically measured protein levels to 80 % of that in normal gland (Huang Sasaki and Kolattukudy unpublished results). These molecular changes were reflected in the wax ester composition. Injection of aza-

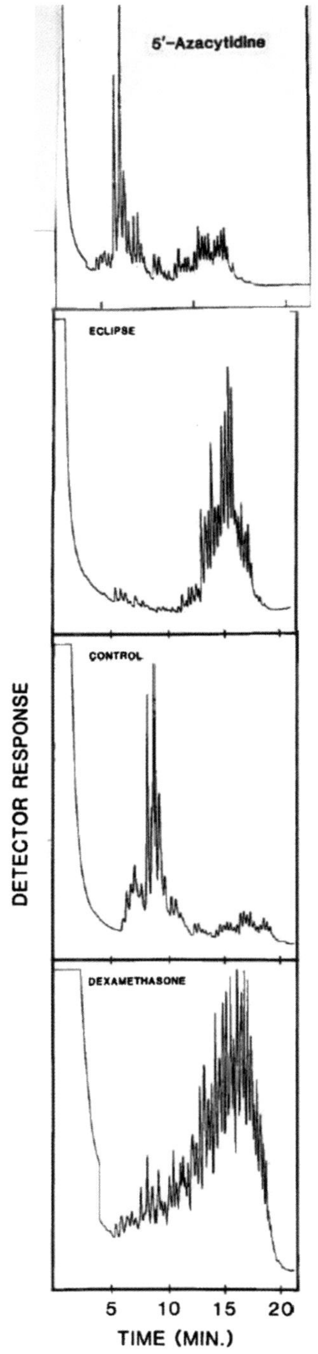

Fig. 11.8 Gas chromatograms showing wax ester distribution in the uropygial glands of male mallard ducks during non-eclipse period (*control*), non-eclipse period but injected with (Dexamethasone) eclipse period (*eclipse*), azacytidine injected during eclipse (*5-Azacytidine*)

cytidine, during eclipse, when the gland produces long-chain waxes as the major secretion products, caused the production of short-chain waxes as the major secretion product, essentially reversing the effects of eclipse (Fig. 11.8). Southern hybridization analysis with methylation sensitive restriction enzymes Msp1 and Hpa2 revealed higher level of methylation of the thioesterase gene in the glands from animals in eclipse and animals out of eclipse but injected with estradiol. Injection of azacytidine that resulted in reversal of eclipse also caused loss of methylation of the thioesterase gene. Thus, during eclipse, estrogen mediates suppression of S-acyl-FAS thioesterase gene expression at the transcriptional level by methylating the thioesterase gene and causes a major shift in wax ester composition. After the eclipse period estrogen level decreases preventing thioesterase gene methylation in the newly differentiating cells in the gland, allowing resumption of production of the thioesterase, thus leading to formation of short-chain wax esters.

DNA methylation, including that caused by estradiol (Zhang and Ho 2011), is usually thought to cause long lasting shut down of gene expression. In the uropygial gland cells that produce wax esters, DNA methylation also probably causes permanent shut down of the thioesterase gene expression. However, as these cells have a short lifetime in the holocrene gland, the newly produced differentiating cells that become the secretory lipid-producing cells can respond to the changes in the hormonal status, and consequent methylation status and thus produce the appropriate lipids. Whether the negative regulation of mammalian sebaceous gland lipid production by estrogen (Smith and Thiboutot 2008) involves DNA methylation remains to be explored.

Female Mallards

Female mallards showed dramatic changes in the uropygial gland lipids during the mating season (Kolattukudy et al. 1987a). Short-chain monoester waxes constitute the major component of the uropygial gland secretion during most of the year. However, with the beginning of the mating season in the middle of March, a polar component appears, and it becomes the dominant and sole component of the secretion through April and May. As the mating season ends in June, monoester waxes become the sole component of the secretion. The polar component consists of diester wax composed of n-C8, n-C10, and n-C12 3-hydroxy fatty acids esterified with n-C16 and n-C18 alcohols and n-C6–C16 even chain fatty acids. These changes are superimposed on the seasonal changes in the chain length composition of monoester waxes. Thus, immediately after the mating season, long-chain wax esters (>C12) dominate until the shorter chain monoester waxes become dominant by the end of August just as in the case of male mallards. These changes in the short-chain content are reflected in the levels of S-acyl-FAS thioesterase just like in the male mallards. The 30 KDa thioesterase virtually disappears during the eclipse period.

Biochemical studies indicated that 3-hydroxy fatty acids used in the diester synthesis is probably produced via hydration of the 2-enoyl-CoA generated by peroxisomal fatty acyl-CoA oxidase (Bohnet et al. 1991). Subcellular fractionation by sucrose and Nycodenz density gradient centrifugation followed by electron microscopic and biochemical examination of the fractions indicated that diester biosynthesis and monoester biosynthesis occur in different compartments in the cell. The diester synthesis occurs in catalase-containing microbodies (peroxisomes), whereas monoester waxes are produced by endoplasmic reticulum. Acyl-CoA reductase required for this synthesis of both monoester and diesters are located in both peroxisomes and endoplasmic reticulum (Zhang and Ho 2011).

Hormonal Regulation of Sebaceous Gland Lipid Biosynthesis

Steroid and thyroxin levels in ducks have attracted much attention (Jallageas et al. 1978; Sharp et al. 1986; Jallageas and Assenmacher 1979). During the mating season estrogen levels peak in female mallards, coinciding with the production of diester waxes as the sole component in the gland secretion. In males also estrogen levels peak during eclipse when the S-acyl-FAS thioesterase is at a minimum. In fact, estradiol injection suppressed S-acyl FAS thioesterase production (Bohnet et al. 1991). Whether the consensus estrogen receptor binding site, found in the first intron of the S-acyl FAS thioesterase gene (Sasaki et al. 1988), is involved in the regulation of expression of the thioesterase gene by estradiol remains to be established. Estradiol injection caused peroxisome proliferation and the production of diesters in the uropygial glands (Bohnet et al. 1991). Thyroxin that can augment the effects of steroid hormones (Mueckler and Pitot 1983; Mueckler et al. 1984), when injected with estradiol, induced peroxisome proliferation in the uropygial glands of both male and female mallards and both produced short-chain diesters that probably act as pheromones. This hormonal treatment induced eclipse molt in males that caused them to look like females. With this hormonally induced molt and production of females pheromones, these males were treated as females by normal untreated males that tried to mount the treated males (Rogers and Kolattukudy, unpublished). Estradiol induction of molting, discovered in ducks, has been applied for pharmacological molt induction in penguins. It appears that penguins in their natural habitat get estrogen prior to molting from their krill diet but in captivity penguins can have difficulty in molting. Administration of estradiol induces molting in such penguins in captivity (Hines et al. 1993).

Peroxisome proliferation activated receptor $\gamma 1$ was found to be the major PPAR induced uniquely in the uropygial gland of duck that responds to estradiol induction but not in the uropygial gland of other species in which estradiol did not induce peroxisome proliferation or diesters synthesis (Ma et al. 1998a). Estrogen treatment of the duck was found to enhance metabolism of arachidonic acid in the uropygial gland. Estradiol treatment induced conversion of arachidonic acid into a prostaglandin D metabolite that was able to activate duck PPAR $\gamma 1$ to the same extent as Δ^{12} prostaglandin J2, and 15–droxy –$\Delta^{12,14}$ prostaglandin J2 the two of the most powerful activators of PPAR γ (Ma et al. 1998b). Thus, estradiol induces the production of the 3-hydroxy fatty acid diester pheromone by inducing the synthesis of a potent activator of PPAR $\gamma 1$ in the duck uropygial gland with the consequent peroxisome proliferation in the gland.

Concluding Remarks

The unusual lipids that coat the surface of animals offer biochemical challenges whose solution could offer novel ways to deal with health issues. Availability of modern analytical technologies, such as GC/LC/MS, allow the elucidation of the complex surface lipidomics. The great advances in the genomic technology and gene knockout technology allowed identification of some of the genes involved and their function in the biosynthesis of some of the surface lipids. Advances are being made in elucidating the signaling molecules and mechanisms involved in maintaining the function of sebaceous glands (McNairn et al. 2013). Progress is also being made in molecular intervention in sebum production to deal with such problems as acne pathology (Sato et al. 2013). Continued advances in the regulation of the sebum biosynthetic process that may lead to novel ways to manipulate the process to yield health benefits.

Acknowledgments I thank Dr. Tatiana Sirakova, Kelly Brussel, Maria Windyga, and Bracken Roberts for assistance in the preparation of the manuscript.

References

Abalain JH, Amet Y, Lecaque D, Secchi J, Daniel JY, Floch HH. Ultrastructural changes in the uropygial gland of the male Japanese quail, Coturnix coturnix, after testosterone treatment. Comparison with the sebaceous gland of the male rat. Cell Tissue Res. 1986;246(2):373–8.

Anderson GJ, Kolattukudy PE. Fatty acid chain elongation by microsomal enzymes from the bovine meibomian gland. Arch Biochem Biophys. 1985;237:177–85.

Bedord CJ, Kolattukudy PE, Rogers L. Isolation and characterization of a tryptic fragment containing the thioesterase segment of fatty acid synthetase from the uropygial gland of goose. Arch Biochem Biophys. 1978;186:139–51.

Biester EM, Hellenbrand J, Gruber J, Hamberg M, Frentzen M. Identification of avian wax synthases. BMC Biochem. 2012;13:4.

Bohnet S, Rogers L, Sasaki G, Kolattukudy PE. Estradiol induces proliferation of peroxisome-like microbodies and the production of 3-hydroxy fatty acid diesters, the female pheromones, in the uropygial glands of male and female mallards. J Biol Chem. 1991;266:9795–804.

Buckner JS, Kolattukudy PE. Lipid biosynthesis in the sebaceous glands: synthesis of multibranched fatty acids from methylmalonyl-Coenzyme A in cell-free preparations from the uropygial gland of goose. Biochemistry. 1975a;14:1774–82.

Buckner JS, Kolattukudy PE. Lipid biosynthesis in sebaceous glands: regulation of the synthesis of n- and branched fatty acids by malonyl-Coenzyme A decarboxylase. Biochemistry. 1975b;14:1768–73.

Buckner JS, Kolattukudy PE. One-step purification and properties of a two-peptide fatty acid synthetase from the uropygial gland of the goose. Biochemistry. 1976a;15:1948–57.

Buckner JS, Kolattukudy PE. Biochemistry of bird waxes. In: Kolattukudy PE, editor. Chemistry and biochemistry of natural waxes. Amsterdam: Elsevier; 1976b. pp. 148–97.

Butovich IA. Lipidomics of human meibomian gland secretions: chemistry, biophysics and physiological role of meibomian lipids. Prog Lipid Res. 2011;50:278–301.

Camera E, Ludovici M, Galante M, Sinagra JL, Picardo M. Comprehensive analysis of the major lipid classes in sebum by rapid resolution high-performance liquid chromatography and electrospray mass spectrometry. J Lipid Res. 2010;51(11):3377–88.

Cheesbrough TM, Kolattukudy PE. Alkane biosynthesis by decarbonylation of aldehydes, a novel biochemical reaction catalyzed by a particulate preparation from *Pisum sativum*. Proc Natl Acad Sci U S A. 1984;81:6613–7.

Cheesbrough TM, Kolattukudy PE. Microsomal preparation from an animal tissue catalyzes release of carbon monoxide from a fatty aldehyde to generate an alkane. J Biol Chem. 1988;263:2738–43.

Cheng JB, Russell DW. Mammalian wax biosynthesis. I. Identification of two fatty acyl-Coenzyme A reductases with different substrate specificities and tissue distributions. J Biol Chem. 2004a;279(36):37789–97.

Cheng JB, Russell DW. Mammalian wax biosynthesis. II. Expression cloning of wax synthase cDNAs encoding a member of the acyltransferase enzyme family. J Biol Chem. 2004b;279(36):37798–807. (Epub 2004 June 27).

Courchesne-Smith C, Jang S-H, Shi Q, DeWille J, Sasaki G, Kolattukudy PE. Cytoplasmic accumulation of a normally mitochondrial malonyl-CoA decarboxylase by the use of an alternate transcription start site. Arch Biochem Biophys. 1992;298:576–86.

Das D, Eser BE, Han J, Sciore A, Marsh EN. Oxygen-independent decarbonylation of aldehydes by cyanobacterial aldehyde decarbonylase: a new reaction of diiron enzymes. Angew Chem Int Ed Engl. 2011;50(31):7148–52.

Dennis MW, Kolattukudy PE. A cobalt-porphyrin enzyme converts a fatty aldehyde to a hydrocarbon and CO. Proc Natl Acad Sci U S A. 1992;89:5306–10.

deRenobales M, Rogers L, Kolattukudy PE. Involvement of a thioesterase in the production of short-chain fatty acids in the uropygial glands of mallard ducks (*Anas platyrhynchos*). Arch Biochem Biophys. 1980;205:464–77.

Downing DT. Mammalian waxes. In: Kolattukudy PE, editor. Chemistry and biochemistry of natural waxes. Amsterdam: Elsevier; 1976. pp. 18–42.

Fernandes ND, Kolattukudy PE. Methylmalonyl coenzyme A selectivity of cloned and expressed acyltransferase and beta-ketoacyl synthase domains of mycocerosic acid synthase from mycobacterium bovis BCG. J Bacteriol. 1997;179:7538–43.

Finn RD, McLaughlin LA, Hughes C, Song C, Henderson CJ, Roland Wolf C. Cytochrome b5 null mouse: a new model for studying inherited skin disorders and the role of unsaturated fatty acids in normal homeostasis. Transgenic Res. 2011;20(3):491–502. doi:10.1007/s11248-010-9426-1. (Epub 2010 July 30).

Foster RJ, Bonsall RF, Poulose AJ, Kolattukudy PE. Interaction of S-acyl fatty acid synthase thioester hydrolase with fatty acid synthase: direct measurement of binding by fluorescence anisotropy. J Biol Chem. 1985a;260:1386–9.

Foster RJ, Poulose AJ, Bonsall RF, Kolattukudy PE. Measurement of distance between the active serine of the thioesterase domain and the pantetheine thiol of fatty acid synthase by fluorescence resonance energy transfer. J Biol Chem. 1985b;260:2826–31.

Friedberg SJ, Greene RC. The enzymatic synthesis of wax in liver. J Biol Chem. 1967;242(2):234–7.

Froese DS, et al. Crystal structures of malonyl-coenzyme a decarboxylase provide insights into its catalytic mechanism and disease-causing mutations. Structure. 2013;21(7):1182–92.

Goodridge AG, Jenrik RA, McDevitt MA, Morris SM Jr, Winberry LK. Malic enzyme and fatty acid synthase in the uropygial gland and liver of embryonic and neonatal ducklings. Tissue-specific regulation of gene expression. Arch Biochem Biophys. 1984;230, 82–92.

Guo L, Zhang X, Zhou D, Okunade AL, Su X. Stereospecificity of fatty acid 2-hydroxylase and differential functions of 2-hydroxy fatty acid enantiomers. J Lipid Res. 2012;53(7):1327–35.

Hardwick JP. Cytochrome P450 omega hydroxylase (CYP4) function in fatty acid metabolism and metabolic diseases. Biochem Pharmacol. 2008;75(12):2263–75.

Hellenbrand J, Biester EM, Gruber J, Hamberg M, Frentzen M. Fatty acyl-CoA reductases of birds. BMC Biochem. 2011;12:64.

Hines R, Kolattukudy PE, Sharkey P. Pharmacological induction of molt and gonadal involution in birds. Proceedings Association of Avian Veterinarians; 1993. pp. 127–34.

Honsho M, Asaoku S, Fujiki Y. Posttranslational regulation of fatty acyl-CoA reductase 1, Far1, controls ether glycerophospholipid synthesis. J Biol Chem. 2010;285(12):8537–42.

Jacob J. Bird waxes. In: Kolattukudy PE, editor. Chemistry and biochemistry of natural waxes. Amsterdam: Elsevier; 1976. pp. 94–141.

Jacobsen E, Billings JK, Frantz RA, Kinney CK, Stewart ME, Downing DT. Age-related changes in sebaceous wax ester secretion rates in men and women. J Invest Dermatol. 1985 Nov;85(5):483–5.

Jallageas M, Assenmacher I. Further evidence for reciprocal interactions between the annual sexual and thyroid cycles in male Peking ducks. Gen Comp Endocrino. 1979;37:44–51.

Jallageas M, Tamisier A, Assenmacher I. A comparative study of the annual cycles in sexual and thyroid function in male Peking ducks (*Anas platyrhynchos*) and teal (*Anas crecca*). Gen Comp Endocrinol. 1978;36:201–10.

Jang SH, Cheesbrough TM, Kolattukudy PE. Molecular cloning, nucleotide sequence, and tissue distribution of malonyl-CoA decarboxylase. J Biol Chem. 1989;264:3500–5.

Kelly EJ, Nakano M, Rohatgi P, Yarov-Yarovoy V, Rettie AE. Finding homes for orphan cytochrome P450s: CYP4V2 and CYP4F22 in disease states. Mol Interv. 2011;11(2):124–32.

Khnykin D, Miner JH, Jahnsen F. Role of fatty acid transporters in epidermis: implications for health and disease. Dermatoendocrinology. 2011;3(2):53–61. doi:10.4161/derm.3.2.14816. (Epub 2011 April 1).

Kihara A. Very long-chain fatty acids: elongation, physiology, and related disorders. J Biochem. 2012;152(5):387–95.

Knop E, Knop N, Millar T, Obata H, Sullivan DA. The international workshop on meibomian gland dysfunction: report of the subcommittee on anatomy, physiology, and pathophysiology of the meibomian gland. IOVS. 2011;52(4):1938–78. (Special Issue).

Kolattukudy PE. Mechanisms of synthesis of waxy esters in broccoli (*Brassica oleracea*). Biochemistry. 1967;6:2705.

Kolattukudy PE. Reduction of fatty acids to alcohols by cell free preparations of *Euglena gracilis*. Biochemistry. 1970;9:1095–102.

Kolattukudy PE. Enzymatic synthesis of fatty alcohols in *Brassica oleracea*. Arch Biochem Biophys. 1971;142:701–9.

Kolattukudy PE. Structure and cell-free synthesis of alkane-1,2-diols of the uropygial gland of white crowned sparrow *Zonotrichia leucophrys*. Biochem Biophys Res Commun. 1972;49:1376–83.

Kolattukudy PE. Avian uropygial (preen) gland. In: Lowenstein JM, editor. Methods in enzymology. Vol. 72. New York: Academic; 1981. pp. 714–20.

Kolattukudy PE. Lipid derived defensive polymers and waxes and their role in plant-microbe interaction. In: Stumpf PK, editor. The biochemistry of plants. Vol. 9. New York: Academic; 1987. pp. 291–314.

Kolattukudy PE, Rogers LM. Acyl-CoA reductase and acyl-CoA fatty alcohol acyl transferase in the microsomal preparation from the bovine meibomian gland. J Lipid Res. 1986;27:404–11.

Kolattukudy PE, Rogers LM. Biosynthesis of 3-hydroxy fatty acids, the pheromone components of female mallard ducks, by cell-free preparation from the uropygial gland. Arch Biochem Biophys. 1987;252:121–9.

Kolattukudy PE, Sawaya WN. Age dependent structural changes in the diol esters of uropygial glands of chicken. Lipids. 1974;9:290–2.

Kolattukudy PE, Rogers LM, Nicolaides N. Biosynthesis of lipids by bovine meibomian glands. Lipids. 1985a;20:468–74.

Kolattukudy PE, Bohnet S, Rogers L. Disappearance of short chain acids from the preen gland wax of mallard ducks during eclipse. J Lipid Res. 1985b;26:989–94.

Kolattukudy PE, Rogers L, Flurkey W. Suppression of a thioesterase gene expression and the disappearance of short chain fatty acids in the preen gland of the mallard duck during eclipse, the period following postnuptial molt. J Biol Chem. 1985c;260:10789–93.

Kolattukudy PE, Bohnet S, Rogers LM. Diesters of 3-hydroxy fatty acids produced by the uropygial glands of female mallards uniquely during the mating season. J Lipid Res. 1987a;28:582–8.

Kolattukudy PE, Rogers LM, Poulose AJ, Jang SH, Kim YS, Cheesbrough TM, Liggitt DH. Developmental pattern of the expression of malonyl-CoA decarboxylase gene and the production of unique lipids in the goose uropygial glands. Arch Biochem Biophys. 1987b;256:446–54.

Kolattukudy PE, Bohnet S, Sasaki G, Rogers L. Developmental changes in the expression of S-acyl fatty acid synthase thioesterase gene and lipid composition in the uropygial gland of mallard ducks (*Anas platyrhynchos*). Arch Biochem Biophys. 1991;284:201–6.
Lardizabal KD, Metz JG, Sakamoto T, Hutton WC, Pollard MR, Lassner MW. Purification of a jojoba embryo wax synthase, cloning of its cDNA, and production of high levels of wax in seeds of transgenic arabidopsis. Plant Physiol. 2000;122(3):645–55.
Lee SY, Tong L. Lipid-containing lubricants for dry eye: a systematic review. Optom Vis Sci. 2012;89(11):1654–61.
Lefèvre C, Bouadjar B, Ferrand V, Tadini G, Mégarbané A, Lathrop M, Prud'homme JF, Fischer J. Mutations in a new cytochrome P450 gene in lamellar ichthyosis type 3. Hum Mol Genet. 2006;15(5):767–76. (Epub 2006 Jan 25).
Lin MH, Hsu FF, Miner JH. Requirement of fatty acid transport protein 4 for development, maturation, and function of sebaceous glands in a mouse model of ichthyosis prematurity syndrome. J Biol Chem. 2013a;288(6):3964–76.
Lin F, Das D, Lin XN, Marsh EN. Aldehyde-forming fatty acyl-CoA reductase from cyanobacteria: expression, purification and characterization of the recombinant enzyme. FEBS J. 2013b;280(19):4773–81.
Lopaschuk GD, Stanley WC. Malonyl-CoA decarboxylase inhibition as a novel approach to treat ischemic heart disease. Cardiovasc Drugs Ther. 2006;20:433–9.
Ma H, Tam QT, Kolattukudy PE. Peroxisome proliferator-activated receptor gamma1 (PPAR-gamma1) as a major PPAR in a tissue in which estrogen induces peroxisome proliferation. FEBS Lett. 1998a;434:394–400.
Ma H, Sprecher HW, Kolattukudy PE. Estrogen-induced production of a peroxisome proliferator-activated receptor (PPAR) ligand in a PPAR γ-expressing tissue. J Biol Chem. 1998b;273:30131–8.
Maier H, Meixner M, Hartmann D, Sandhoff R, Wang-Eckhardt L, Zöller I, Gieselmann V, Eckhardt M. Normal fur development and sebum production depends on fatty acid 2-hydroxylase expression in sebaceous glands. J Biol Chem. 2011;286(29):25922–34.
McNairn AJ, et al. TGFβ signaling regulates lipogenesis in human sebaceous glands cells. BMC Dermatol. 2013;13:2.
Menon GK, Menon J. Avian epidermal lipids: functional considerations and relationship to feathering. Amer Zool. 2000;40(4):540–52.
Miklaszewska M, Kawiński A, Banaś A. Detailed characterization of the substrate specificity of mouse wax synthase. Acta Biochim Pol. 2013;60(2):209–15.
Moto K, Yoshiga T, Yamamoto M, Takahashi S, Okano K, Ando T, Nakata T, Matsumoto S. Pheromone gland-specific fatty-acyl reductase of the silkmoth, Bombyx mori. Proc Natl Acad Sci U S A. 2003;100(16):9156–61.
Mueckler MM, Pitot HC. Transcriptional control of ornithine aminotransferase synthesis in rat kidney by estrogen and thyroid hormone. J Biol Chem. 1983;268:1781–4.
Mueckler MM, Moran S, Pitot HC. Transcriptional control of ornithine aminotransferase synthesis in rat kidney by estrogen and thyroid hormone. J Biol Chem. 1984;259:2302–5.
Nazzaro-Porro M, Passi S, Boniforti L, Belsito F. Effects of aging on fatty acids in skin surface lipids. J Invest Dermatol. 1979;73(1):112–7.
Nicolaides N. Skin lipids: their biochemical uniqueness. Science. 1974;186(4158):19–26.
Nicolaides N, Santos EC. The di- and triesters of the lipids of steer and human meibomian glands. Lipids. 1985;20(7):454–67.
Pappas A. Epidermal surface lipids. Dermantoendocrinology. 2009;1(2):72–6.
Poulose AJ, Rogers L, Cheesbrough TM, Kolattukudy PE. Cloning and sequencing of the cDNA for S-acyl fatty acid synthase thioesterase from the uropygial gland of mallard duck. J Biol Chem. 1985;260:15953–8.
Proksch E, Jensen J. Skin as an organ of protection. In: Goldsmith LA, et al. editors. Fitzpatrick's dermatology in general medicine. New York: McGraw-Hill, 2012.
Qiu Y, Tittiger C, Wicker-Thomas C, Le Goff G, Young S, Wajnberg E, Fricaux T, Taquet N, Blomquist GJ, Feyereisen R. An insect-specific P450 oxidative decarbonylase for cuticular hydrocarbon biosynthesis. Proc Natl Acad Sci U S A. 2012;109(37):14858–63.

Rainwater DL, Kolattukudy PE. Fatty acid biosynthesis in mycobacterium tuberculosis var. bovis bacillus calmette-guerin: purification and characterization of a novel fatty acid synthase, mycocerosic acid synthase, which elongates n-fatty acyl-CoA with methylmalonyl-CoA. J Biol Chem. 1985;260:616–23.
Rogers L, Kolattukudy PE. Purification of S-acyl fatty acid synthase thioester hydrolase by affinity chromatography with fatty acid synthase attached to sepharose. Anal Biochem. 1984;137:444–8.
Rogers L, Kolattukudy PE, deRenobales M. Purification and characterization of an S-acyl fatty acid synthase thioester hydrolase which modifies the product specificity of fatty acid synthase in the uropygial gland of mallard. J Biol Chem. 1982;257:880–6.
Salibian A, Montalti D. Physiological and biochemical aspects of the avian uropygial gland. Braz J Biol. 2009;69(2):437–46.
Sasaki GC, Cheesbrough V, Kolattukudy PE. Nucleotide sequence of the S-acyl fatty acid synthase thioesterase gene and its tissue specific expression. DNA. 1988;7:449–57.
Sato T, Akimoto N, Kitamura K, Kurihara H, Hayashi N, Ito A. Adapalene suppresses sebum accumulation via the inhibition of triacylglycerol biosynthesis and perilipin expression in differentiated hamster sebocytes in vitro. J Dermatol Sci. 2013;70(3):204–10.
Sawaya WN, Kolattukudy PE. Structure and biosynthesis of diesters of alkane-2,3-diols of the uropygial glands of ring-necked pheasants. Biochemistry. 1972;11:4398–406.
Sawaya WN, Kolattukudy PE. Enzymatic esterification of alkane-2,3-diols by the microsomes of the uropygial glands of ring-necked pheasants (*Phasianus colchicus*). Arch Biochem Biophys. 1973;157:309–19.
Schirmer A, Rude MA, Li X, Popova E, del Cardayre SB. Microbial biosynthesis of alkanes. Science. 2010;329(5991):559–62.
Schirra F, Suzuki T, Richards SM, Jensen RV, Liu M, Lombardi MJ, Rowley P, Treister NS, Sullivan DA. Androgen control of gene expression in the mouse meibomian gland. Invest Ophthalmol Vis Sci. 2005;46(10):3666–75.
Schmuth M, Ortegon AM, Mao-Qiang M, Elias PM, Feingold KR, Stahl A. Differential expression of fatty acid transport proteins in epidermis and skin appendages. J Invest Dermatol. 2005;125(6):1174–81.
Sharp PJ, Klandorf H, McNeilly AS. Plasma prolactin, thyroxine, triiodothyronine, testosterone, and luteinizing hormone during a photoinduced reproductive cycle in mallard drakes. J Exp Zool. 1986;238:409–13.
Sirakova TD, Deb C, Daniel J, Singh HD, Maamar H, Dubey VS, Kolattukudy PE. Wax ester synthesis is required for mycobacterium tuberculosis to enter in vitro dormancy. PLoS ONE. 2012;7(12):e51641.
Smith KR, Thiboutot DM. Thematic review series: skin lipids. Sebaceous gland lipids: friend or foe? J Lipid Res. 2008;49(2):271–81.
Stark K, Törmä H, Oliw EH. Co-localization of COX-2, CYP4F8, and mPGES-1 in epidermis with prominent expression of CYP4F8 mRNA in psoriatic lesions. Prostaglandins Other Lipid Mediat. 2006;79(1–2):114–25. (Epub 2006 Jan 27).
Strauss JS, Pochi PE, Downing DT. Skin lipids and acne. Annu Rev Med. 1975;26:27–32.
Tang BY, Hansen IA. Synthesis of 2,3-diols in chicken uropygial glands. Comp Biochem Physiol B. 1976;54(4):483–8.
Thiboutot D, Jabara S, McAllister JM, Sivarajah A, Gilliland K, Cong Z, Clawson G. Human skin is a steroidogenic tissue: steroidogenic enzymes and cofactors are expressed in epidermis, normal sebocytes, and an immortalized sebocyte cell line (SEB-1). J Invest Dermatol. 2003;120(6):905–14.
Turkish AR, Sturley SL. The genetics of neutral lipid biosynthesis: an evolutionary perspective. Am J Physiol Endocrinol Metab. 2009;297(1):E19–27.
Turkish AR, Henneberry AL, Cromley D, Padamsee M, Oelkers P, Bazzi H, Christiano AM, Billheimer JT, Sturley SL. Identification of two novel human acyl-CoA wax alcohol acyltransferases: members of the diacylglycerol acyltransferase 2 (DGAT2) gene superfamily. J Biol Chem. 2005;280(15):14755–64. (Epub 2005 Jan 25).

Uchida Y. The role of fatty acid elongation in epidermal structure and function. Dermatoendocrinology. 2011;3(2):65–9.

Ussher JR, Lopaschuk GD. The malonyl CoA axis as a potential target for treating ischaemic heart disease. Cardiovasc Res. 2008;79:259–68.

Vasireddy V, Uchida Y, Salem N Jr, Kim SY, Mandal MN, Reddy GB, Bodepudi R, Alderson NL, Brown JC, Hama H, Dlugosz A, Elias PM, Holleran WM, Ayyagari R. Loss of functional ELOVL4 depletes very long-chain fatty acids (> or = C28) and the unique omega-O-acylceramides in skin leading to neonatal death. Hum Mol Genet. 2007;16(5):471–82. (Epub 2007 Jan 5).

Vioque J, Kolattukudy PE. Resolution and purification of an aldehyde generating and an alcohol generating fatty acyl CoA reductase from pea leaves (*Pisum sativum* L.). Arch Biochem Biophys. 1997;340:64–72.

Wang X, Kolattukudy PE. Solubilization, purification and characterization of fatty acyl-CoA reductase from duck uropygial gland. Biochem Biophys Res Commun. 1995a;208:210–5.

Wang X, Kolattukudy PE. Solubilization and purification of aldehyde-generating fatty acyl-CoA reductase from green alga Botryococcus braunii. FEBS Lett. 1995b;370:15–8.

Westerberg R, Tvrdik P, Undén AB, Månsson JE, Norlén L, Jakobsson A, Holleran WH, Elias PM, Asadi A, Flodby P, Toftgård R, Capecchi MR, Jacobsson A. Role for ELOVL3 and fatty acid chain length in development of hair and skin function. J Biol Chem. 2004;279(7):5621–9. (Epub 2003 Oct 27).

Wróbel A, Seltmann H, Fimmel S, Müller-Decker K, Tsukada M, Bogdanoff B, Mandt N, Blume-Peytavi U, Orfanos CE, Zouboulis CC. Differentiation and apoptosis in human immortalized sebocytes. J Invest Dermatol. 2003;120(2):175–81.

Yen CL, Monetti M, Burri BJ, Farese RV Jr. The triacylglycerol synthesis enzyme DGAT1 also catalyzes the synthesis of diacylglycerols, waxes, and retinyl esters. J Lipid Res. 2005a;46(7):1502–11.

Yen CL, Brown CH 4th, Monetti M, Farese RV Jr. A human skin multifunctional O-acyltransferase that catalyzes the synthesis of acylglycerols, waxes, and retinyl esters. J Lipid Res. 2005b;46(11):2388–97.

Yoder JA, Denlinger DL, Dennis MW, Kolattukudy PE. Enhancement of diapausing flesh fly puparia with additional hydrocarbons and evidence for alkane biosynthesis by a decarbonylation mechanism. Insect Biochem Mol Biol. 1992;22:237–43.

Zhang X, Ho SM. Epigenetics meets endocrinology. J Mol Endocrinol. 2011;46(1):R11–32.

Zouboilis et al. Establishment and characterization of an immortalized human sebaceous gland cell line (SZ95). J Invest Dermatol. 1999;113(6):1011–20.

Chapter 12
Squalene Chemistry and Biology

Emanuela Camera, Monica Ottaviani and Mauro Picardo

Core Messages

- Squalene is an abundant compound of human sebum that is implicated in defensive mechanisms and skin disorders not observed in other mammalian species.
- Squalene is a double-edged sword as it covers protective functions and, upon environmental stresses and pathological conditions, yields oxidized and breakdown compounds associated with detrimental effects on skin.
- Squalene level is a marker of sebaceous gland activity and density; it is found altered in skin disorders involving skin surface lipids modifications.
- Preservation of physiological concentrations of squalene offers a strategy for the management of acne and prevention of skin photo-damages.

Abbreviations

CE	Cholesterol esters
FA	Fatty acids
SG	Sebaceous gland
SSL	Skin surface lipids
TG	Triglycerides
WE	Wax esters

E. Camera (✉) · M. Ottaviani · M. Picardo
Laboratory of Cutaneous Physiopathology and Integrated Center of Metabolomics Research, San Gallicano Dermatological Institute, IRCCS, Via Elio Chianesi 53, 00144 Rome, Italy
e-mail: camera@ifo.it

A. Pappas (ed.), *Lipids and Skin Health,* DOI 10.1007/978-3-319-09943-9_12

Introduction

Human sebum is renowned as a unique mixture of lipids. Although the complexity of human sebum composition is suggestive of the multiplicity of sebum functions, squalene deserves a special focus due to its extraordinary abundance on human skin. Noteworthy, squalene is the main standalone and unsaponifiable component in sebum. The majority of sebaceous lipid components are members of lipid families, such as fatty acids (FA), triglycerides (TG), wax esters (WE), and cholesterol esters (CE). In particular, TG together with free FA account for the 40–60 % of the sebum weight, followed by WE (19–26 %) and squalene (10–15 %). Although minimal amounts of cholesterol are measurable in sebum, its exact origin is still unclear. It is likely that epidermal cholesterol or membrane remnants are the sources of cholesterol in sebum. Apart from components found ubiquitously in human tissues and biofluids such as TG, free FA, and CE, the sebaceous gland (SG) shares the synthesis of wax esters with the apocrine and the meibomian glands, whereas the high-yield production of squalene seems to be specific to fully differentiated sebocytes in the SG.

Squalene Chemistry and Biosynthetic Pathways

Squalene (2, 6, 10, 15, 19, 23-hexamethyltetracosa-2, 6, 10, 14, 18, 22-hexaene) is an unsaturated aliphatic hydrocarbon with 30 carbon atoms and six unconjugated double bonds with trans geometry (Fig. 12.1). The structure, with its six methyl branches, is characteristic of triterpenoid compounds, partly shared with carotenoids, Q10 ubiquinone, and tocopherols. Squalene is the biochemical precursor of the whole family of steroids being the intermediate product in the biosynthetic pathway leading to cholesterol. Enzymes involved in the pathways of biosynthesis of squalene have been found and characterized in the human SG and cultured sebocyte cells. Lipid synthesis and packaging into lipid droplets occur consistently with the differentiation of sebocytes. Cholesterol and squalene share the first steps of their biosynthesis in the SG, similarly to other tissues that manufacture lipids. The first step of the biosynthetic pathway involves the HMG-CoA synthase enzyme that catalyzes the condensation of acetyl-CoA with acetoacetyl-CoA yielding 3-hydroxy-3-methylglutaryl-coenzyme A (HMG-CoA). Expression and modulation of HMG-CoA synthase have been assessed in the SG (Rosignoli et al. 2003). The following step is catalyzed by the HMG-CoA reductase, which is the rate limiting enzyme for the de novo cholesterol synthesis. This enzyme has proven to be expressed and active in the human SG (Smythe et al. 1998). This is accounted in the mevalonate pathway which is critical in the synthesis of terpenoids. The symmetry of the squalene backbone is due to head-to-head condensation of two molecules of farnesyl pyrophosphate, with reduction by nicotinamide adenine dinucleotide phosphate (NADPH) catalyzed by the squalene synthase enzyme. Activity, expression, and transcript levels of squalene synthase have been determined in

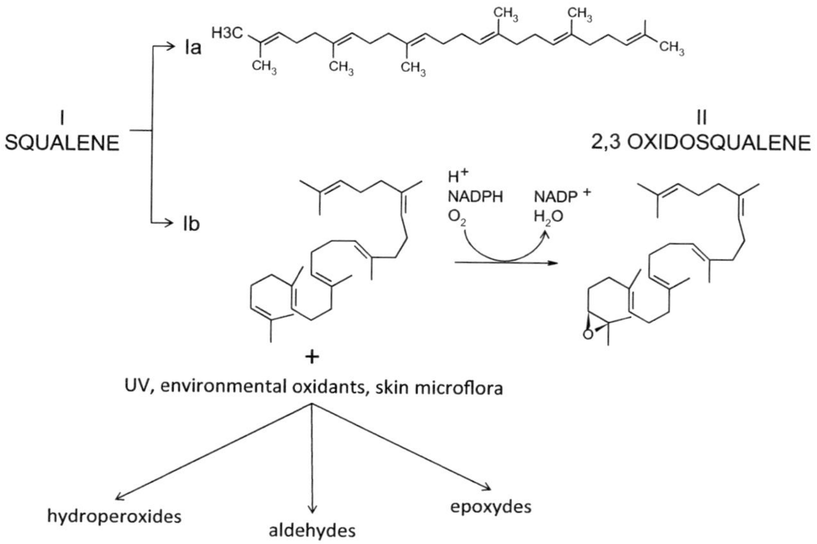

Fig. 12.1 Linear (*Ia*) and folded (*Ib*) structures of squalene (*I*) and structure of 2,3-oxidosqualene (*II*, squalene epoxide). Squalene is transformed into several oxygenated derivatives and breakdown compounds by different stressors (*Ultra violet* (*UV*), *environmental oxidants*, *skin microflora*). Enzymatic oxygenation of squalene to 2,3-oxidosqualene is catalyzed by squalene monooxygenase enzyme, consuming a mole of NADPH and one of oxygen

sebocytes in culture (Pelle et al. 2008). The final step that introduces squalene into the steroid structure is the oxidation of one of the terminal double bonds. The intermediate 2,3-squalene oxide undergoes enzyme-catalyzed cyclization to yield lanosterol, which is further transformed into cholesterol and the downstream steroids. Nevertheless, cholesterol itself does not take part in the sebogenesis as its synthesis is interrupted at the level of squalene, which instead accumulates in the lipid droplets. NADPH is implicated in the synthesis of cholesterol from squalene, however other factors seem to determine a low cholesterol/squalene ratio and are to be searched among those regulating the activity of squalene epoxidase and downstream enzymes (Downie and Kealey 1998). Detection of low levels of 2,3-squalene oxide in sebum might be due to a partial activity of the squalene monooxygenase or to nonenzymatic squalene oxidation. Nevertheless, a role played by enzymes from the cutaneous microflora in the processing of human squalene has been little investigated.

The lipogenic activity of the SG is finely regulated and depends partly on lipid signaling. Similarly to other lipogenic tissues the SG expresses liver X receptors (LXRs), which are members of nuclear receptor superfamily (Hong et al. 2008; Russell et al. 2007). LXRs staining has been detected within the nucleus of sebocytes. LXRs are key players of the lipid homeostasis acting as sensors of cholesterol levels. Upon binding to oxysterols, LXRs form dimers with the retinoid X receptor alpha (RXRα) and binds consensus sequences within promoters of target genes that regulate cholesterol and phospholipids efflux (Schultz et al. 2000). The expression of sterol regulatory element binding protein 1 (SREBP-1) is under direct control of LXRs following heterodimerization with RXRα (Swinnen et al. 1997). Detection of LXRα and LXRβ receptors in cultured sebocytes has led to investigate LXR agonists for their effects on cell proliferation and sebogenesis (Russell et al. 2007). Sterol regulatory element binding proteins (SREBPs) affect the expression of genes for both fatty acid and cholesterol synthesis. In particular, SREBP-1 is increased in SEB-1 human sebocyte in culture following stimulation with insulin-like growth factor (IGF-1; Smith et al. 2006). *FAS* gene is another target of the LXRs (Joseph et al. 2002; Hong et al. 2008).

The observation that incubation of the isolated SG in vitro reflects the relative rates of lipid biosynthesis as they occur in vivo, at least in their initial stage, has allowed studying metabolic pathways of sebogenesis (Downing et al. 1977). Experiments based on the incorporation of radiolabeled acetate, have demonstrated that the SG possesses the capacity to synthesize lipids de novo (Summerly and Woodbury 1971; Guy et al. 1999). Based on the patterns of lipids synthesized from different exogenous substrates, it has been suggested that NADPH was a limiting and directing factor of the sebaceous lipids biosynthesis in isolated SG (Middleton et al. 1988; Downie and Kealey 1998). Exogenous radiolabeled glucose, a good source of NADPH, favors TG synthesis, whereas exogenous radiolabeled acetate, a poor source of NADPH, is in favor of squalene. TG synthesis, unlike squalene synthesis, is heavily NADPH dependent (Middleton et al. 1988). This has suggested that the glandular level of NADPH is the limiting and driving factor leading to the squalene/TG ratio typical of sebum. Thus, sources of NADPH, such as glucose and glutamine, favor TG synthesis. Supplying sebocytes with acetate, which is a poor source of NADPH, results in a diminished abundance of TG, whereas squalene biosynthesis is favored. The synthesis of cholesterol from squalene is also NADPH dependent, raising the question whether the high amount of squalene in sebum would be a NADPH driven process (Ridden et al. 1990). Endogenous glycogen is a potential lipogenic substrate. Following glycogen breakdown, glycerophosphate is as important as NADPH in governing the sebaceous lipogenesis. Squalene synthesis is favored over TG when acetate and glutamine are the substrates, independently on the glycogen status of the gland (Downie and Kealey 1998).

Levels of Squalene in Skin Surface Lipids

Squalene accounts for 10–15% of the total sebum lipids. (Kellum 1967; Downing et al. 1969, 1977; Greene et al. 1970; Stewart and Downing 1985). The SG has likely evolved a mechanism that halts the cholesterol biosynthesis and warrants the accumulation of squalene. On the other hand, the anaerobic environment of the SG might limit the access of molecular oxygen required for the further processing of squalene (Smith et al. 2008). The high content of squalene together with an exceptional variety and abundance of WE distinguishes the sebaceous lipid secretion

Table 12.1 Changes in surface lipid composition with age. (Modified from Ramasastry et al. (1970). Reported values are the weight/weight percentage of lipids in sebum)

Age range	Subjects	Free FA	TG	WE	Cholesterol	CE	Squalene
1 month–2 years	7	20.8	38.4	17.6	3.7	10.3	9.4
2–4 years	8	22.9	49.6	8.0	4.2	8.9	6.2
4–8 years	13	15.9	45.6	6.9	7.2	14.6	7.7
8–10 years	8	17.8	47.4	17.8	3.2	5.7	8.3
10–15 years	7	18.8	42.9	23.6	1.8	4.2	8.4
18–40 years	17	16.4	41.0	25.0	1.4	2.1	12.0
Correlations with squalene		−0.5172	−0.7510	0.7707	−0.5746	−0.5597	1

FA fatty acids, *TG* triglycerides, *WE* wax esters, *CE* cholesterol esters

from other lipid compartments. Noteworthy, elevated levels of squalene occur only in human sebaceous secretion (Nikkari 1974a). It can be speculated that squalene together with WE conveys the major skin protective purpose of sebum thought to be limitation of dehydration. The SG is fully functional at birth. The maternal androgens and the endogenous steroid synthesis stimulate the SG resulting in the secretion of the vernix caseosa, which appears starting from the third trimester of gestation. Composition of sebum has been evaluated from birth to adulthood. Although interindividual variations of squalene and other sebaceous lipids, reported as weight/weight percentage of skin surface lipids, is very high, average values show consistency in a wide range of ages (Table 12.1). Measures of sebum composition from infancy to adulthood demonstrated that varying the rate of sebaceous lipogenesis does not necessarily change the composition of the lipid mixture being synthesized (Table 12.1). In particular, squalene concentrations appear not to correlate with development from infancy to adulthood (Ramasastry et al. 1970). Upon elaboration of the early data on the sebum composition, it can be inferred that squalene and TG proportions are inversely correlated. This observation possibly indicates that factors regulating the squalene synthesis are negative modulators of TG (Table 12.1). This inverse relationship is consistent with the different dependency on NADPH of the respective biosynthetic pathways as discussed earlier (Middleton et al. 1988; Downie and Kealey 1998). Further evidence on the influence of substrate availability on the sebum composition arises from studies in humans. Changes in the concentration of sebaceous lipids on the skin surface in normal weight and obese human subjects resulting from fasting, demonstrated that squalene is the component that is preserved under caloric restriction. Starvation lowers the total sebum excretion rate and decreases the percentage of WE and TG. In contrast, the proportion of squalene increases upon starvation (Pochi et al. 1970; Downing et al. 1972). Thus, the reduction of sebum excretion occurring upon caloric restriction is due to relented synthesis of all sebum constituent except squalene. Estimate of the delay between the synthesis of sebum and its excretion to skin surface time came from the observation that the maximum rate of increase of squalene concentration in the skin surface lipids occurred at the eighth day after the fast was started. Together these evidences support the hypothesis that during caloric restriction or starvation the biosynthetic pathways leading to FA are shut down to preserve squalene biosynthesis. The authors concluded that alterations of nutritional status could be detectable by analyzing the skin surface lipids. The concentration of squalene in SSL is also indicative of the uneven distribution of the SG in different body sites. In particular, the squalene/cholesterol ratio is a useful marker of the proportion between sebaceous secretion and epidermal lipids. Forehead, chest, and back are in the order the body sites with the highest squalene/cholesterol ratio. Whereas lower values of squalene/cholesterol ratio were obtained from anatomical sites with less SG density, e.g., legs (Michael-Jubeli et al. 2011).

Interaction Between Blood and Sebaceous Lipids

Squalene is measurable in blood, likely as a result of a lipid leakage from the cholesterol biosynthetic processes in metabolic tissues and from dietary sources (Rajaratnam et al. 1999). Squalene as the intermediate of cholesterol synthesis is included in surrogate markers of cholesterol metabolism and absorption. Early studies have demonstrated that squalene is accumulated in lipid bodies of human adipose cells (Tilvis et al. 1982). Significance of squalene stores in the adipose tissue has not been clarified. Shark liver oil is the richest source of squalene; however, there are several botanical sources as well, including rice bran, wheat germ, olive oil, palm oil, cottonseed oil, and rapeseed oil. Squalene is found abundant also in yeast lipid extracts. Squalene is absorbed from dietary sources, transported mostly in association with very-low-density lipoprotein (VLDL) and low-density lipoprotein (LDL), and distributed to the periphery (Simonen et al. 2007). Ingestion of squalene-rich dietary oils results into increased serum levels of squalene and conversion into cholesterol. (Strandberg et al. 1990; Miettinen and Vanhanen 1994). Serum concentration in healthy woman has been observed to range from about 1–3 μM and to be in correlation with cholesterol levels and the lipoprotein distribution. It remains to be elucidated how the blood levels of squalene are connected with the cholesterol absorption and metabolism through the conversion to lanosterol (Rajaratnam et al. 1999). The same group concluded that elevated blood levels of squalene were associated with higher risks of cardiovascular diseases in women. (Rajaratnam et al. 2000). Serum levels of squalene have been demonstrated to be associated with the prevalence of the metabolic syndrome (Gylling et al. 2007).

There is evidence that the activity of lipogenetic enzymes in the SG is regulated by exogenous lipids (Smythe et al. 1998). Nevertheless, the influence of circulating lipids on the lipogenic activity of the SG awaits further investigations. Cholesterol enters the cells by the LDL receptor-mediated endocytosis causing inhibition of HMG-CoA reductase transcription and translation and stimulation of the enzyme degradation (Goldstein and Brown 1990). As HMG-CoA reductase is regulated by the cholesterol levels in the SG environment, cholesterolemia could have some impact on the squalene levels in the sebum. Conflicting indications have been provided on the blood–skin interaction concerning cholesterol. In humans, 26–46 % of skin surface cholesterol is derived from the bloodstream (Nikkari et al.

1974b, 1975). However, the amount of skin cholesterol has no apparent relationship with hypercholesterolemic or normocholesterolemic conditions of subjects (Bhattacharyya et al. 1972). Nevertheless, the SG responds to higher levels of lipoproteins by lowering the lipidogenic capacity through inhibition of the HMG-CoA reductase activity; this evidence indicating intracellular delivery of exogenous cholesterol occurs through interaction with the LDL receptor (Smythe et al. 1998). Interestingly, the enzymatic activity of HMG-CoA reductase varies considerably in the apparently healthy skin of acne-unaffected individuals (CV = 30–46 %). Moreover, the declined sebum production occurring with age can be at least in part ascribable to the lower activity of HMG-CoA reductase in older subjects.

Functions of Squalene in Sebum

Multiple roles have been attributed to squalene, although the most specific and prominent one has not been recognized as yet. The typically high squalene concentration found in sebum is reached in fully differentiated and mature SG. As such, squalene is considered the biological marker of sebocyte differentiation. Altogether the SSL participate to the barrier function of skin in many respects. In particular, squalene covers several functions that range from the immunomodulatory, to the regulation of the skin microflora, to contrasting UV damages and skin ageing (Picardo et al. 1991a; Dennis and Shibamoto 1989; Chiba et al. 1999). As soon as the SG becomes fully active and excretes sebum, *Proprionibacterium acnes* (*P. acnes*) finds the favorable conditions for the colonization of the pilosebaceous follicle. With the limited knowledge accumulated so far, it has been concluded that *P. acnes* uses the sebaceous FA as nutrients upon hydrolysis of TG operated by its own lipases. Noteworthy, *P. acnes* colonization causes the incretion of photosensitizers, i.e., porphyrins, that participate in the photo-oxidative processes leading to lipid degradation, mainly affecting squalene integrity (Saint-Leger et al. 1986a). Alteration of SSL, both in quantity and quality, has been found to be involved in different skin diseases. Levels of SSL are affected in a number of skin disorders. Marked modifications of SSL have been reported in atopic dermatitis, psoriasis, and seborrheic dermatitis (Picardo et al. 1991b; Passi et al. 1991; Zouboulis 2010). The impact of these skin disorders on the SSL composition has been reviewed (De Luca and Valacchi 2010). Apart from compositional alterations of ceramides, which are pivotal in the homeaostasis of the skin barrier (Masukawa et al. 2008), subjects affected from atopic dermatitis present significantly lower levels of lipids of the sebaceous type, including squalene. Correspondingly, levels of other lipid species are increased, indicating that the lipid proportions in sebum are impaired in conditions of epidermal barrier disruption. Notably, systemic alterations of main lipophilic and enzymatic antioxidants are frequently associated with the above diseases (Passi et al. 1991).

In acne, about 30 % of the wet weight of follicular casts is formed by lipid matter, wherein squalene accounts for the 20 % of total lipids (Nordstrom et al. 1986).

Although squalene has been addressed as a comedogenic factor, per se it is not an inflammatory mediator in acne, suggested by the observation that intradermal injection of squalene induces no inflammatory response in receiving subjects (Puhvel and Sakamoto 1977). In contrast, substances contained in sebum rather than squalene per se stimulate comedogenesis, providing the first clue that qualitative changes played a role (Kligman and Katz 1968). Sebum outflow is a route of excretion of lipophilic antioxidants to the skin surface. Vitamin E (α-tocopherol) and coenzyme Q10 are cosecreted with squalene in the skin surface lipids. Levels of α-tocopherol directly correlate with those of squalene, indicating a reciprocal influence in the mechanisms regulating their secretion and degradation by environmental factors (Thiele et al. 1999; De Luca and Valacchi 2010). The lipophilic antioxidants prevent squalene from excessive photooxidation, as they act as scavengers of free radicals and reactive oxygen species in the lipid phase (Passi et al. 2002). In spite of the low levels that exogenous antioxidants reach in sebum, they exhibit a synergic inhibition of the UV induced effects that results in the rescue of squalene depletion.

Squalene Oxidation Products and Skin Disorders

Squalene is regarded as the major lipid undergoing peroxidation in sebum, based on its chemical structure and its concentration in the SSL. Because of the elevated number of unsaturated sites, squalene represents a sensitive target of peroxidation reactions, wherein squalene monohydroperoxide is the primary oxidation product (Kohno et al. 1995). Due to the symmetry of the molecule, six possible positional isomers of monohydroperoxide derivatives are produced. Further oxidation of monohydroperoxides generates by-products bearing two or more –OOH groups (Mountfort et al. 2007). Oxidative reactions involving squalene yields other by-products, mainly squalene epoxides, also characterized by positional isomerism (De Luca et al. 1996; Mountfort et al. 2007). Furthermore, 2,3-squalene oxide can be formed enzymatically by mammalian or fungal squalene monoxigenases (Fig. 12.1). Several in vitro studies indicate that squalene oxidation can be initiated by light exposure and both UVA and UVB radiations. Squalene is very efficient in scavenging singlet oxygen, which is generated upon excitation of photosensitizing molecules, which enhance the yield and the kinetics of oxidation by-products formation (Ohsawa et al. 1984; Dennis and Shibamoto 1989; Kohno et al. 1995; Matsuo et al. 1983; Ryu et al. 2009). Breakdown products of oxidized squalene expand the chemical variety of the oxidation by-products. Transfarnesal and short chain reactive aldehydes, in particular formaldehyde and malonyl dialdehyde complete the wide range of by-products generated by oxidative degradation of squalene. (De Luca et al. 1996; Dennis and Shibamoto 1989; Yeo and Shibamoto 1992; Wei and Shibamoto 2007). It has been demonstrated that in physiological conditions, UVA induces oxidative degradation of squalene at a much higher extent than UVB (Ryu et al. 2009). Early reports of squalene peroxidation upon UVB were likely ascribable to the minimal UVA doses contaminating the UVB sources (Ekanayake

Mudiyanselage et al. 2003). UV exposure of the skin produces marked squalene degradation especially when compared to the other unsaturated lipid fractions such as cholesterol, sebum TG, and free FA. Consistent with this, squalene can be considered as integral part of the SSL antioxidant defense providing skin lipids protection along with α-tocopherol and ubiquinone that are, at least in part, cosecreted and cosynthesized with squalene by the SG, respectively (Thiele et al. 1999; Packer and Valacchi 2002). However, it has been demonstrated that squalene oxidation takes place in vivo only after α-tocopherol and ubiquinone are depleted at a significant extent indicating that antioxidants are the first line targets of UV irradiation (Podda et al. 1998; Thiele et al. 2001; Shvedova et al. 2001). Squalene oxidation products generated by skin exposure to different peroxidative stimuli such as UV irradiation or microbial peroxidizing metabolism may be considered as important physiological mediators of the biological effects exerted on the skin by oxidative stimuli (Picardo et al. 1991c). For example, the discoloration observed in pityriasis versicolor, has been associated with the squalene oxidation (Nazzaro-Porro et al. 1986). Peroxidated squalene induces wrinkles and roughness in hairless mice, indicating that it partly delivers the pro-ageing effects of UV light (Chiba et al. 1999). Early studies have suggested that by-products of squalene oxidation are involved in the onset and progression of acne vulgaris (Leyden 1995). The lipid peroxidation hypothesis of acne pathogenesis has been unraveled in several studies revealing that localized free radical damage and peroxides might be involved in initiating the inflammatory reactions. Lipid peroxidation is evident in acne, wherein localized free radical damage and peroxides might be involved in initiating the damaging inflammatory reactions. Particularly, squalene shows enhanced comedogenicity when oxidized (Mills et al. 1978). Both squalene and its oxidized metabolites have been found at much higher levels in acne vs. healthy controls (Mills et al. 1978; Saint-Leger et al. 1986a, b; Hanaoka et al. 1971; Cotterill et al. 1972; Pappas et al. 2009). The mechanisms of squalene oxidation by-products formation in acne remain to be elucidated. Nevertheless, they are one of the factors of the pathological functions exerted by sebum. *P. acnes* metabolism produces coproporphyrin which has been demonstrated to be an efficient singlet oxygen generator upon UVA irradiation. This observation could provide a link between P. acnes colonization and increased oxidative damages of SSL in acne (Arakane et al. 1996). Application of UV-irradiated squalene onto cultured keratinocytes elicits proinflammatory effects, partly mediated by the increased expression of inflammatory cytokines and cell hyperproliferation (Ottaviani et al. 2006). Thus, inflammatory manifestations in acne are likely secondary events triggered by peroxidated lipids. It has been observed that the accumulation of lipoperoxides in the comedones correlates with the severity of the lesions and biomarkers of inflammation and cell hyperproliferation. Higher levels of lipid peroxides and IL-1α within the comedones were associated with the obstruction of the infundibulum observed at different stages of comedones formation, from microcomedones to inflammatory open or close comedones (Tochio et al. 2009). Specifically, by-products of squalene derived from oxidative processes have been related to comedones formation. The topical application of squalene monohydroperoxide has been proven comedogenic in animal models (Motoyoshi 1983;

Chiba et al. 2000). In addition, the study of intra-comedonal lipids of acne patients indicated the presence of high amount of polar lipids that appeared mainly derived from squalene peroxidation (Saint-Leger et al. 1986b). Further mechanisms through which oxidized squalene causes detrimental effects are centered on the depletion of antioxidant defense systems (Ikeno et al. 2011). In vitro study on fibroblast demonstrated that glutathione is an important endogenous antioxidant implicated in the protection of skin cells from oxidative damage induced by squalene monohydroperoxide. Increased sensitivity to squalene peroxides along with a more pronounced comedogenic response have been observed upon glutathione depletion (Chiba et al. 2001). The pathogenetic role of squalene peroxidation in acne points out to antioxidants as suitable means to attenuate comedogenicity (Bowe et al. 2012). In particular, evidence has been provided on the value of oral vitamins A and E in attenuating lipid peroxidation in acne (Ayres and Mihan 1978).

Potential Clinical Uses of Squalene

Several benefits deriving from the clinical use of squalene are suggested by the available experimental evidence. Squalene administration can be advantageous in the modulation of cholesterol metabolism, in the detoxification of xenobiotics, to reinforce the antioxidant defense system and immune functions. However, there is a lack of controlled studies that would support benefits of squalene administration in the management of skin disorders. A considerable amount of squalene is absorbed from dietary sources. Direct application of squalene to the skin overcomes the extensive biotransformation that squalene can undergo when administered systemically. Nevertheless, studies are necessary to evaluate the impact on skin metabolism and properties of systemic or topical squalene.

Summary

Squalene is an endogenous triterpene that takes a large share of the several functions carried out by sebum. Its biochemical features, including abundance in the SSL and transformation into oxygenated by-products, have proven to play a key role in the skin response to endogenous and environmental stressors. In physiological conditions, sebum can be regarded as a vehicle that transports outward lipid molecules, among these ones squalene represents a first line defense system of direct and indirect effects of solar light. However, marked degradation of squalene produces a wide range of derivatives that deteriorates the skin functions. Squalene by-products have demonstrated cytotoxic, immunogenic, and hyperproliferative effects. Squalene shows altered levels in conditions of disrupted barrier function, such as psoriasis and atopic dermatitis, and in the major disorder of the SG, such as acne. Nevertheless, inconsistency between squalene levels and acne severity degrees calls for further investigation of the factors that actually regulate its concentration at the skin level. Modulating squalene concentration in SSL, either endogenously or topically, and preventing its oxidation appear to be beneficial in the management of inflammatory skin diseases and for the improvement of skin health.

Acknowledgments The authors would like to acknowledge the support received by the Italian Ministry of Health through the grant RF-2010-2316435.

References

Arakane K, Ryu A, Hayashi C, Masunaga T, Shinmoto K, Mashiko S, Nagano T, Hirobe M. Singlet oxygen (1 delta g) generation from coproporphyrin in propionibacterium acnes on irradiation. Biochem Biophys Res Commun. 1996;223(3):578–82.

Ayres S, Mihan R. Acne vulgaris and lipid peroxidation: new concepts in pathogenesis and treatment. Int J Dermatol. 1978;17:305–7.

Bhattacharyya AK, Connor WE, Spector AA. Excretion of sterols from the skin of normal and hypercholesterolemic humans. Implications for sterol balance studies. J Clin Invest. 1972;51:2060–70.

Bowe WP, Patel N, Logan AC. Acne vulgaris: the role of oxidative stress and the potential therapeutic value of local and systemic antioxidants. J Drugs Dermatol. 2012;11(6):742–6. (Review).

Chiba K, Sone T, Kawakami K, Onoue M. Skin roughness and wrinkle formation induced by repeated application of squalene-monohydroperoxide to the hairless mouse. Exp Dermatol. 1999;8(6):471–9.

Chiba K, Yoshizawa K, Makino I, Kawakami K, Onoue M. Comedogenicity of squalene monohydroperoxide in the skin after topical application. J Toxicol Sci. 2000;25(2):77–83.

Chiba K, Yoshizawa K, Makino I, Kawakami K, Onoue M. Changes in the levels of glutathione after cellular and cutaneous damage induced by squalene monohydroperoxide. J Biochem Mol Toxicol. 2001;15(3):150–8.

Cotterill JA, Cunliffe WJ, Williamson B, Bulusu L. Further observations on the pathogenesis of acne. Br Med J. 1972;3:444–6.

De Luca C, Valacchi G. Surface lipids as multifunctional mediators of skin responses to environmental stimuli. Mediators Inflamm. 2010;2010:321494. doi:10.1155/2010/321494.

De Luca C, Picardo M, Breathnach A, Passi S. Lipoperoxidase activity of pityrosporum: characterisation of by-products and possible role in pityriasis versicolor. Exp Dermatol. 1996;5(1):49–56.

Dennis KJ, Shibamoto T. Production of malonaldehyde from squalene, a major skin surface lipid, during UV-irradiation. Photochem Photobiol. 1989;49(5):711–6.

Downie MM, Kealey T. Lipogenesis in the human sebaceous gland: glycogen and glycerophosphate are substrates for the synthesis of sebum lipids. J Invest Dermatol. 1998;111(2):199–205.

Downing DT, Strauss JS, Pochi PE. Variability in the chemical composition of human skin surface lipids. J Invest Dermatol. 1969;53(5):322–7.

Downing DT, Strauss JS, Pochi PE. Changes in skin surface lipid composition induced by severe caloric restriction in man. Am J Clin Nutr. 1972;25(4):365–7.

Downing DT, Strauss JS, Norton LA, Pochi PE, Stewart ME. The time course of lipid formation in human sebaceous glands. J Invest Dermatol. 1977;69(4):407–12.

Ekanayake Mudiyanselage S, Hamburger M, Elsner P, Thiele JJ. Ultraviolet A induces generation of squalene monohydroperoxide isomers in human sebum and skin surface lipids in vitro and in vivo. J Invest Dermatol. 2003;120(6):915–22.

Goldstein JL, Brown MS. Regulation of the mevalonate pathway. Nature. 1990;343:425–30.

Greene RS, Downing DT, Pochi PE, Strauss JS. Anatomical variation in the amount and composition of human skin surface lipid. J Invest Dermatol. 1970;54(3):240–7.

Guy R, Downie M, Kealey T. The organ maintained human sebaceous gland. Exp Dermatol. 1999;8(4):315–7.
Gylling H, Hallikainen M, Kolehmainen M, Toppinen L, Pihlajamäki J, Mykkänen H, Agren JJ, Rauramaa R, Laakso M, Miettinen TA. Cholesterol synthesis prevails over absorption in metabolic syndrome. Transl Res. 2007;149(6):310–6.
Hanaoka H, Ohkido M, Hattori Y, Maruta T, Arai T. Reexamination of the sebaceous function with relation to squalene. Jpn J Dermatol. 1971;81:103.
Hong I, Lee MH, Na TY, Zouboulis CC, Lee MO. LXRalpha enhances lipid synthesis in SZ95 sebocytes. J Invest Dermatol. 2008;128(5):1266–72.
Ikeno H, Tochio T, Tanaka H, Nakata S. Decrease in glutathione may be involved in pathogenesis of acne vulgaris. J Cosmet Dermatol. 2011;10(3):240–4.
Joseph SB, Laffitte BA, Patel PH, Watson MA, Matsukuma KE, Walczak R, Collins JL, Osborne TF, Tontonoz P. Direct and indirect mechanisms for regulation of fatty acid synthase gene expression by liver X receptors. J Biol Chem. 2002;277(13):11019–25.
Kellum RE. Human sebaceous gland lipids. Analysis by thin-layer chromatography. Arch Dermatol. 1967;95(2):218–20.
Kligman AM, Katz AG. Pathogenesis of acne vulgaris. I. Comedogenic properties of human sebum in external ear canal of the rabbit. Arch Dermatol. 1968;98(1):53–7.
Kohno Y, Egawa Y, Itoh S, Nagaoka S, Takahashi M, Mukai K. Kinetic study of quenching reaction of singlet oxygen and scavenging reaction of free radical by squalene in n-butanol. Biochim Biophys Acta. 1995;1256(1):52–6.
Leyden JJ. New understandings of the pathogenesis of acne. J Am Acad Dermatol. 1995;32(5 Pt 3):S15–25. (Review).
Masukawa Y, Narita H, Shimizu E, Kondo N, Sugai Y, Oba T, Homma R, Ishikawa J, Takagi Y, Kitahara T, Takema Y, Kita K. Characterization of overall ceramide species in human stratum corneum. J Lipid Res. 2008;49(7):1466–76.
Matsuo I, Yoshino K, Ohkido M. Mechanism of skin surface lipid peroxidation. Curr Probl Dermatol. 1983;11:135–43.
Michael-Jubeli R, Bleton J, Baillet-Guffroy A. High-temperature gas chromatography-mass spectrometry for skin surface lipids profiling. J Lipid Res. 2011;52(1):143–51.
Middleton B, Birdi I, Heffron M, Marsden JR. The substrate determines the rate and pattern of neutral lipid synthesized by isolated human sebaceous glands. FEBS Lett. 1988;231(1):59–61.
Miettinen TA, Vanhanen H. Serum concentration and metabolism of cholesterol during rapeseed oil and squalene feeding. Am J Clin Nutr. 1994;59(2):356–63.
Mills OH, Porte M, Kligman AM. Enhancement of comedogenic substances by ultraviolet radiation. Br J Dermatol. 1978;98:145–50.
Motoyoshi K. Enhanced comedo formation in rabbit ear skin by squalene and oleic acid peroxides. Br J Dermatol. 1983;109(2):191–8.
Mountfort KA, Bronstein H, Archer N, Jickells SM. Identification of oxidation products of squalene in solution and in latent fingerprints by ESI-MS and LC/APCI-MS. Anal Chem. 2007;79(7):2650–7.
Nazzaro-Porro M, Passi S, Picardo M, Mercantini R, Breathnach AS. Lipoxygenase activity of pityrosporum in vitro and in vivo. J Invest Dermatol. 1986;87(1):108–12.
Nikkari T. Comparative chemistry of sebum. J Invest Dermatol. 1974;62(3):257–67. (Review).
Nikkari T, Schreibman PH, Ahrens EH Jr. In vivo studies of sterol and squalene secretion by human skin. J Lipid Res. 1974;15(6):563–73.
Nikkari T, Schreibman PH, Ahrens EH Jr. Isotope kinetics of human skin cholesterol secretion. J Exp Med. 1975;141(3):620–34.
Nordstrom KM, Labows JN, McGinley KJ, Leyden JJ. Characterization of wax esters, triglycerides, and free fatty acids of follicular casts. J Invest Dermatol. 1986;86(6):700–5.
Ohsawa K, Watanabe T, Matsukawa R, Yoshimura Y, Imaeda K. The possible role of squalene and its peroxide of the sebum in the occurrence of sunburn and protection from the damage caused by U.V. irradiation. J Toxicol Sci. 1984;9(2):151–9.

Ottaviani M, Alestas T, Flori E, Mastrofrancesco A, Zouboulis CC, Picardo M. Peroxidated squalene induces the production of inflammatory mediators in HaCaT keratinocytes: a possible role in acne vulgaris. J Invest Dermatol. 2006;126(11):2430–7.

Packer L, Valacchi G. Antioxidants and the response of skin to oxidative stress: vitamin E as a key indicator. Skin Pharmacol Appl Skin Physiol. 2002;15(5):282–90. (Review).

Pappas A, Johnsen S, Liu JC, Eisinger M. Sebum analysis of individuals with and without acne. Dermatoendocrinol. 2009;1(3):157–61.

Passi S, Picardo M, Morrone A, De Luca C, Ippolito F. Skin surface lipids in HIV sero-positive and HIV sero-negative patients affected with seborrheic dermatitis. J Dermatol Sci. 1991;2:84–91.

Passi S, De PitÃ O, Puddu P, Littarru GP. Lipophilic antioxidants in human sebum and aging. Free Radic Res. 2002;36(4):471–7.

Pelle E, McCarthy J, Seltmann H, Huang X, Mammone T, Zouboulis CC, Maes D. Identification of histamine receptors and reduction of squalene levels by an antihistamine in sebocytes. J Invest Dermatol. 2008;128(5):1280–5.

Picardo M, Zompetta C, De Luca C, Cirone M, Faggioni A, Nazzaro-Porro M, Passi S, Prota G. Role of skin surface lipids in UV-induced epidermal cell changes. Arch Dermatol Res. 1991a;283(3):191–7.

Picardo M, Passi S, De Luca C, Morrone A, Bartoli F, Ippolito F. Skin surface lipids in patients affected with atopic dermatitis. In: Czernielewski JM, editor. Immunological and pharmacological aspects of atopic and contact eczema. Pharmacology and the skin. Vol. 4. Basel: Karger; 1991b. pp. 173–4.

Picardo M, Zompetta C, De Luca C, Amantea A, Faggioni A, Nazzaro-Porro M, Passi S. Squalene peroxides may contribute to ultraviolet light-induced immunological effects. Photodermatol Photoimmunol Photomed. 1991c;8(3):105–10.

Pochi PE, Downing DT, Strauss JS. Sebaceous gland response in man to prolonged total caloric deprivation. J Invest Dermatol. 1970;55(5):303–9.

Podda M, Traber MG, Weber C, Yan LJ, Packer L. UV-irradiation depletes antioxidants and causes oxidative damage in a model of human skin. Free Radic Biol Med. 1998;24(1):55–65.

Puhvel SM, Sakamoto M. An in vivo evaluation of the inflammatory effect of purified comedonal components in human skin. J Invest Dermatol. 1977;69(4):401–6.

Rajaratnam RA, Gylling H, Miettinen TA. Serum squalene in postmenopausal women without and with coronary artery disease. Atherosclerosis. 1999;146(1):61–4.

Rajaratnam RA, Gylling H, Miettinen TA. Independent association of serum squalene and noncholesterol sterols with coronary artery disease in postmenopausal women. J Am Coll Cardiol. 2000;35(5):1185–91.

Ramasastry P, Downing DT, Pochi PE, Strauss JS. Chemical composition of human skin surface lipids from birth to puberty. J Invest Dermatol. 1970;54(2):139–44.

Ridden J, Ferguson D, Kealey T. Organ maintenance of human sebaceous glands: in vitro effects of 13-cis retinoic acid and testosterone. J Cell Sci. 1990;95(Pt 1):125–36.

Rosignoli C, Nicolas JC, Jomard A, Michel S. Involvement of the SREBP pathway in the mode of action of androgens in sebaceous glands in vivo. Exp Dermatol. 2003;12(4):480–9.

Russell LE, Harrison WJ, Bahta AW, Zouboulis CC, Burrin JM, Philpott MP. Characterization of liver X receptor expression and function in human skin and the pilosebaceous unit. Exp Dermatol. 2007;16(10):844–52.

Ryu A, Arakane K, Koide C, Arai H, Nagano T. Squalene as a target molecule in skin hyperpigmentation caused by singlet oxygen. Biol Pharm Bull. 2009;32(9):1504–9.

Saint-Leger D, Bague A, Cohen E, Chivot M. A possible role for squalene in the pathogenesis of acne. I. In vitro study of squalene oxidation. Br J Dermatol. 1986a;114:535–42.

Saint-Leger D, Bague A, Lefebvre E, Cohen E, Chivot M. A possible role for squalene in the pathogenesis of acne. II. In vivo study of squalene oxides in skin surface and intra-comedonal lipids of acne patients. Br J Dermatol. 1986b;114(5):543–52.

Schultz JR, Tu H, Luk A, Repa JJ, Medina JC, Li L, Schwendner S, Wang S, Thoolen M, Mangelsdorf DJ, Lustig KD, Shan B. Role of LXRs in control of lipogenesis. Gene Dev. 2000;14(22):2831–8.

Shvedova AA, Tyurina YY, Tyurin VA, Kikuchi Y, Kagan VE, Quinn PJ. Quantitative analysis of phospholipid peroxidation and antioxidant protection in live human epidermal keratinocytes. Biosci Rep. 2001;21(1):33–43.

Simonen PP, Gylling H, Miettinen TA. The distribution of squalene and non-cholesterol sterols in lipoproteins in type 2 diabetes. Atherosclerosis. 2007;194(1):222–9.

Smith TM, Cong Z, Gilliland KL, Clawson GA, Thiboutot DM. Insulin-like growth factor-1 induces lipid production in human SEB-1 sebocytes via sterol response element-binding protein-1. J Invest Dermatol. 2006;126(6):1226–32.

Smith TM, Gilliland K, Clawson GA, Thiboutot D. IGF-1 induces SREBP-1 expression and lipogenesis in SEB-1 sebocytes via activation of the phosphoinositide 3-kinase/Akt pathway. J Invest Dermatol. 2008;128(5):1286–93.

Smythe CD, Greenall M, Kealey T. The activity of HMG-CoA reductase and acetyl-CoA carboxylase in human apocrine sweat glands, sebaceous glands, and hair follicles is regulated by phosphorylation and by exogenous cholesterol. J Invest Dermatol. 1998;111:139–48.

Stewart ME, Downing DT. Measurement of sebum secretion rates in young children. J Invest Dermatol. 1985;84(1):59–61.

Strandberg TE, Tilvis RS, Miettinen TA. Metabolic variables of cholesterol during squalene feeding in humans: comparison with cholestyramine treatment. J Lipid Res. 1990;31(9):1637–43.

Summerly R, Woodbury S. The in vitro incorporation of 14 C-acetate into the isolated sebaceous glands and appendage-freed epidermis of human skin. A technique for the study of lipid synthesis in the isolated sebaceous gland. Br J Dermatol. 1971;85(5):424–31.

Swinnen JV, Ulrix W, Heyns W, Verhoeven G. Coordinate regulation of lipogenic gene expression by androgens: evidence for a cascade mechanism involving sterol regulatory element binding proteins. Proc Natl Acad Sci U S A. 1997;94(24):12975–80.

Thiele JJ, Weber SU, Packer L. Sebaceous gland secretion is a major physiologic route of vitamin E delivery to skin. J Invest Dermatol. 1999;113(6):1006–10.

Thiele JJ, Schroeter C, Hsieh SN, Podda M, Packer L. The antioxidant network of the stratum corneum. Curr Probl Dermatol. 2001;29:26–42. (Review).

Tilvis R, Kovanen PT, Miettinen TA. Metabolism of squalene in human fat cells. Demonstration of a two-pool system. J Biol Chem. 1982;257(17):10300–5.

Tochio T, Tanaka H, Nakata S, Ikeno H. Accumulation of lipid peroxide in the content of comedones may be involved in the progression of comedogenesis and inflammatory changes in comedones. J Cosmet Dermatol. 2009;8(2):152–8.

Wei A, Shibamoto T. Antioxidant activities of essential oil mixtures toward skin lipid squalene oxidized by UV irradiation. Cutan Ocul Toxicol. 2007;26(3):227–33.

Yeo HC, Shibamoto T. Formation of formaldehyde and malonaldehyde by photooxidation of squalene. Lipids. 1992;27(1):50–3.

Zouboulis CC. The sebaceous gland. Hautarzt. 2010;61:467–77.

Part V
Subcutaneous Fat

Chapter 13
Adipose Tissue and Fat Cell Biology

Jan Kopecky

Core Messages

- Adipose tissue is present in most vertebrates and is comprised of several fat depots around the body; composed from fat cells called adipocytes.
- Other cell types present are preadipocytes, endothelial cells, pericytes and innate immune cells.
- Primarily two types of adipose tissue are recognized, white and brown adipose tissue (*WAT* and *BAT*, respectively), which have distinct and opposite functions, energy storage (in WAT) and energy dissipation (in BAT).
- The remarkable plasticity of the adipose tissue reveals the key role of adipose tissue in buffering metabolic energy under physiological as well as disease conditions.
- Endocrine functions of adipose tissue represent another fundamental function of the fat tissue.
- Fat depot- and gender-specific feature of adipose tissue are of importance for skin health, as well as for development of obesity-associated diseases.

Introduction

Adipose tissue is present in most vertebrates, including mammals, birds, reptiles, amphibians and most of the fish. This tissue is formed from mesoderm during embryonic development and is comprised of several fat depots around the body. It is composed from adipocytes, the cells forming the majority of the tissue when studied in histological sections, as well as several other cell types, namely preadipocytes, endothelial cells, pericytes and innate immune cells (Schipper et al. 2012; Olefsky and Glass 2010). In all mammals including humans, two types of adipose

J. Kopecky (✉)
Department of Adipose Tissue Biology, Institute of Physiology of the Academy of Sciences of the Czech Republic, v.v.i., Videnska 1083, 14220 Prague 4, Czech Republic
E-mail: kopecky@biomed.cas.cz

A. Pappas (ed.), *Lipids and Skin Health*, DOI 10.1007/978-3-319-09943-9_13

tissue are recognized, white and brown adipose tissue (*WAT* and *BAT*, respectively), which are engaged in two major distinct and opposite functions: energy storage (in WAT) and energy dissipation (in BAT). White adipocytes are spherical unilocular cells, filled by a large lipid droplet, with all cellular organelles packed in a thin cytoplasmic rim in the vicinity of plasma membrane. Brown adipocytes are polygonal multilocular cells, with cytoplasmic space filled with typical mitochondria containing many cristae (Cinti 1999, 2012). Brown adipocytes are marked by the presence of mitochondrial uncoupling protein-1 (*UCP1*), which mediates cold- and diet-inducible non-shivering thermogenesis that is adrenergically regulated (Rothwell and Stock 1981; Nedergaard et al. 2001; Cannon and Nedergaard 2004).

Differentiation of Adipocytes and Ontogenic Development of Adipose Tissue

Adipocytes can be differentiated *in vitro* from preadipocytes isolated from adipose tissue collagenase digests. Using this technique, as well as using genetically modified mice to study adipogenesis *in situ*, complex interplay between extracellular (hormonal and neural) and intracellular factors, namely intracellular signalling pathways and specific transcriptional mechanisms involved in differentiation of fat cells was described (reviewed in Rodriguez et al. 2004; Farmer 2006; Kajimura et al. 2010; Algire et al. 2013; Beranger et al. 2013).

During ontogenesis, formation of individual fat depots is tightly linked to development of adipose tissue microcirculation since adipocytes are probably derived from vascular endothelial cells (reviewed in Cinti 1999, 2002; Algire et al. 2013; Crandall et al. 1997). Ontogenesis of adrenergic innervation of adipose tissue, which is of great importance for its development (Cinti 2012; Beranger et al. 2013; Lafontan and Berlan 1993) and for the control of its metabolic functions (Cannon and Nedergaard 2004; Lafontan and Langin 2009; Sponarova et al. 2005), is not so well characterized (Cinti 1999, 2002). During early ontogenesis, development of BAT occurs earlier than that of WAT (Cinti 2002) and differences in the timing of BAT recruitment during perinatal development exist among different mammalian species (Cannon and Nedergaard 2004; see Section "Gross Anatomy and Function of BAT").

In rodents, the existence of several fat cell lineages was uncovered, which underlie the differentiation of precursor cells into: (i) classical multilocular brown adipocytes, which are closely related to myocytes, and which are responsible for the bulk of the adaptive UCP1-mediated thermogenesis; (ii) classical unilocular white adipocytes lacking UCP1, which are fully competent for ATP synthesis by oxidative phosphorylation (*OXPHOS*; Martin and Denton 1970; Flachs et al. 2011; Pietilainen et al. 2008); and (iii) so called 'brite adipocytes' (Walden et al. 2012; Petrovic et al. 2010), which are also named 'beige cells' (reviewed in Algire et al. 2013; Wu et al. 2012). 'Brite cells' are interspersed in specific WAT, they show high inducibility of the UCP1-linked thermogenic programme, and their abundance correlates negatively with the propensity to obesity while being under genetic control (Guerra et al. 1998). Besides the sympathetic stimulation, synthetic ligands of peroxisome proliferator-activated receptor-γ (*PPARγ*; reviewed in Ohno et al. 2012; Pardo et al. 2011; Wilson-Fritch et al. 2004), as well as several hormonal factors could induce 'brite cells' in rodents (Fisher et al. 2012; Bordicchia et al. 2012; Hondares et al. 2010; and reviewed in Bonet et al. 2013). However, in spite of the fact that markers discriminating brown adipocytes from 'brite cells' have been identified in rodents (Walden et al. 2012; Wu et al. 2012; Bostrom et al. 2012; Gburcik et al. 2012), the existence of 'brite cells' in humans remains to be unequivocally established (Wu et al. 2012; Villarroya 2012). In fact, inducibility of at least some of the UCP1-containing cells in WAT depots probably reflects reversible transdifferentiation of white to brown adipocytes (Frontini and Cinti 2010; Frontini et al. 2013; Rosenwald et al. 2013), bringing another level of complexity to the identification of markers characteristic of cell lineages engaged in adipose tissue plasticity. Reflecting the transdifferentiation theory, the fact that both brown and white adipocytes with opposing functions can co-exist in the same fat depot, and the inherent plasticity of various types of adipose tissue, Cinti together with some other researchers (Cinti 2002, 2012; Smorlesi et al. 2012) are coining the term 'adipose organ', which comprises all fat depots in the body. Abundance of white and brown adipocytes, respectively, in different fat depots depends on the anatomical location of the depot, on physiological conditions, and on the species (Cinti 1999, 2012). Most importantly, results from the group of Cinti (2012) also indicate that milk-secreting glands formed during pregnancy and lactation in mice originate from subcutaneous WAT and that milk-secreting epithelial cells are formed by transdifferentiation of adipocytes. The whole process could reverse during the post-lactation period. Human studies in this respect are needed.

In humans, growth of WAT could occur by both increasing the number of adipocytes (hyperplasia) and by enlargement of the size of existing adipocytes (hypertrophy). The first process takes place during childhood and adolescence, while obesity in adults is predominantly associated with hypertrophy of fat cells, with increases in fat cell number appearing only in morbidly obese subject (reviewed in Kissebah and Krakower 1994; Arner et al. 2011; Spalding et al. 2008; see also Section "White Adipose Tissue").

Brown Adipose Tissue

Brown adipose tissue is a thermogenic organ, which exists only in mammalians. Thus, BAT was found in all mammals studied, except for few (like pigs and kangaroos; reviewed in Cinti 2012; Cannon and Nedergaard 2004; Frontini and Cinti 2010). Despite of its relatively low mass, BAT could contribute to more than one half of all oxygen consumed in cold-adapted animals exposed to cold (Cannon and Nedergaard 2004).

Gross Anatomy and Function of BAT

In newborns, BAT plays an important role in adaptation to extrauterine life. In small animal species like in rodents, BAT is located in relatively well surrounded fat depots, which are reddish in colour (due to high content of mitochondrial cytochromes) and are associated with the major vessels (aorta and its branches). This enables for convection of BAT-generated heat to vital organ *via* circulation. The most prominent BAT depot is located in the interscapular region closely under the skin. All BAT depots are rich in adrenergic nerves and they are well vascularized. Brown fat continues to be present in smaller animal species during all their life, reflecting the relatively high energy requirements for thermogenesis in small size warm-blooded organisms (Porter and Brand 1993) when exposed to cold. Adaptation to cold results in dramatic induction of growth of BAT and of its thermogenic capacity, which is marked by UCP1 content of the tissue (Nedergaard and Cannon 2013). In hamsters and several other species, thermogenesis in BAT is essential for arousal from hibernation (Cannon and Nedergaard 2004).

In humans, BAT starts to develop at the beginning of the last trimester of pregnancy and in neonates, it is present in similar anatomical locations as in rodents (Houstek et al. 1993b; Merklin 1974). However, human BAT is less brownish-coloured (reflecting presence of more white adipocytes in the fat depots) to be unequivocally macroscopically identified during dissection (our observation). Thus, identification of BAT depots in humans depends in large on the detection of brown adipocytes at the histological level by their typical multilocular appearance (Frontini and Cinti 2010; Merklin 1974) and their UCP1 content (Frontini and Cinti 2010). Prematurely born babies suffer from insufficient development of BAT, which is manifested as low thermogenic capacity of these newborns (Cannon and Nedergaard 2004). Human BAT was believed to disappear during aging. However, some older studies demonstrated presence of brown adipocytes and/or UCP1 gene expression even in adult humans (Kortelainen et al. 1993; Oberkofler et al. 1997), and recent discovery of functional BAT in adult humans (Zingaretti et al. 2009; Virtanen et al. 2013) has led to a revival of the strategy (Rothwell and Stock 1979; Himms-Hagen 1983; Kopecky et al. 1995) to counteract obesity and associated metabolic disorders by inducing UCP1-mediated thermogenesis (Algire et al. 2013; Wu et al. 2013a).

The prevailing opinion is that UCP1-mediated thermogenesis in BAT is the only mechanism responsible for cold- and diet-induced adrenergically regulated thermogenesis (Rothwell and Stock 1981; Nedergaard et al. 2001; Cannon and Nedergaard 2004). However, examination of some older studies (Thurlby and Ellis 1986; Ma and Foster 1989) casts doubts on the unique role of the UCP1-mediated thermogenesis. That UCP1-independent energy expenditure in WAT could be induced using dietary and pharmacological interventions, and help to reduce adiposity and improve health is supported by recent evidence in UCP1-ablated mice (Meyer et al. 2010) and by some other studies (reviewed in Flachs et al. 2013; see Section "Dietary and Pharmacological Interventions").

Metabolic Features of BAT

Energy-dissipating function of UCP1 located in the inner mitochondrial membrane depends on its protonophoric activity, which is activated by fatty acids (*FA*) in response to β-adrenergic stimulation of lipolysis and allows for a full unmasking of mitochondrial oxidative capacity without concomitant synthesis of ATP (for review, see Krauss et al. 2005). In fact, the content of mitochondrial ATP synthase in classical brown adipocytes is extremely low (Houstek et al. 1995; Andersson et al. 1997), and this feature is observed already at the time of perinatal recruitment of BAT thermogenesis both in rodents (Houstek et al. 1988, 1990) and humans (Houstek et al. 1993a). Adrenergic stimulation of BAT not only activates UCP1-mediated proton leak but also leads to complex metabolic adaptations (for reviews on BAT metabolism see Cannon and Nedergaard 2004), including mitochondrial biogenesis, aimed at increasing fuel supply and oxidation. Thus, lipolysis of intracellular triacylglycerols (*TAG*) stores is accelerated and the uptake of FA derived from blood-born lipoproteins is increased due to the action of lipoprotein lipase (*LPL*) that is rapidly recruited during cold exposure (Carneheim et al. 1984). Although glucose is not a major substrate for BAT thermogenesis, glucose utilization by activated BAT increases dramatically (Marette and Bukowiecki 1989) as does lactate production. Glycolysis supplies ATP when its production *via* OXPHOS is attenuated, and also makes pyruvate available for the synthesis of oxaloacetate, the condensing partner for acetyl-CoA formed from β-oxidation of FA, thus ensuring a continuous supply of citric acid cycle intermediates (anaplerosis; Cannon and Nedergaard 1979). High expression of the muscle type carnitine palmitoyl transferase-1 (Brown et al. 1997) allows for rapid β-oxidation of FA (Nedergaard and Cannon 1979), while high activity of mitochondrial glycerol-3-phosphate dehydrogenase supports TAG synthesis and controls cytoplasmic NADH levels (Mracek et al. 2013). Longer exposure to cold (days) leads to increased lipogenesis, which is necessary for sustained thermogenesis. Triiodothyronine locally produced from thyroxine in brown adipocytes (Christoffolete et al. 2004) amplifies adrenergic stimulation of UCP1 gene transcription (Bianco et al. 1988), helps to ensure an optimal balance between lipolysis and lipogenesis (Christoffolete et al. 2004), and contributes to systemic triiodothyronine levels (Silva and Matthews 1984).

White Adipose Tissue

White fat, as the most plastic among the metabolically relevant tissues, could be responsible for 5–60 % of total body weight (Kissebah and Krakower 1994; Lee et al. 2013). Fat mass reflects energy balance, however, adipocyte number is very static in adult humans and independent on fluctuation in body weight, even in response to massive weight loss, indicating that adipocyte number is set during childhood and adolescence (Kissebah and Krakower 1994); approximately 10 % of fat cells are renewed annually in adult humans (Spalding et al. 2008).

Gross Anatomy and Functions of WAT

WAT includes several anatomical depots, which are similar in mice and in humans (Cinti 2012; Lee et al. 2013), except for the absence of epididymal WAT in the men (Frontini and Cinti 2010). Intra-abdominal (also called visceral) fat includes intraperitoneal fat (omental depot close to stomach, mesenteric depot along the intestine, and epiploic fat along the colon) and retroperitoneal fat around the kidney; for details in humans, see Lee et al. 2013). For long, accumulation of visceral fat, which characterizes upper body obesity, is known to correlate with metabolic syndrome (Despres and Lemieux 2006). This is typical for men, who in contrast to women, accumulate fat centrally in both subcutaneous and visceral depots (Enzi et al. 1986), and they do not increase accumulation of subcutaneous fat until morbidly obese (Kissebah and Krakower 1994). That sex hormones are important in regulation of body fat distribution is supported by the observation that increased androgenic tonus under various pathological conditions, including, e.g. polycystic ovary syndrome, is associated with upper body obesity and increased incidence of metabolic syndrome, as is also the case of post-menopausal women, facing decreased levels of estrogens, relative increase of androgenic tonus and redistribution of body fat to support development of abdominal obesity (Kissebah and Krakower 1994).

Subcutaneous WAT is of special interest for dermatologists. As also recalled in the position paper published in collaboration of several experts in the field (Klein et al. 2007), subcutaneous WAT has important role in the perceptions of women's beauty while the subcutaneous fat lobules are also involved in the manifestation of cellulite. However, the important functions of subcutaneous fat reflect its involvement in the mechanical and thermal insulation and in energy storage (see also Section "Metabolic Features of WAT"). Moreover, subcutaneous WAT, forms a functional unit with the skin components (Ryan and Curri 1989), and there is also a wild range of other recently emerging biological functions of subcutaneous fat (see below). Therefore, cosmetic surgery based on lipectomy of subcutaneous fat could impose health risk, which is difficult to judge (Klein et al. 2004), and which should not be ignored.

Major depots of subcutaneous fat in human are located in abdominal, gluteal and femoral regions (for details, see Lee et al. 2013). As could be revealed using computerized tomography, a layer of connective tissue (Scarpa's fascia) separates deep from superficial subcutaneous fat, which exert distinct metabolic properties and correlate independently with metabolic complications of obesity, while the deep subcutaneous layer exerts a stronger link to insulin resistance (Markman and Barton 1987; Kelley et al. 2000). Compared with visceral fat, human subcutaneous fat could even exert beneficial effects on metabolic disorders associated with obesity (Frayn 2000, 2005; Unger 2003). Accordingly, women, who tend to accumulate more subcutaneous fat (lower body obesity) than men, are more resistant than men to adverse metabolic disorders linked to obesity.

Contradictory data exist with respect to fat depot-specific involvement in weight loss in response to fasting or physical exercise (Sponarova et al. 2005; Kissebah and Krakower 1994). However, most of the data suggests preferential loss of visceral than subcutaneous fat, in association with beneficial effects on metabolic syndrome (Chaston and Dixon 2008). In contrast, starvation-induced weight loss in mice is reflected more by reduction of the weight of subcutaneous as compared with that of epididymal fat (Sponarova et al. 2005). Relatively high resistance to reduction of adiposity in the epididymal fat in mice correlates with a larger size of adipocytes in gonadal WAT as compared with the subcutaneous fat in mice (Medrikova et al. 2011), and also with smaller fat cells in visceral fat than subcutaneous fat in women, while in men, the size of adipocytes in both fat depots is similar (reviewed in Kissebah and Krakower 1994; Lee et al. 2013).

Concerning energy expenditure, the role of WAT metabolism itself is usually neglected. However, it is not insignificant. Thus, the contribution of WAT to resting metabolic rate in lean adult humans is close to 5% and it doubles in obesity (reviewed in Bottcher and Furst 1997), while in adult mice reared at 20 °C the total oxidative capacity of WAT represents ~30–50% of that in BAT (Kopecky et al. 1996); see Section "Metabolic Features of WAT". In both humans (Hallgren et al. 1989) and rodents (Hallgren et al. 1984), oxygen consumption in WAT cells declines with age and it is negatively correlated with obesity in humans (Hallgren et al. 1989). Decreased white fat cell thermogenesis in obese humans was also found using direct microcalorimetry (Bottcher and Furst 1997). Accordingly, mitochondrial content in WAT is relatively low in genetically obese and insulin-resistant mice (reviewed in Flachs et al. 2013). Moreover, in monozygotic twins discordant for obesity, copy number of mitochondrial DNA in subcutaneous WAT was reduced by 47% in obese co-twins, in association with reduced expression of the genes for mitochondrial proteins (Pietilainen et al. 2008). Very recently, it has been reported (Vosselman et al. 2012) that systemic non-selective β-adrenergic stimulation in humans, that increases energy expenditure to the same extent as cold exposure, does not activate BAT, and the increase energy expenditure could be explained only in part by the activation of muscle non-shivering thermogenesis (for review on muscle thermogenesis, see Thurlby and Ellis 1986; Simonsen et al. 1992; Mejsnar and Pacha 1983). These data support the notion that also adrenergically-stimulated energy expenditure in WAT may, though to a relatively small extent and independent of UCP1, influence total energy balance (see also Simonsen et al. 1992; see Section "Metabolic Features of WAT").

WAT has a major role in the control of systemic FA levels (see Section "Metabolic Features of WAT"). Accordingly, both hypertrophy and atrophy (Gavrilova et al. 2000) of WAT are associated with abnormal deposits of lipids in other tissues and lipotoxic damage of insulin signalling. As originally suggested by Danforth (2000), the inability to recruit new small white adipocytes while the older ones are full of lipids leads to diabetes. Larger adipocytes are less insulin-sensitive than smaller ones, and small adipocytes can store more extra fat (reviewed in Medrikova et al. 2011). In both humans (Kissebah and Krakower 1994) and mice (Medrikova et al. 2011), females compared with males demonstrate increased capacity for adipocyte enlargement under obesogenic conditions, and lower risk of development of metabolic disorders associated with obesity.

The key role of WAT in glucose homeostasis (Heilbronn et al. 2004; Virtue and Vidal-Puig 2008) is supported by the recent experimental evidence of anti-diabetic effects of WAT-specific up-regulation of PPARγ (Sugii et al. 2009). Consistently with the Danfroth's hypothesis, PPARγ agonists thiazolidinediones (*TZDs*) could improve insulin sensitivity while inducing adipogenic programme and formation of small adipocytes, especially in subcutaneous rather than in visceral fat (Sears et al. 2009; Fonseca 2003).

Adipose-Derived Peptide and Protein Hormones

Through its endocrine functions, WAT is involved in the regulation of glucose and energy homeostasis, as well as immune response while affecting development of both infectious diseases and cancer. Adipose-derived protein factors, which are released into circulation and exert endocrine effects are collectively called adipokines and could be regarded as a subclass of cytokines. Two major adipokines are leptin and adiponectin, but there are probably 100s of other adipokines and cytokines produced in adipose tissue with some of them exerting only autocrine and paracrine, but not systemic effects (reviewed in Trujillo and Scherer 2006; Sell et al. 2012). In contrast to WAT, secretory functions of BAT are poorly characterized, which provides an opportunity for exciting studies in future.

The seminal discovery of leptin as WAT-derived signalling factor for body weight homeostasis was published in 1994 by the group of Friedman (Zhang et al. 1994), and later studies demonstrated multiple biological effects of leptin, including its beneficial effects on wound healing through modulation of leptin receptors in adipocytes, fibroblasts and keratinocytes in subcutaneous WAT (reviewed in Klein et al. 2007). Adiponectin, as the most abundant transcript in human fat and a prominent blood protein, was cloned independently by Scherer et al. (1995) and Hu et al. (1996). Adiponectin is a marker of insulin sensitivity and cardiovascular health (Turer and Scherer 2012). Circulating levels of both these adipokines are higher in women than in men (reviewed in Trujillo and Scherer 2006). Recently, it has been also demonstrated by the group of Scherer that a new adipokines, endotrophin, could be involved in the known link between obesity and cancer, especially in the development of breast cancer, the most common malignancy found in women (Park and Scherer 2012). Endotrophin is a cleavage product of collagen VIα3, the major component of extracellular matrix in WAT, and its inhibition confers cisplatin sensitivity to tumours; such inhibition could be achieved using TZDs (Park et al. 2013). Increased extracellular matrix formation resulting in tissue fibrosis is associated with low-grade inflammation of WAT, which occurs in obesity (Schipper et al. 2012; Olefsky and Glass 2010; Lumeng et al. 2007; Khan et al. 2009; see Section "Local Lipid Mediators in Regulation of WAT Metabolism").

In addition to adipocytes, their precursors, pericytes and endothelial cells, as well as innate immune cells including macrophages, mast cells, neutrophils or eosinophils can be found in WAT (Schipper et al. 2012; Olefsky and Glass 2010).

Among them, adipose-tissue macrophages, major antigen-presenting cells, are functionally and numerically dominant and exert both immune and housekeeping functions. Macrophages release either pro-inflammatory (IL-6, TNF-α, IFN-γ) or anti-inflammatory (IL-4, IL-10, IL-13) cytokines, respectively (Gonzalez-Periz and Claria 2010). Obesity is associated with low-grade inflammation of WAT and infiltration of the tissue with pro-inflammatory macrophages (*M1* phenotype), which produce pro-inflammatory cytokines (Schipper et al. 2012; Olefsky and Glass 2010; Lumeng et al. 2007; Khan et al. 2009). M1 macrophages release cytokines, which block insulin action in adipocytes *via* TNF-α pathway, promote extracellular matrix remodelling and fibrosis of WAT, recruit other immune cells such as mast cells, B-cells or T-cells, and in histological sections, they form crown-like-structures that encircle necrotic adipocytes (Murano et al. 2008). Also other molecules like lipopolysaccharide or palmitic acid promote M1-polarization *via* TLR4 pathway (Saberi et al. 2009; Suganami et al. 2007), and TLR antagonists including EPA or the adiponectin paralog C1q/TNF-related protein 3 (Kopp et al. 2010), an endogenous LPS antagonist in adipose tissue, might be slowing down the inflammation. WAT macrophages in lean mice show a less-inflammatory properties (*M2* phenotype; Galli et al. 2011). All the macrophage-released mediators affect also adipocytes, which themselves produce all kinds of adipokines with pro- or anti-inflammatory properties (monocyte chemoattractant protein-1, resistin, adiponectin, etc.; Schipper et al. 2012).

Immune responses in subcutaneous WAT, including production of various adipokines and cytokines (see above) are associated with similar processes ongoing in keratinocytes and form an important part of the innate immune system, which protects organism against invading pathogens at the level of the skin (reviewed in Klein et al. 2007). In addition, adipocytes may also modulate the antigen-presenting function of lymph node dendritic cells by changing FA composition around the nodes enclosed in WAT (Klein et al. 2007; Mattacks et al. 2004; Sadler et al. 2005).

Local Lipid Mediators in Regulation of WAT Metabolism

Besides the adipokines and cytokines, a large family of endogenous lipid mediators, eicosanoids, derived from 20 carbon polyunsaturated fatty acids (*PUFA*), mainly from arachidonic acid (*AA*; 20:4 *n*-6) and eicosapentaenoic acid (*EPA*; 20:5 *n*-3; Gonzalez-Periz and Claria 2010) are present in WAT. Eicosanoids are potent local mediators of signal transduction and modulate the inflammatory response. In general, AA-derived eicosanoids promote inflammation, while EPA-derived eicosanoids act anti-inflammatory. Although immune cells are the main producers of eicosanoids, also adipocytes synthetize prostanoids (PGE2, PGI2, PGF2α) and leukotrienes and express eicosanoid receptors. Thus, members of both adipocyte and immune cell lineages are able to communicate using lipid mediators as membrane or nuclear receptor ligands. As reviewed in (Flachs et al. 2013), prostanoids are important for differentiation of adipocytes, can regulate lipolysis in an autocrine and paracrine manner through PGE2 or even shift the differentiation of defined mesenchymal progenitors towards a brown adipocyte phenotype; PGE2 could induce UCP1 in white adipocytes and stimulate thermogenesis in BAT; cyclopentenone prostaglandins (e.g. 15d-PGJ2) act as PPARγ ligands and modulate transcription of genes including those involved in the production of adipokines linked to inflammation. Discovery of pro-resolving and novel anti-inflammatory lipid mediators called resolvins (E-resolvins and D-resolvins), protectins, and maresins, which are derived from *n*-3 PUFA docosahexaenoic acid (*DHA*; 22:6 *n*-3) and EPA, opened a new field concerning the active resolution of WAT inflammation (White et al. 2010; Titos et al. 2011; Weylandt et al. 2012).

Dietary interventions may affect inflammatory status and metabolism of WAT. High intake of linoleic acid (18:2 *n*-6), the precursor of AA, may result in increased tonus of endocannabinoid system, impairment of energy metabolism in WAT and it could support development of obesity in affluent societies (reviewed in Flachs et al. 2013). Dietary interventions using EPA and DHA counteract the obesogenic effect of endocannabinoids and promote formation of anti-inflammatory eicosanoids and other lipid mediators (reviewed in Flachs et al. 2013; see above). In addition, *n*-3 PUFA in combination with calorie restriction could exert additive effect in modulation of levels of lipid mediators in WAT, to resolve inflammation and even modulate metabolic properties of adipocytes through PPARs (Flachs et al. 2011). In fact, changes in the levels of lipid mediators may exert stronger effects than those exerted by pharmacological treatments. Thus, e.g. resolvins RvD1 and RvD2 counteract both local adipokine production and monocyte accumulation in inflamed WAT with a 300-fold higher efficiency as compared with TZD rosiglitazone (Claria et al. 2012).

It has been demonstrated that obesity-associated macrophage infiltration of WAT in mice is one order of magnitude higher in gonadal as compared with subcutaneous WAT, and that this inflammatory response is much lower in female as compared with male mice. Furthermore, female mice showed a relatively weak link between adipocyte hypertrophy and various metabolic disorders linked to obesity (Medrikova et al. 2011; see Section "Gross Anatomy and Functions of WAT"). Accordingly, the obesity-associated WAT inflammation resulted in mitochondrial dysfunction in epididymal but not in periovarian WAT in the rat (Amengual-Cladera et al. 2012). Therefore, it has to be learned how mouse strain-, fat depot- and gender-specific differences in macrophage infiltration of WAT could affect metabolism of adipocytes.

In addition to the role of endotrophin in the link between obesity and cancer (see above), it has been also shown that especially PGE2, its receptors, and cyclooxygenase-2 pathway are involved in increased aromatase activation leading to breast cancer (Subbaramaiah et al. 2011, 2012; Chen et al. 2007; Hahn et al. 2006). On the contrary, growth of breast cancer cells was decreased by *n*-3 PUFA incorporated in plasma membranes, thus creating a pool for the synthesis of anti-inflammatory mediators (Schley et al. 2007).

Metabolic Features of WAT

Intrinsic metabolism of WAT is of key importance for the whole organism with respect to storage of energy in TAG during postprandial state, as well as the mobilization of energy reserves during fasting or exercise. The introduction of *in situ* microdialysis technique by Lonnroth et al. (1987) was of the key importance for characterization of metabolic features of subcutaneous WAT in humans. Excellent reviews on adipose tissue metabolism have been published (e.g. Lafontan 2008; Girousse and Langin 2012).

Conservation of metabolic energy in TAG reflects uptake of FA from plasma by the action of LPL and *de novo* lipogenesis of FA from products of glycolysis. LPL is secreted from adipocytes to the lumen of capillaries by a mechanism, which is up-regulated in the fed state (reviewed in Wang and Eckel 2009). Transport of FA released from lipoproteins into adipocytes is mediated by specific transporter proteins (Lafontan 2008).

WAT is an important site of *de novo* FA synthesis, mainly in rodents, however, even in humans WAT may account for up to 40% of whole-body lipogenesis (Chascione et al. 1987). Formation of acetyl-CoA for FA synthesis depends on citrate exported from mitochondria into cytosol, where it is cleaved to generate acetyl-CoA and oxaloacetate. Cytosolic oxaloacetate is reduced to malate by NADH, and the malate is subsequently decarboxylated to pyruvate in a reaction that also generates NADPH to be used during the synthesis of FA. Pyruvate returns to the mitochondrion, where it can be converted to oxaloacetate in a reaction that requires ATP and is catalyzed by pyruvate carboxylase. This enzyme is a key component of the 'pyruvate cycle' and its activity is about three-fold higher in WAT than in the liver (reviewed in Flachs et al. 2013).

TAG production requires a continuous supply of glycerol-3-phosphate to esterify FA. Glycerol-3-phosphate can be formed from: (i) glucose *via* glycolytic pathway, (ii) glycerol, which is converted to glycerol-3-phosphate by glycerol kinase, and (iii) precursors other than glucose and glycerol (such as pyruvate, lactate and amino acids), by glyceroneogenesis, which forms phosphoenolpyruvate *via* the mitochondrial dicarboxylic shuttle and subsequently produces glycerol-3-phosphate by a partial reversion of glycolysis. While glycerol kinase is essential for glycerol-3-phosphate formation in BAT (Festuccia et al. 2003), glyceroneogenesis is the dominant pathway for TAG-glycerol synthesis in WAT (Nye et al. 2008; reviewed in Flachs et al. 2013).

Mitochondrial ATP production is required both for *de novo* FA synthesis as well as for re-esterification of FA. In white adipocytes, similarly as in most other cells, most of ATP is formed by mitochondrial OXPHOS (Martin and Denton 1970; Flachs et al. 2011). Mitochondria are located in the thin cytoplasmic rim of adipocytes (Cinti 1999). In rodents, gonadal adipocytes contain more mitochondria than adipocytes in subcutaneous WAT (Deveaud et al. 2004; Sackmann-Sala et al. 2012). Similarly, in obese humans, mitochondrial DNA copy number per mg tissue is higher in visceral than in abdominal subcutaneous WAT (Kraunsoe et al. 2010).

Although mitochondrial respiration normalized to cell number or mtDNA is lower in visceral WAT compared to subcutaneous WAT due to smaller adipocytes in the former fat depot, the visceral fat is in this respect more active than the subcutaneous WAT (Kraunsoe et al. 2010). In obese state mitochondria in WAT (Kusminski and Scherer 2012), similarly to mitochondria in other tissues (Koves et al. 2008), are probably not able to cope with increasing demands for FA oxidation, resulting in incomplete β-oxidation, which is also associated with a stress of endoplasmic reticulum (Kusminski and Scherer 2012). Altered redox state, increased mitochondrial ROS formation and accumulation of the products of incomplete β-oxidation (Koves et al. 2008) leads to a deterioration of insulin sensitivity (Muoio and Neufer 2012).

Lipolysis of TAG in adipocytes is under complex hormonal control and it is mediated by several co-operating enzymes (reviewed in Girousse and Langin 2012), i.e. desnutrin/adipose triglyceride lipase (*ATGL*), an enzyme catalyzing the initial step in TAG hydrolysis and rate-limiting hydrolysis of TAG (Haemmerle et al. 2006), and the hormone-sensitive lipase (*HSL*), which catalyzes the hydrolysis of TAG into diacylglycerols and diacylglycerols into monoacylglycerols, and monoglyceride lipase. Lipolysis of TAG in white adipocytes is associated with re-esterification of a part (30–90 % at basal state, and 10–20 % at stimulated state) of lipolyzed FA back into adipose TAG, i.e. futile *TAG/FA* cycling (reviewed in Flachs et al. 2013; Bezaire et al. 2009; Campbell et al. 1992; Hanson and Reshef 2003; Reshef et al. 2003). As for the first time suggested by Newsholme and Crabtree (see ref. Newsholme and Crabtree 1970), substrate cycles could provide a mechanism of variable sensitivity for controlling the flux through a metabolic pathway (see Brooks 2013). Also in WAT, an extensive rate of futile FA recycling probably allows for a fine and fast tuning of these opposite metabolic fluxes in response to metabolic demands (Kalderon et al. 2000; Large et al. 2004), and it is essential for buffering of plasma FA levels (Nye et al. 2008; Campbell et al. 1992; Cadoudal et al. 2008). Thus, TAG/FA cycle in white adipocytes has the key role in metabolic flexibility of adipocytes, as well as the whole organism (reviewed in Flachs et al. 2013).

Complex modulation of lipolysis in white adipocytes (Ahmadian et al. 2011; Daval et al. 2005; Djouder et al. 2010) and fine tuning of TAG/FA cycle is probably supported by AMP-activated protein kinase through independent regulation of the activities of the key enzymes involved (Lafontan and Langin 2009; Flachs et al. 2013). It has been also recently suggested that FA liberated from TAG intracellularly are released from the adipocyte and immediately taken back by FA translocase/CD36-mediated mechanism to impose an additional level of regulation on lipolysis (Zhou et al. 2012). Higher activity of TAG/FA cycle in human visceral WAT (Marin et al. 1992) corresponds with a higher mitochondrial content and OXPHOS activity in this fat depot as compared with subcutaneous WAT.

In the absence of the stimulation of TAG/FA cycle, oxidation of FA is relatively slow. For instance, in normal fed rats only 0.2 % of endogenous FA in white adipocytes were oxidized while 50.1 % were released and 49.7 % re-esterified (Wang et al. 2003). However, the amount of FA disposed through oxidative pathway increased about 1.5-fold with the stimulation of lipolysis and TAG/FA cycle by fasting (Wang et al. 2003), which could be explained by the increased need for ATP production to support *de novo* FA synthesis and FA re-esterification. It is to be inferred that TAG/FA cycle in WAT could be responsible for about 2–3 % of resting metabolic rate in moderately obese humans (Flachs et al. 2013; see also Section "Gross Anatomy and Functions of WAT").

Similarly as in the heart (Haemmerle et al. 2011), also in adipose tissue (Mottillo et al. 2012; Li et al. 2005) ATGL and HSL-mediated lipolysis could generate ligands for nuclear receptors such as PPARα or PPARδ, which control mitochondrial FA oxidation and OXPHOS activity (Haemmerle et al. 2011; Li et al. 2005), as well as expression of UCP1 in BAT (Li et al. 2005). As suggested by Granneman et al. (2005), the induction of FA catabolism by this mechanism in WAT is important for limiting pro-inflammatory signaling during chronic lipolytic stimulation .

Dietary and Pharmacological Interventions

Important biological functions of adipose tissue, and namely their disturbances underlying development of obesity-associated disorders, point to dietary and pharmacological interventions targeting either BAT or WAT or both tissues as a part of complex treatment strategies. Such interventions may result in decreased accumulation of body fat reflecting modulation of metabolic fluxes in adipocytes and in secondary amelioration of obesity-associated disorders, and they may also exert beneficial effects by a selective modulation of specific biological functions of adipose tissue.

Various pharmacological and transgenic interventions could counteract obesity in rodents, e.g. pharmacological stimulation of PPARα (Hondares et al. 2011), forced lipolysis (Yehuda-Shnaidman et al. 2010; Jaworski et al. 2009), leptin treatment (Wang et al. 1999), or overexpression of PGC-1α (Puigserver and Spiegelman 2003) by activating the whole program of UCP1-mediated thermogenesis and mitochondrial uncoupling within adipocytes, including the induction of 'brite cells' (reviewed in Bonet et al. 2013). However, in most cases, the changes in lipid accumulation in WAT are also associated with the induction of UCP1-mediated thermogenesis in classical BAT depots (reviewed in Bonet et al. 2013), i.e. the process, which is probably much more relevant with respect to changes of whole body energy metabolism.

Reflecting the large mass of WAT, even small changes in its metabolism or secretory functions could elicit important systemic effects. This is well documented by the physiological changes in the rates of energy storage and FA release in the tissue (see Section "Metabolic Features of WAT") in association with the changes in the activity of the leptin signaling (see Section "Adipose-Derived Peptide and Protein Hormones"), reflecting whole body energy balance and activity of neuro hormonal signalling to adipose tissue. However, even some micronutrients may dramatically affect WAT. Thus, e.g. dietary *n*-3 PUFA, which in humans act anti-inflammatory (Calder 2011) while lowering cardiovascular morbidity and possibly also the incidence of type 2 diabetes (Nettleton and Katz 2005; Kromhout et al. 2011), could reduce low-grade inflammation in WAT in mice (Todoric et al. 2006; Kuda et al. 2009) and induce adiponectin in both mice (Flachs et al. 2006) and in humans (Wu et al. 2013b). We have also demonstrated UCP1-independent induction of mitochondrial FA oxidation in WAT in dietary-obese mice (Flachs et al. 2011, 2005), reflecting probably increased activity of TAG/FA cycling in adipocytes (Janovska et al. 2013), which could result in reduction of adiposity, especially if combined with calorie restriction (Flachs et al. 2011; reviewed in Flachs et al. 2013).

A question remains, however, how important the induction of energy dissipation in white adipocytes could be with respect to the regulation of body fat stores. Useful information can be drawn from the obesity-resistant phenotype of transgenic mice expressing UCP1 gene from the aP2 gene promoter, with an enhanced expression of UCP1 gene in BAT and ectopic expression of UCP1 gene in WAT (aP2-*Ucp1* mice; ref. Kopecky et al. 1995). Interestingly, obesity resistance of these mice likely results from enhanced energy expenditure in white adipocytes located in subcutaneous rather than visceral fat (for review see Klaus et al. 2011). Furthermore, UCP1-ablated mice are resistant to obesity in cold but not at thermoneutral temperature. This striking induction of obesity resistance could involve increased energy expenditure in both visceral and subcutaneous WAT, as suggested by the elevated cytochrome oxidase activity of the tissue (reviewed in Meyer et al. 2010). It could be speculated that under conditions of increased sympathetic stimulation and increased lipolysis in WAT during cold exposure, TAG/FA cycle could be activated and contribute to the increase in energy expenditure.

Low-grade inflammation of WAT could be reduced (Kuda et al. 2009) and induction of adiponectin could be achieved (Kim et al. 2004) in response to PPARγ agonists TZDs. That TZDs could improve cisplatin sensitivity to tumours like breast carcinomas by modulating secretory functions of WAT (Park et al. 2013; see Section "Adipose-Derived Peptide and Protein Hormones") could be a part of the complex anti-inflammatory effect of TZDs. Also mitochondrial insufficiency in white adipocytes of obese mice could be normalized by the treatment using TZDs (Wilson-Fritch et al. 2004). TZDs support catabolism of branched-chain amino acids in WAT (Hsiao et al. 2011), which may contribute significantly to the insulin-sensitizing effect of TZDs in obesity (Newgard 2012). TZDs also induce selectively in WAT expression of genes involved in glyceroneogenesis (see Cadoudal et al. 2008; Guan et al. 2002).Thus, concomitantly with the induction of mitochondrial oxidative capacity, TZDs could stimulate glycerol incorporation into TAG, futile FA re-esterification cycle and thus reduce FA release from adipocytes (Nye et al. 2008). The complex modulation of adipocyte metabolism by PPARγ agonists documents further the tight link between mitochondrial OXPHOS and TAG/FA cycle in white adipocytes.

Conclusions

Impressive plasticity of adipose tissue with respect to its weight, metabolic phenotype of adipocytes and tissue secretory functions, reflects the importance of adipose tissue under physiological conditions, as well as during development of both obesity and adipose tissue atrophy and associated metabolic disorders. Brown adipocytes endowed with a high capacity to produce heat by uncoupled mitochondrial OXPHOS are essential for non-shivering thermogenesis and possibly also for the control of energy balance, while white adipocyte exerting high capacity for TAG/FA cycling, which is dependent on ATP formed by mitochondrial OXPHOS are essential for regulatable energy storage as well as highly flexible control of blood lipid levels. Endocrine functions of adipose tissue represent an integral part of the regulation of glucose homeostasis, innate immunity, and affect development of specific forms of cancer. Fat depot- and gender-specific feature of adipose tissue are of importance for skin health, as well as for development of obesity-associated diseases. Novel therapeutical strategies based on specific modulation of processes ongoing in adipose tissue are likely to be developed.

Acknowledgements Supported by the EU FP7 project DIABAT (HEALTH-F2-2011-278373), and by the Czech Science Foundation (13-00871S).

References

Ahmadian M, Abbott MJ, Tang T, Hudak CS, Kim Y, Bruss M, Hellerstein MK, Lee HY, Samuel VT, Shulman GI, Wang Y, Duncan RE, Kang C, Sul HS. Desnutrin/ATGL is regulated by AMPK and is required for a brown adipose phenotype. Cell Metab. 2011;13:739–48.

Algire C, Medrikova D, Herzig S. White and brown adipose stem cells: from signaling to clinical implications. Biochim Biophys Acta. 2013;1831:896–904.

Amengual-Cladera E, Llado I, Proenza AM, Gianotti M. High-fat diet feeding induces a depot-dependent response on the pro-inflammatory state and mitochondrial function of gonadal white adipose tissue. Br J Nutr. 2012;109(3):1–12.

Andersson U, Houstek J, Cannon B. ATP synthase subunit c expression: physiological regulation of the P1 and P2 genes. Biochem J. 1997;323:379–85.

Arner P, Bernard S, Salehpour M, Possnert G, Liebl J, Steier P, Buchholz BA, Eriksson M, Arner E, Hauner H, Skurk T, Ryden M, Frayn KN, Spalding KL. Dynamics of human adipose lipid turnover in health and metabolic disease. Nature. 2011;478:110–3.

Beranger GE, Karbiener M, Barquissau V, Pisani DF, Scheideler M, Langin D, Amri EZ. In vitro brown and "brite"/"beige" adipogenesis: human cellular models and molecular aspects. Biochim Biophys Acta. 2013;1831:905–14.

Bezaire V, Mairal A, Ribet C, Lefort C, Girousse A, Jocken J, Laurencikiene J, Anesia R, Rodriguez AM, Ryden M, Stenson BM, Dani C, Ailhaud G, Arner P, Langin D. Contribution of adipose triglyceride lipase and hormone-sensitive lipase to lipolysis in hMADS adipocytes. J Biol Chem. 2009;284:18282–91.

Bianco AC, Sheng XY, Silva JE. Triiodothyronine amplifies norepinephrine stimulation of uncoupling protein gene transcription by a mechanism not requiring protein synthesis. J Biol Chem. 1988;263:18168–75.

Bonet ML, Oliver P, Palou A. Pharmacological and nutritional agents promoting browning of white adipose tissue. Biochim Biophys Acta. 2013;1831:969–85.

Bordicchia M, Liu D, Amri EZ, Ailhaud G, Dessi-Fulgheri P, Zhang C, Takahashi N, Sarzani R, Collins S. Cardiac natriuretic peptides act via p38 MAPK to induce the brown fat thermogenic program in mouse and human adipocytes. J Clin Invest. 2012;122:1022–36.

Bostrom P, Wu J, Jedrychowski MP, Korde A, Ye L, Lo JC, Rasbach KA, Bostrom EA, Choi JH, Long JZ, Kajimura S, Zingaretti MC, Vind BF, Tu H, Cinti S, Hojlund K, Gygi SP, Spiegelman BM. A PGC1-alpha-dependent myokine that drives brown-fat-like development of white fat and thermogenesis. Nature. 2012;481:463–8.

Bottcher H, Furst P. Decreased white fat cell thermogenesis in obese individuals. Int J Obes. 1997;21:439–44.

Brooks BJ. Studies on the triglyceride—fatty acid cycle. Oxford: Merton College; 2013.

Brown NF, Hill JK, Esser V, Kirkland JL, Corkey BE, Foster DW, McGarry JD. Mouse white adipocytes and 3T3-L1 cells display an anomalous pattern of carnitine palmitoyltransferase (CPT) I isoform expression during differentiation. Inter-tissue and inter-species expression of CPT I and CPT II enzymes. Biochem J. 1997;327(Pt 1):225–31.

Cadoudal T, Distel E, Durant S, Fouque F, Blouin JM, Collinet M, Bortoli S, Forest C, Benelli C. Pyruvate dehydrogenase kinase 4: regulation by thiazolidinediones and implication in glyceroneogenesis in adipose tissue. Diabetes. 2008;57:2272–9.

Calder PC. Fatty acids and inflammation: the cutting edge between food and pharma. Eur J Pharmacol. 2011;668:S50–8.

Campbell PJ, Carlson MG, Hill JO, Nurjhan N. Regulation of free fatty acid metabolism by insulin in humans: role of lipolysis and reesterification. Am J Physiol. 1992;263:E1063–9.

Cannon B, Nedergaard J. The physiological role of pyruvate carboxylation in hamster brown adipose tissue. Eur J Biochem. 1979;94:419–26.

Cannon B, Nedergaard J. Brown adipose tissue: function and physiological significance. Physiol Rev. 2004;84:277–359.

Carneheim C, Nedergaard J, Cannon B. Beta-adrenergic stimulation of lipoprotein lipase in rat brown adipose tissue during acclimation to cold. Am J Physiol. 1984;246:E327–33.

Chascione C, Elwyn DH, Davila M, Gil KM, Askanazi J, Kinney JM. Effect of carbohydrate intake on de novo lipogenesis in human adipose tissue. Am J Physiol. 1987;253:E664–9.

Chaston TB, Dixon JB. Factors associated with percent change in visceral versus subcutaneous abdominal fat during weight loss: findings from a systematic review. Int J Obes (Lond). 2008;32:619–28.

Chen D, Reierstad S, Lin Z, Lu M, Brooks C, Li N, Innes J, Bulun SE. Prostaglandin E(2) induces breast cancer related aromatase promoters via activation of p38 and c-Jun NH(2)-terminal kinase in adipose fibroblasts. Cancer Res. 2007;67:8914–22.

Christoffolete MA, Linardi CC, de Jesus L, Ebina KN, Carvalho SD, Ribeiro MO, Rabelo R, Curcio C, Martins L, Kimura ET, Bianco AC. Mice with targeted disruption of the Dio2 gene have cold-induced overexpression of the uncoupling protein 1 gene but fail to increase brown adipose tissue lipogenesis and adaptive thermogenesis. Diabetes. 2004;53:577–84.

Cinti S. The adipose organ. Milano: Editrice Kurtis; 1999.

Cinti S. Adipocyte differentiation and transdifferentiation: plasticity of the adipose organ1. J Endocrinol Invest. 2002;25:823–35.

Cinti S. The adipose organ at a glance. Dis Model Mech. 2012;5:588–94.

Claria J, Dalli J, Yacoubian S, Gao F, Serhan CN. Resolvin D1 and resolvin D2 govern local inflammatory tone in obese fat. J Immunol. 2012;189:2597–605.

Crandall DL, Hausman GJ, Kral JG. A review of the microcirculation of adipose tissue: anatomic, metabolic, and angiogenic perspectives. Microcirculation. 1997;4:211–32.

Danforth E Jr. Failure of adipocyte differentiation causes type II diabetes mellitus? Nat Genet. 2000;26:13.

Daval M, Diot-Dupuy F, Bazin R, Hainault I, Viollet B, Vaulont S, Hajduch E, Ferre P, Foufelle F. Anti-lipolytic action of AMP-activated protein kinase in rodent adipocytes. J Biol Chem. 2005;280:25250–7.

Despres JP, Lemieux I. Abdominal obesity and metabolic syndrome. Nature. 2006;444:881–7.

Deveaud C, Beauvoit B, Salin B, Schaeffer J, Rigoulet M. Regional differences in oxidative capacity of rat white adipose tissue are linked to the mitochondrial content of mature adipocytes. Mol Cell Biochem. 2004;267:157–66.

Djouder N, Tuerk RD, Suter M, Salvioni P, Thali RF, Scholz R, Vaahtomeri K, Auchli Y, Rechsteiner H, Brunisholz RA, Viollet B, Makela TP, Wallimann T, Neumann D, Krek W. PKA phosphorylates and inactivates AMPKalpha to promote efficient lipolysis. EMBO J. 2010;29:469–81.

Enzi G, Gasparo M, Biondetti PR, Fiore D, Semisa M, Zurlo F. Subcutaneous and visceral fat distribution according to sex, age, and overweight, evaluated by computed tomography. Am J Clin Nutr. 1986;44:739–46.

Farmer SR. Transcriptional control of adipocyte formation. Cell Metab. 2006;4:263–73.

Festuccia WTL, Guerra-Sa R, Kawashita NH, Garofalo MAR, Evangelista EA, Rodrigues V, Kettelhut IC, Migliorini RH. Expression of glycerokinase in brown adipose tissue is stimulated by the sympathetic nervous system. Am J Physiol Regul Integr Comp Physiol. 2003;284:R1536–41.

Fisher FM, Kleiner S, Douris N, Fox EC, Mepani RJ, Verdeguer F, Wu J, Kharitonenkov A, Flier JS, Maratos-Flier E, Spiegelman BM. FGF21 regulates PGC-1alpha and browning of white adipose tissues in adaptive thermogenesis. Genes Dev. 2012;26:271–81.

Flachs P, Horakova O, Brauner P, Rossmeisl M, Pecina P, Franssen-van Hal NL, Ruzickova J, Sponarova J, Drahota Z, Vlcek C, Keijer J, Houstek J, Kopecky J. Polyunsaturated fatty acids of marine origin upregulate mitochondrial biogenesis and induce beta-oxidation in white fat. Diabetologia. 2005;48:2365–75.

Flachs P, Mohamed-Ali V, Horakova O, Rossmeisl M, Hosseinzadeh-Attar MJ, Hensler M, Ruzickova J, Kopecky J. Polyunsaturated fatty acids of marine origin induce adiponectin in mice fed high-fat diet. Diabetologia. 2006;49:394–7.

Flachs P, Ruhl R, Hensler M, Janovsk P, Zouhar P, Kus V, Macek JZ, Papp E, Kuda O, Svobodova M, Rossmeisl M, Tsenov G, Mohamed-Ali V, Kopecky J. Synergistic induction of lipid catabolism and anti-inflammatory lipids in white fat of dietary obese mice in response to calorie restriction and n-3 fatty acids. Diabetologia. 2011;54:2626–38.

Flachs P, Rossmeisl M, Kuda O, Kopecky J. Stimulation of mitochondrial oxidative capacity in white fat independent of UCP1: a key to lean phenotype. Biochim Biophys Acta. 2013;1831:986–1003.

Fonseca V. Effect of thiazolidinediones on body weight in patients with diabetes mellitus. Am J Med. 2003;115(Suppl 8A):42–8S.

Frayn KN. Visceral fat and insulin resistance–causative or correlative? Br J Nutr. 2000;83(Suppl 1):S71–7.

Frayn KN. Obesity and metabolic disease: is adipose tissue the culprit? Proc Nutr Soc. 2005;64:7–13.

Frontini A, Cinti S. Distribution and development of brown adipocytes in the murine and human adipose organ. Cell Metab. 2010;11:253–6.

Frontini A, Vitali A, Perugini J, Murano I, Romiti C, Ricquier D, Guerrieri M, Cinti S. White-to-brown transdifferentiation of omental adipocytes in patients affected by pheochromocytoma. Biochim Biophys Acta. 2013;1831:950–9.

Galli SJ, Borregaard N, Wynn TA. Phenotypic and functional plasticity of cells of innate immunity: macrophages, mast cells and neutrophils. Nat Immunol. 2011;12:1035–44.

Gavrilova O, Marcus-Samuels B, Graham D, Kim JK, Shulman GI, Castle AL, Vinson C, Eckhaus M, Reitman ML. Surgical implantation of adipose tissue reverses diabetes in lipoatrophic mice. J Clin Invest. 2000;105:271–8.

Gburcik V, Cawthorn WP, Nedergaard J, Timmons JA, Cannon B. An essential role for Tbx15 in the differentiation of brown and "brite" but not white adipocytes. Am J Physiol Endocrinol Metab. 2012;303(8):E1053–60.

Girousse A, Langin D. Adipocyte lipases and lipid droplet-associated proteins: insight from transgenic mouse models. Int J Obes (Lond). 2012;36:581–94.

Gonzalez-Periz A, Claria J. Resolution of adipose tissue inflammation. Sci World J. 2010;10:832–56.

Granneman JG, Li P, Zhu Z, Lu Y. Metabolic and cellular plasticity in white adipose tissue I: effects of beta3-adrenergic receptor activation. Am J Physiol Endocrinol Metab. 2005;289:E608–16.

Guan HP, Li Y, Jensen MV, Newgard CB, Steppan CM, Lazar MA. A futile metabolic cycle activated in adipocytes by antidiabetic agents. Nat Med. 2002;8:1122–8.

Guerra C, Koza RA, Yamashita H, King KW, Kozak LP. Emergence of brown adipocytes in white fat in mice is under genetic control. Effects on body weight and adiposity. J Clin Invest. 1998;102:412–20.

Haemmerle G, Lass A, Zimmermann R, Gorkiewicz G, Meyer C, Rozman J, Heldmaier G, Maier R, Theussl C, Eder S, Kratky D, Wagner EF, Klingenspor M, Hoefler G, Zechner R. Defective lipolysis and altered energy metabolism in mice lacking adipose triglyceride lipase. Science. 2006;312:734–7.

Haemmerle G, Moustafa T, Woelkart G, Buttner S, Schmidt A, van de Weijer T, Hesselink M, Jaeger D, Kienesberger PC, Zierler K, Schreiber R, Eichmann T, Kolb D, Kotzbeck P, Schweiger M, Kumari M, Eder S, Schoiswohl G, Wongsiriroj N, Pollak NM, Radner FP, Preiss-Landl K, Kolbe T, Rulicke T, Pieske B, Trauner M, Lass A, Zimmermann R, Hoefler G, Cinti S, Kershaw EE, Schrauwen P, Madeo F, Mayer B, Zechner R. ATGL-mediated fat catabolism regulates cardiac mitochondrial function via PPAR-alpha and PGC-1. Nat Med. 2011;17:1076–85.

Hahn T, Alvarez I, Kobie JJ, Ramanathapuram L, Dial S, Fulton A, Besselsen D, Walker E, Akporiaye ET. Short-term dietary administration of celecoxib enhances the efficacy of tumor lysate-pulsed dendritic cell vaccines in treating murine breast cancer. Int J Cancer. 2006;118:2220–31.

Hallgren P, Raddatz E, Bergh CH, Kucera P, Sjostrom L. Oxygen consumption in collagenase-liberated rat adipocytes in relation to cell size and age. Metabolism. 1984;33:897–900.

Hallgren P, Sjostrom L, Hedlund H, Lundell L, Olbe L. Influence of age, fat cell weight, and obesity on O_2 consumption of human adipose tissue. Am J Physiol. 1989;256:E467–74.

Hanson RW, Reshef L. Glyceroneogenesis revisited. Biochimie. 2003;85:1199–205.

Heilbronn L, Smith SR, Ravussin E. Failure of fat cell proliferation, mitochondrial function and fat oxidation results in ectopic fat storage, insulin resistance and type II diabetes mellitus. Int J Obes Relat Metab Disord. 2004;28(Suppl 4):S12–21.

Himms-Hagen J. Role of thermogenesis in brown adipose tissue in the regulation of energy balance. In: Hollenberg CH, Roncardi DAK, editors. The adipose and obesity: cellular and molecular mechanisms. New York: Raven Press; 1983. pp. 259–70.

Hondares E, Rosell M, Gonzalez FJ, Giralt M, Iglesias R, Villarroya F. Hepatic FGF21 expression is induced at birth via PPARalpha in response to milk intake and contributes to thermogenic activation of neonatal brown fat. Cell Metab. 2010;11:206–12.

Hondares E, Rosell M, Diaz-Delfin J, Olmos Y, Monsalve M, Iglesias R, Villarroya F, Giralt M. Peroxisome proliferator-activated receptor alpha (PPARalpha) induces PPARgamma coactivator 1alpha (PGC-1alpha) gene expression and contributes to thermogenic activation of brown fat: involvement of PRDM16. J Biol Chem. 2011;286:43112–22.

Houstek J, Kopecky J, Rychter Z, Soukup T. Uncoupling protein in embryonic brown adipose tissue—existence of nonthermogenic and thermogenic mitochondria. Biochim Biophys Acta. 1988;935:19–25.

Houstek J, Janikova D, Bednar J, Kopecky J, Sebestian J, Soukup T. Postnatal appearance of uncoupling protein and formation of thermogenic mitochondria in hamster brown adipose tissue. Biochim Biophys Acta. 1990;1015:441–9.

Houstek J, Vizek K, Pavelka S, Kopecky J, Krejcova E, Hermanska E, Cermakova S. Type II iodothyronine 5′-deiodinase and uncoupling protein in brown adipose tissue of human newborns. J Clin Endocrinol Metab. 1993a;77:382–7.

Houstek J, Vizek K, Pavelka S, Kopecky J, Krejcova E, Hermanska J, Cermakova M. Development of type II iodothyronine deiodinase and thermogenic function of brown fat in human newborns. Glasgow: IUPS; 1993b.

Houstek J, Andersson U, Tvrdik P, Nedergaard J, Cannon B. The expression of subunit c correlates with and thus may limit the biosynthesis of the mitochondrial F0 F1-ATPase in brown adipose tissue. J Biol Chem. 1995;270:7689–94.

Hsiao G, Chapman J, Ofrecio JM, Wilkes J, Resnik JL, Thapar D, Subramaniam S, Sears DD. Multi-tissue, selective PPARgamma modulation of insulin sensitivity and metabolic pathways in obese rats. Am J Physiol Endocrinol Metab. 2011;300:E164–74.

Hu E, Liang P, Spiegelman BM. AdipoQ is a novel adipose-specific gene dysregulated in obesity. J Biol Chem. 1996;271:10697–703.

Janovska P, Flachs P, Kazdova L, Kopecky J. Anti-obesity effect of n-3 polyunsaturated fatty acids in mice fed high-fat diet is independent of cold-induced thermogenesis. Physiol Res. 2013;62:153–61.

Jaworski K, Ahmadian M, Duncan RE, Sarkadi-Nagy E, Varady KA, Hellerstein MK, Lee HY, Samuel VT, Shulman GI, Kim KH, de Val S, Kang C, Sul HS. AdPLA ablation increases lipolysis and prevents obesity induced by high-fat feeding or leptin deficiency. Nat Med. 2009;15:159–68.

Kajimura S, Seale P, Spiegelman BM. Transcriptional control of brown fat development. Cell Metab. 2010;11:257–62.

Kalderon B, Mayorek N, Berry E, Zevit N, Bar-Tana J. Fatty acid cycling in the fasting rat. Am J Physiol Endocrinol Metab. 2000;279:E221–7.

Kelley DE, Thaete FL, Troost F, Huwe T, Goodpaster BH. Subdivisions of subcutaneous abdominal adipose tissue and insulin resistance. Am J Physiol Endocrinol Metab. 2000;278:E941–8.

Khan T, Muise ES, Iyengar P, Wang ZV, Chandalia M, Abate N, Zhang BB, Bonaldo P, Chua S, Scherer PE. Metabolic dysregulation and adipose tissue fibrosis: role of collagen VI. Mol Cell Biol. 2009;29:1575–91.

Kim H, Haluzik M, Gavrilova O, Yakar S, Portas J, Sun H, Pajvani UB, Scherer PE, LeRoith D. Thiazolidinediones improve insulin sensitivity in adipose tissue and reduce the hyperlipidaemia without affecting the hyperglycaemia in a transgenic model of type 2 diabetes. Diabetologia. 2004;47:2215–25.

Kissebah AH, Krakower GR. Regional adiposity and morbidity. Physiol Rev. 1994;74:761–811.

Klaus S, Keipert S, Rossmeisl M, Kopecky J. Augmenting energy expenditure by mitochondrial uncoupling: a role of AMP-activated protein kinase. Genes Nutr. 2011;7:369–86.

Klein S, Fontana L, Young VL, Coggan AR, Kilo C, Patterson BW, Mohammed BS. Absence of an effect of liposuction on insulin action and risk factors for coronary heart disease. N Engl J Med. 2004;350:2549–57.

Klein J, Permana PA, Owecki M, Chaldakov GN, Bohm M, Hausman G, Lapiere CM, Atanassova P, Sowinski J, Fasshauer M, Hausman DB, Maquoi E, Tonchev AB, Peneva VN, Vlachanov KP, Fiore M, Aloe L, Slominski A, Reardon CL, Ryan TJ, Pond CM, Ryan TJ. What are subcutaneous adipocytes really good for? Exp Dermatol. 2007;16:45–70.

Kopecky J, Clarke G, Enerback S, Spiegelman B, Kozak LP. Expression of the mitochondrial uncoupling protein gene from the aP2 gene promoter prevents genetic obesity. J Clin Invest. 1995;96:2914–23.

Kopecky J, Rossmeisl M, Hodny Z, Syrovy I, Horakova M, Kolarova P. Reduction of dietary obesity in the aP2-Ucp transgenic mice: mechanism and adipose tissue morphology. Am J Physiol. 1996;270:E776–86.

Kopp A, Bala M, Buechler C, Falk W, Gross P, Neumeier M, Scholmerich J, Schaffler A. C1q/TNF-related protein-3 represents a novel and endogenous lipopolysaccharide antagonist of the adipose tissue. Endocrinology. 2010;151:5267–78.

Kortelainen ML, Pelletier G, Ricquier D, Bukowiecki LJ. Immunohistochemical detection of human brown adipose tissue uncoupling protein in an autopsy series. J Histochem Cytochem. 1993;41:759–64.

Koves TR, Ussher JR, Noland RC, Slentz D, Mosedale M, Ilkayeva O, Bain J, Stevens R, Dyck JR, Newgard CB, Lopaschuk GD, Muoio DM. Mitochondrial overload and incomplete fatty acid oxidation contribute to skeletal muscle insulin resistance. Cell Metab. 2008;7:45–56.

Kraunsoe R, Boushel R, Hansen CN, Schjerling P, Qvortrup K, Stockel M, Mikines KJ, Dela F. Mitochondrial respiration in subcutaneous and visceral adipose tissue from patients with morbid obesity. J Physiol. 2010;588:2023–32.
Krauss S, Zhang CY, Lowell BB. The mitochondrial uncoupling-protein homologues. Nat Rev Mol Cell Biol. 2005;6:248–61.
Kromhout D, Geleijnse JM, de GJ, Oude Griep LM, Mulder BJ, de Boer MJ, Deckers JW, Boersma E, Zock PL, Giltay EJ. n-3 fatty acids, ventricular arrhythmia-related events, and fatal myocardial infarction in postmyocardial infarction patients with diabetes. Diabetes Care. 2011;34:2515–20.
Kuda O, Jelenik T, Jilkova Z, Flachs P, Rossmeisl M, Hensler M, Kazdova L, Ogston N, Baranowski M, Gorski J, Janovska P, Kus V, Polak J, Mohamed-Ali V, Burcelin R, Cinti S, Bryhn M, Kopecky J. n-3 Fatty acids and rosiglitazone improve insulin sensitivity through additive stimulatory effects on muscle glycogen synthesis in mice fed a high-fat diet. Diabetologia. 2009;52:941–51.
Kusminski CM, Scherer PE. Mitochondrial dysfunction in white adipose tissue. Trends Endocrinol Metab. 2012;23:435–43.
Lafontan M. Advances in adipose tissue metabolism. Int J Obes (Lond). 2008;32(Suppl 7):S39–51.
Lafontan M, Berlan M. Fat cell adrenergic receptors and the control of white and brown fat cell function. J Lipid Res. 1993;34:1057–91.
Lafontan M, Langin D. Lipolysis and lipid mobilization in human adipose tissue. Prog Lipid Res. 2009;48:275–97.
Large V, Peroni O, Letexier D, Ray H, Beylot M. Metabolism of lipids in human white adipocyte. Diabetes Metab. 2004;30:294–309.
Lee MJ, Wu Y, Fried SK. Adipose tissue heterogeneity: implication of depot differences in adipose tissue for obesity complications. Mol Aspects Med. 2013;34(1):1–11. doi:10.1016/j.mam.2012.10.001. Epub 2012 Oct 13.
Li P, Zhu Z, Lu Y, Granneman JG. Metabolic and cellular plasticity in white adipose tissue II: role of peroxisome proliferator-activated receptor-alpha. Am J Physiol Endocrinol Metab. 2005;289:E617–26.
Lonnroth P, Jansson PA, Smith U. A microdialysis method allowing characterization of intercellular water space in humans. Am J Physiol. 1987;253:E228–31.
Lumeng CN, Bodzin JL, Saltiel AR. Obesity induces a phenotypic switch in adipose tissue macrophage polarization. J Clin Invest. 2007;117:175–84.
Ma SW, Foster DO. Brown adipose tissue, liver, and diet-induced thermogenesis in cafeteria diet-fed rats. Can J Physiol Pharmacol. 1989;67:376–81.
Marette A, Bukowiecki LJ. Stimulation of glucose transport by insulin and norepinephrine in isolated rat brown adipocytes. Am J Physiol. 1989;257:C714–21.
Marin P, Andersson B, Ottosson M, Olbe L, Chowdhury B, Kvist H, Holm G, Sjostrom L, Bjorntorp P. The morphology and metabolism of intraabdominal adipose tissue in men. Metabolism. 1992;41:1242–8.
Markman B, Barton FE Jr. Anatomy of the subcutaneous tissue of the trunk and lower extremity. Plast Reconstr Surg. 1987;80:248–54.
Martin BR, Denton RM. The intracellular localization of enzymes in white-adipose-tissue fat-cells and permeability properties of fat-cell mitochondria. Biochem J. 1970;117:861–77.
Mattacks CA, Sadler D, Pond CM. Site-specific differences in fatty acid composition of dendritic cells and associated adipose tissue in popliteal depot, mesentery, and omentum and their modulation by chronic inflammation and dietary lipids. Lymphat Res Biol. 2004;2:107–29.
Medrikova D, Macek JZ, Bardova K, Janovsk P, Rossmeisl M, Kopecky J. Sex differences during the course of diet-induced obesity in mice: adipose tissue expandability and glycemic control. Int J Obes. 2011;36:262–72.
Mejsnar J, Pacha J. Thermogenesis due to noradrenaline in muscles under different rates of perfusion. Pflugers Arch. 1983;397:149–51.
Merklin RJ. Growth and distribution of human fetal brown fat. Anat Rec. 1974;178:637–45.

Meyer CW, Willershauser M, Jastroch M, Rourke BC, Fromme T, Oelkrug R, Heldmaier G, Klingenspor M. Adaptive thermogenesis and thermal conductance in wild-type and UCP1-KO mice. Am J Physiol Regul Integr Comp Physiol. 2010;299:R1396–406.

Mottillo EP, Bloch AE, Leff T, Granneman JG. Lipolytic products activate peroxisome proliferator-activated receptor (PPAR) alpha and delta in brown adipocytes to match fatty acid oxidation with supply. J Biol Chem. 2012;287:25038–48.

Mracek T, Drahota Z, Houstek J. The function and the role of the mitochondrial glycerol-3-phosphate dehydrogenase in mammalian tissues. Biochim Biophys Acta. 2013;1827:401–10.

Muoio DM, Neufer PD. Lipid-induced mitochondrial stress and insulin action in muscle. Cell Metab. 2012;15:595–605.

Murano I, Barbatelli G, Parisani V, Latini C, Muzzonigro G, Castellucci M, Cinti S. Dead adipocytes, detected as crown-like structures, are prevalent in visceral fat depots of genetically obese mice. J Lipid Res. 2008;49:1562–8.

Nedergaard J, Cannon B. Overview–preparation and properties of mitochondria from different sources. Methods Enzymol. 1979;55:3–28.

Nedergaard J, Cannon B. UCP1 mRNA does not produce heat. Biochim Biophys Acta. 2013;1831:943–9.

Nedergaard J, Golozoubova V, Matthias A, Asadi A, Jacobsson A, Cannon B. UCP1: the only protein able to mediate adaptive non-shivering thermogenesis and metabolic inefficiency. Biochim Biophys Acta. 2001;1504:82–106.

Nettleton JA, Katz R. n-3 long-chain polyunsaturated fatty acids in type 2 diabetes: a review. J Am Diet Assoc. 2005;105:428–40.

Newgard CB. Interplay between lipids and branched-chain amino acids in development of insulin resistance. Cell Metab. 2012;15:606–14.

Newsholme EA, Crabtree B. The role of fructose-1,6-diphosphatase in the regulation of glycolysis in skeletal muscle. FEBS Lett. 1970;7:195–8.

Nye C, Kim J, Kalhan SC, Hanson RW. Reassessing triglyceride synthesis in adipose tissue. Trends Endocrinol Metab. 2008;19:356–61.

Oberkofler H, Dallinger G, Liu YM, Hell E, Krempler F, Patsch W. Uncoupling protein gene: quantification of expression levels in adipose tissues of obese and non-obese humans. J Lipid Res. 1997;38:2125–33.

Ohno H, Shinoda K, Spiegelman BM, Kajimura S. PPARgamma agonists induce a white-to-brown fat conversion through stabilization of PRDM16 protein. Cell Metab. 2012;15:395–404.

Olefsky JM, Glass CK. Macrophages, inflammation, and insulin resistance. Annu Rev Physiol. 2010;72:219–46.

Pardo R, Enguix N, Lasheras J, Feliu JE, Kralli A, Villena JA. Rosiglitazone-induced mitochondrial biogenesis in white adipose tissue is independent of peroxisome proliferator-activated receptor gamma coactivator-1alpha. PLoS ONE. 2011;6:e26989.

Park J, Scherer PE. Adipocyte-derived endotrophin promotes malignant tumor progression. J Clin Invest. 2012;122:4243–56.

Park J, Morley TS, Scherer PE. Inhibition of endotrophin, a cleavage product of collagen VI, confers cisplatin sensitivity to tumours. EMBO Mol Med. 2013;5:935–48.

Petrovic N, Walden TB, Shabalina IG, Timmons JA, Cannon B, Nedergaard J. Chronic peroxisome proliferator-activated receptor gamma (PPARgamma) activation of epididymally derived white adipocyte cultures reveals a population of thermogenically competent, UCP1-containing adipocytes molecularly distinct from classic brown adipocytes. J Biol Chem. 2010;285:7153–64.

Pietilainen KH, Naukkarinen J, Rissanen A, Saharinen J, Ellonen P, Keranen H, Suomalainen A, Gotz A, Suortti T, Yki-Jarvinen H, Oresic M, Kaprio J, Peltonen L. Global transcript profiles of fat in monozygotic twins discordant for BMI: pathways behind acquired obesity. PLoS Med. 2008;5:e51.

Porter RK, Brand MD. Body mass dependence of H + leak in mitochondria and its relevance to metabolic rate. Nature. 1993;362:628–30.

Puigserver P, Spiegelman BM. Peroxisome proliferator-activated receptor-gamma coactivator 1 alpha (PGC-1 alpha): transcriptional coactivator and metabolic regulator. Endocr Rev. 2003;24:78–90.
Reshef L, Olswang Y, Cassuto H, Blum B, Croniger CM, Kalhan SC, Tilghman SM, Hanson RW. Glyceroneogenesis and the triglyceride/fatty acid cycle. J Biol Chem. 2003;278:30413–6.
Rodriguez AM, Elabd C, Delteil F, Astier J, Vernochet C, Saint-Marc P, Guesnet J, Guezennec A, Amri EZ, Dani C, Ailhaud G. Adipocyte differentiation of multipotent cells established from human adipose tissue. Biochem Biophys Res Commun. 2004;315:255–63.
Rosenwald M, Perdikari A, Rulicke T, Wolfrum C. Bi-directional interconversion of brite and white adipocytes. Nat Cell Biol. 2013;15:659–67.
Rothwell NJ, Stock MJ. A role for brown adipose tissue in diet-induced thermogenesis. Nature. 1979;281:31–5.
Rothwell NJ, Stock MJ. Influence of noradrenaline on blood flow to brown adipose tissue in rats exhibiting diet-induced thermogenesis. Pflugers Arch. 1981;389:237–42.
Ryan TJ, Curri SP. Cutaneous adipose tissue. Philadelphia: Lippincott; 1989.
Saberi M, Woods NB, de LC, Schenk S, Lu JC, Bandyopadhyay G, Verma IM, Olefsky JM. Hematopoietic cell-specific deletion of toll-like receptor 4 ameliorates hepatic and adipose tissue insulin resistance in high-fat-fed mice. Cell Metab. 2009;10:419–29.
Sackmann-Sala L, Berryman DE, Munn RD, Lubbers ER, Kopchick JJ. Heterogeneity among white adipose tissue depots in male C57BL/6J mice. Obesity (Silver Spring). 2012;20:101–11.
Sadler D, Mattacks CA, Pond CM. Changes in adipocytes and dendritic cells in lymph node containing adipose depots during and after many weeks of mild inflammation. J Anat. 2005;207:769–81.
Scherer PE, Williams S, Fogliano M, Baldini G, Lodish HF. A novel serum protein similar to C1q, produced exclusively in adipocytes. J Biol Chem. 1995;270:26746–9.
Schipper HS, Prakken B, Kalkhoven E, Boes M. Adipose tissue-resident immune cells: key players in immunometabolism. Trends Endocrinol Metab. 2012;23:407–15.
Schley PD, Brindley DN, Field CJ. (n-3) PUFA alter raft lipid composition and decrease epidermal growth factor receptor levels in lipid rafts of human breast cancer cells. J Nutr. 2007;137:548–53.
Sears DD, Hsiao G, Hsiao A, Yu JG, Courtney CH, Ofrecio JM, Chapman J, Subramaniam S. Mechanisms of human insulin resistance and thiazolidinedione-mediated insulin sensitization. Proc Natl Acad Sci U S A. 2009;106:18745–50.
Sell H, Habich C, Eckel J. Adaptive immunity in obesity and insulin resistance. Nat Rev Endocrinol. 2012;8:709–16.
Silva JE, Matthews P. Thyroid hormone metabolism and the source of plasma triiodothyronine in 2-week-old rats: effect of thyroid status. Endocrinology. 1984;114:2394–405.
Simonsen L, Bulow J, Madsen J, Christensen NJ. Thermogenic response to epinephrine in the forearm and abdominal subcutaneous adipose tissue. Am J Physiol. 1992;263:E850–5.
Smorlesi A, Frontini A, Giordano A, Cinti S. The adipose organ: white-brown adipocyte plasticity and metabolic inflammation. Obes Rev. 2012;13(Suppl 2):83–96.
Spalding KL, Arner E, Westermark PO, Bernard S, Buchholz BA, Bergmann O, Blomqvist L, Hoffstedt J, Naslund E, Britton T, Concha H, Hassan M, Ryden M, Frisen J, Arner P. Dynamics of fat cell turnover in humans. Nature. 2008;453:783–7.
Sponarova J, Mustard KJ, Horakova O, Flachs P, Rossmeisl M, Brauner P, Bardova K, Thomason-Hughes M, Braunerova R, Janovska P, Hardie DG, Kopecky J. Involvement of AMP-activated protein kinase in fat depot-specific metabolic changes during starvation. FEBS Lett. 2005;579:6105–10.
Subbaramaiah K, Hudis CA, Dannenberg AJ. The prostaglandin transporter regulates adipogenesis and aromatase transcription. Cancer Prev Res (Phila). 2011;4:194–206.
Subbaramaiah K, Morris PG, Zhou XK, Morrow M, Du B, Giri D, Kopelovich L, Hudis CA, Dannenberg AJ. Increased levels of COX-2 and prostaglandin E2 contribute to elevated aromatase expression in inflamed breast tissue of obese women. Cancer Discov. 2012;2:356–65.

Suganami T, Tanimoto-Koyama K, Nishida J, Itoh M, Yuan X, Mizuarai S, Kotani H, Yamaoka S, Miyake K, Aoe S, Kamei Y, Ogawa Y. Role of the toll-like receptor 4/NF-kappaB pathway in saturated fatty acid-induced inflammatory changes in the interaction between adipocytes and macrophages. Arterioscler Thromb Vasc Biol. 2007;27:84–91.

Sugii S, Olson P, Sears DD, Saberi M, Atkins AR, Barish GD, Hong SH, Castro GL, Yin YQ, Nelson MC, Hsiao G, Greaves DR, Downes M, Yu RT, Olefsky JM, Evans RM. PPARgamma activation in adipocytes is sufficient for systemic insulin sensitization. Proc Natl Acad Sci U S A. 2009;106:22504–9.

Thurlby PL, Ellis RD. Differences between the effects of noradrenaline and the beta-adrenoceptor agonist BRL 28410 in brown adipose tissue and hind limb of the anaesthetized rat. Can J Physiol Pharmacol. 1986;64:1111–4.

Titos E, Rius B, Gonzalez-Periz A, Lopez-Vicario C, Moran-Salvador E, Martinez-Clemente M, Arroyo V, Claria J. Resolvin D1 and its precursor docosahexaenoic acid promote resolution of adipose tissue inflammation by eliciting macrophage polarization toward an M2-like phenotype. J Immunol. 2011;187:5408–18.

Todoric J, Loffler M, Huber J, Bilban M, Reimers M, Kadl A, Zeyda M, Waldhausl W, Stulnig TM. Adipose tissue inflammation induced by high-fat diet in obese diabetic mice is prevented by n-3 polyunsaturated fatty acids. Diabetologia. 2006;49:2109–19.

Trujillo ME, Scherer PE. Adipose tissue-derived factors: impact on health and disease. Endocr Rev. 2006;27:762–78.

Turer AT, Scherer PE. Adiponectin: mechanistic insights and clinical implications. Diabetologia. 2012;55:2319–26.

Unger RH. The physiology of cellular liporegulation. Annu Rev Physiol. 2003;65:333–47.

Villarroya F. Irisin, turning up the heat. Cell Metab. 2012;15:277–8.

Virtanen KA, van Marken Lichtenbelt WD, Nuutila P. Brown adipose tissue functions in humans. Biochim Biophys Acta. 2013;1831:1004–8.

Virtue S, Vidal-Puig A. It's not how fat you are, it's what you do with it that counts. PLoS Biol. 2008;6:e237.

Vosselman MJ, van der Lans AA, Brans B, Wierts R, van Baak MA, Schrauwen P, van Marken Lichtenbelt WD. Systemic beta-adrenergic stimulation of thermogenesis is not accompanied by brown adipose tissue activity in humans. Diabetes. 2012;61:3106–13.

Walden TB, Hansen IR, Timmons JA, Cannon B, Nedergaard J. Recruited vs. nonrecruited molecular signatures of brown, "brite," and white adipose tissues. Am J Physiol Endocrinol Metab. 2012;302:E19–31.

Wang H, Eckel RH. Lipoprotein lipase: from gene to obesity. Am J Physiol Endocrinol Metab. 2009;297:E271–88.

Wang MY, Lee Y, Unger RH. Novel form of lipolysis induced by leptin. J Biol Chem. 1999;274:17541–4.

Wang T, Zang Y, Ling W, Corkey BE, Guo W. Metabolic partitioning of endogenous fatty acid in adipocytes. Obes Res. 2003;11:880–7.

Weylandt KH, Chiu CY, Gomolka B, Waechter SF, Wiedenmann B. Omega-3 fatty acids and their lipid mediators: towards an understanding of resolvin and protectin formation Omega-3 fatty acids and their resolvin/protectin mediators. Prostaglandins Other Lipid Mediat. 2012;97:73–82.

White PJ, Arita M, Taguchi R, Kang JX, Marette A. Transgenic restoration of long-chain n-3 fatty acids in insulin target tissues improves resolution capacity and alleviates obesity-linked inflammation and insulin resistance in high fat-fed mice. Diabetes. 2010;59:3066–73.

Wilson-Fritch L, Nicoloro S, Chouinard M, Lazar MA, Chui PC, Leszyk J, Straubhaar J, Czech MP, Corvera S. Mitochondrial remodeling in adipose tissue associated with obesity and treatment with rosiglitazone. J Clin Invest. 2004;114:1281–9.

Wu J, Bostrom P, Sparks LM, Ye L, Choi JH, Giang AH, Khandekar M, Virtanen KA, Nuutila P, Schaart G, Huang K, Tu H, van Marken Lichtenbelt WD, Hoeks J, Enerback S, Schrauwen P, Spiegelman BM. Beige adipocytes are a distinct type of thermogenic fat cell in mouse and human. Cell. 2012;150:366–76.

Wu J, Cohen P, Spiegelman BM. Adaptive thermogenesis in adipocytes: is beige the new brown? Genes Dev. 2013a;27:234–50.
Wu JH, Cahill LE, Mozaffarian D. Effect of fish oil on circulating adiponectin: a systematic review and meta-analysis of randomized controlled trials. J Clin Endocrinol Metab. 2013b;98:2451–9.
Yehuda-Shnaidman E, Buehrer B, Pi J, Kumar N, Collins S. Acute stimulation of white adipocyte respiration by PKA-induced lipolysis. Diabetes. 2010;59:2474–83.
Zhang Y, Proenca R, Maffei M, Barone M, Leopold L, Friedman JM. Positional cloning of the mouse obese gene and its human homologue. Nature. 1994;372:425–32.
Zhou D, Samovski D, Okunade AL, Stahl PD, Abumrad NA, Su X. CD36 level and trafficking are determinants of lipolysis in adipocytes. FASEB J. 2012;26:4733–42.
Zingaretti MC, Crosta F, Vitali A, Guerrieri M, Frontini A, Cannon B, Nedergaard J, Cinti S. The presence of UCP1 demonstrates that metabolically active adipose tissue in the neck of adult humans truly represents brown adipose tissue. FASEB J. 2009;23:3113–20.

Chapter 14
Facial Subcutaneous Adipocytes

Apostolos Pappas and Su-Hyoun Chon

> **Core Messages**
> - Restoring facial subcutaneous fat is of great interest for the antiaging therapy.
> - Human facial preadipocytes demonstrated a superior adipogenic capacity compared to abdominal preadipocytes, in vitro.
> - In vitro the facial preadipocytes also retained their ability to differentiate through higher passages compared to the abdominal preadipocytes.
> - Facial preadipocytes demonstrated a reduced lipolysis response after exposure to isoproterenol; which was consistent with an approximately 60 % lower expression of the β_2-adrenergic receptor compared to abdominal fat cells.
> - Human facial preadipocytes have different functional and gene expression profiles from abdominal preadipocytes in culture. These differences may have implications for development of treatments for facial fat loss during aging.

Introduction

Hypodermis or the subcutaneous layer is the third and forgotten layer of the skin and is mainly composed of fat cells, the adipocytes, which are grouped together in lobules separated by connective tissue. Although it is believed that this layer provides padding or acts as energy reserve and also provides thermoregulation, it is almost unexplored; especially on how this layer communicates with the dermis and its cell types.

An article with similar content has been sent for publication to the Adipocytes Journal.

A. Pappas (✉) · S.-H. Chon
The Johnson & Johnson Skin Research Center, CPPW, a division of Johnson & Johnson Consumer Companies, Inc., Skillman, NJ 08558, USA
e-mail: APappas@its.jnj.com

A. Pappas (ed.), *Lipids and Skin Health*, DOI 10.1007/978-3-319-09943-9_14

Facial fat loss is one of the common manifestations of aging (Bucky and Kanchwala 2007; Coleman et al. 2009). Autologous fat grafting has become popular to restore facial volume loss in addition to other facial fillers (Bucky and Kanchwala 2007). The enrichment of adipocyte-derived stem/stromal cells (ASC) in fat grafting has been actively considered (Beeson et al. 2011; Cohen and Mailey 2012). Factors that induce endogenous facial preadipocyte differentiation may eventually work to reverse facial fat loss and potentially be considered as alternative treatments to facial fillers and antiaging treatments. The development of such technologies will lead to novel powerful antiaging tools that would target the recessed hypodermis. Nevertheless, facial fat cell isolation and characterization has never been attempted in the past therefore it is not known if facial adipocyte differentiation is inhibited by factors induced by aging or whether certain pathways fundamental for adipocyte differentiation are lost.

Preadipocytes are composing 15–50% of cells in adipose tissues where they mainly reside although small portions are transferred from the circulating pool (Tchkonia et al. 2010). They tend to populate in the vascular smooth muscle cells and pericytes compartment of the adipose vasculature, but this does not happen in the vasculature of other tissues (Tang et al. 2008). As known, the main function of preadipocytes is producing mature fat cells, therefore to deposit the excessive energy to benign storage format in our body. Fat cell differentiation is initiated by several signaling cascades activated by fatty acids and glucocorticoids (Tchkonia et al. 2010). The transcriptional factors peroxisome proliferator-activated receptor gamma (PPAR-γ) and C/EBP-α are key in governing the adipogenesis process (Tchkonia et al. 2010). The capability of proliferation and differentiation of preadipocytes has known to be reduced with aging and this reduction is more profound in the subcutaneous fat region than the omental depot (Sepe et al. 2011). The decreased expression of PPAR-γ and C/EBP-α in preadipocytes from old subjects has been reported to be one of the primary causes (Karagiannides et al. 2001; Sepe et al. 2011). Also increased inflammation and insulin resistance in preadipocytes from old animal studies have been observed and suggested as a mechanistic understanding for the reduced lipogenic activities in preadipocytes from aged subjects (Karagiannides et al. 2001; Sepe et al. 2011). The rate of these changes in adipocytes seems to be different in various depots, contributing metabolic complication and redistribution of body fat during aging (Sepe et al. 2011; Tchkonia et al. 2010) It has been reported that preadipocytes from various body sites possess depot-specific characteristics (Cartwright et al. 2007). For example, PPAR-γ agonists induced lipid droplet formation and induced a greater increase in glycerol 3-phosphate dehydrogenase activity in subcutaneous preadipocytes compared with omental preadipocytes;despite the similar levels of expression of the PPAR-γ protein (Adams et al. 1997). Also, it has shown that insulin mediated antilipolytic activity is higher in the subcutaneous compared to visceral depot. On the other hand catecholamine induced lipolytic activities were lower in subcutaneous than visceral region (Arner 1995; Bolinder et al. 1983; Mauriege et al. 1987). Interestingly, the lipolytic rates from the subcutaneous fat differ between body parts, for example, abdominal subcutaneous fat tissue has higher lipolytic rate than femoral/gluteal region (McQuaid et al. 2010). In addition to heterogeneous metabolic activities, researchers have also reported a distinct gene expression profiles among different fat depots (Gehrke et al. 2013; Gesta et al. 2006). For example, Gesta et al. have shown the difference in gene expression pattern between visceral and subcutaneous fat cells (mature adipocytes as well as preadipocytes) from mice and human by microarray analysis. In particular they found that developmental genes such as homeobox (*HOX*) genes are one of the most profoundly changed, suggesting a new function of these genes in body fat distribution (Gesta et al. 2006). Among a variety of genes those that they have metabolic indication are phospholipase A2, group 2A (PLA2G2A), fibroblast growth factor 10 (FGF10), and insulin-like growth factor binding protein 5 (IGFBP5); these were found to be regulated depot specifically between abdominal compared to gluteal fat (Gehrke et al. 2013). Differential fatty acid composition was also reported, showing abdominal fat tissue more saturated fatty acids in abdominal fat than the gluteal region (Gehrke et al. 2013).

Characterization of Facial Preadipocytes

In preliminary experiments more facial preadipocytes (three different preparations) exhibited distinct differentiated characteristics (lipid droplets, Oil Red O accumulation) in the presence of differentiating media (IBMX, insulin, dexamethasone, and rosiglitazone) than any of the abdominal preadipocyte preparations (Fig. 14.1).

The ability to differentiate was maintained through multiple passages longer in facial preadipocytes than abdominal preadipocytes (Fig. 14.2). It was determined that facial preadipocytes may be induced to differentiate at a passage greater than 10. The morphological and molecular characteristics at passage 11 were similar to those after the third passage (Fig. 14.3a). In Fig. 14.3b, quantitative polymerase chain reaction (qPCR) analysis demonstrated that the expression of the major adipocyte markers: fatty acid binding protein 4 (FABP4), PPAR-γ, and fatty acid synthase (FAS; glucose transporter 4 (GLUT4), G-protein coupled receptor 81 (GPR81), leptin, and adiponectin as well but not shown) are similarly and significantly induced during differentiation of both facial and abdominal preadipocytes. The late-passage (passage 11) facial preadipocytes did not have significantly different expression levels of the adipogenic or the lipolytic genes tested (more than 20 genes), and these expression levels were similar to those of differentiated early-passage (passage 3–5) abdominal cells.

On day 14 or 15 after adding the differentiation media, preadipocyte cultures became fully differentiated adipocytes, as determined by morphology and adipocyte marker gene expression (FABP4, GLUT4, and PPAR-γ). To test the functional response of these cells, passage 3–5 facial and abdominal preadipocytes were treated with 2 μM of isoproterenol, a β_2-adrenergic receptor agonist (Fig. 14.4). After 5 h, lipolytic rate measured by glycerol release was twofold greater in abdominal than facial adipocytes (Fig. 14.4a). After 1 week of exposure to isoproterenol, facial preadipocytes retained more lipid droplets than abdominally derived cells (Fig. 14.4b). Histological examination revealed more lipid droplets of greater diameter in the facial cells than in the abdominal adipocytes.

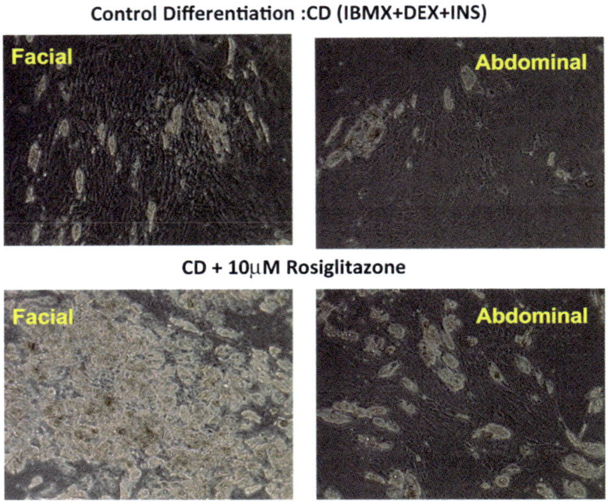

Fig. 14.1 Facial preadipocytes are more responsive to rosiglitazone-containing differentiation compositions, relative to abdominal preadipocytes. Human facial and abdominal preadipocytes in control media (*CD*, insulin, dexamethasone, and *IBMX*), or *CD* plus 10 μM rosiglitazone (*CDR*). Microscopic images of the cells were taken on day 14 after plating. The pictures are representative of three individual experiments. *DEX* dexamethasone, *INS* insulin

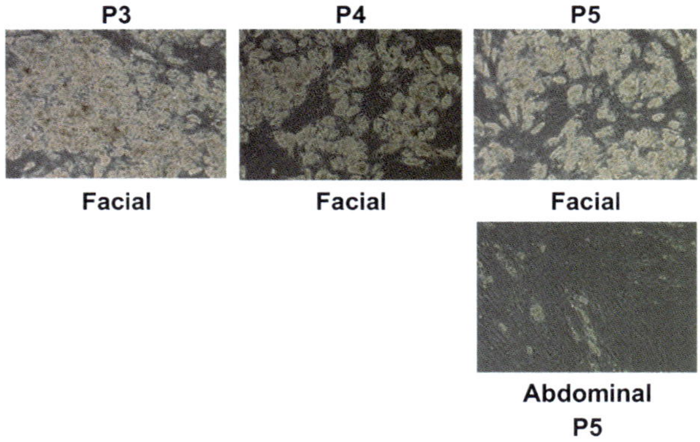

Fig. 14.2 Facial preadipocytes retain their ability to differentiate through later subpassages than abdominal preadipocytes. Human facial and abdominal preadipocytes at the indicated subpassages were induced to differentiate by *CDR*. Images of the cells were taken on day 14 after plating. The pictures are representative of three individual experiments. *CDR* control differentiation media + 10 μM rosiglitazone, *P* passage number

An array of genes associated with lipolysis was then tested to examine the possible differences in expression levels between facial and abdominal cells (Fig. 14.4c). Among the 25 genes tested, the expression of the β_2-adrenergic receptor (ADRB2) was significantly different between facial and the abdominal adipocytes. The lower expression levels of ADRB2 from facial adipocytes compared with abdominal adipocytes was consistent with our finding of reduced lipolytic rate in facial fat cells (Fig. 14.4a), suggesting the underlying mechanism.

The differences noted above in expression of the β_2-adrenergic receptors in differentiated preadipocytes from abdominal and facial sources were identified using cultured adipocytes. Further the differences in gene expression observed between these tissue sites also existed in vivo. The gene profiles using tissue obtained from

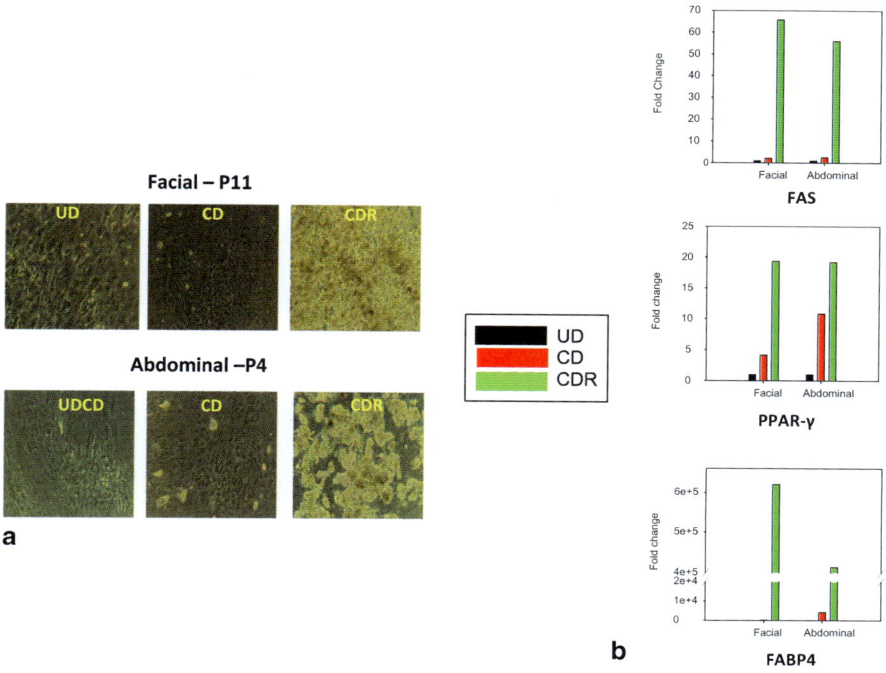

Fig. 14.3 Morphology and expression of lipid gene markers in facial and abdominal preadipocytes: effect of passage number. Facial and abdominal preadipocytes were grown for 14 days in the *CD* media or *CDR* media. **a** Images are representative of four individual experiments. **b** Gene expression was evaluated by qPCR and normalized by expression of the *RPL13A* gene. *UD* undifferentiated (cultured in growth medium), *CD* control differentiation, *CDR* *CD*+10 μM rosiglitazone, *FABP4* fatty acid binding protein 4, *FAS* fatty acid synthase, *P* passage number, *PPAR-γ* peroxisome proliferator-activated receptor gamma, qPCR quantitative polymerase chain reaction

age-matched individuals (45–60 years old) undergoing either facelift or abdominoplasty demonstrated that most gene markers were expressed at similar levels in tissues from both sites. As shown in Fig. 14.5a, significantly less expression of FAS, CPT2, and GLUT4 in facial versus abdominal adipose tissue was observed. Although β_2-adrenergic receptor expression levels was lower in facial versus abdominal adipose tissue, the difference was not statistically significant. The expression the β_3-adrenergic receptor was detected in abdominal adipose tissue, but was completely undetectable in the facial adipose tissue (Fig. 14.5b).

Microarray analysis identified unknown genes that were regulated depot specifically; in five human abdominal adipocytes and four facial. Depot specific patterns of gene expression were observed by principle component analysis (PCA), abdominal samples were separated by facial samples in general, based on the overall expression profiles (Fig. 14.6a). Total of 283 probes differentially regulated between the two depots were identified (Fig. 14.6b). Around 126 probe sets were significantly

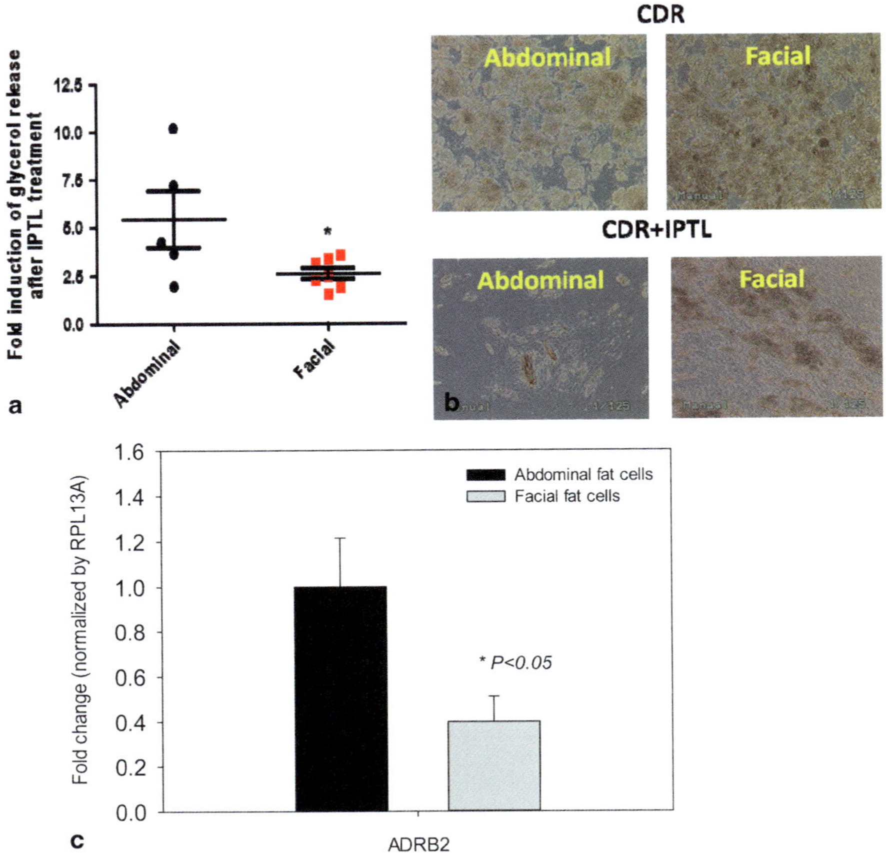

Fig. 14.4 β-adrenergic receptor expression and responsiveness to isoproterenol in differentiated facial and abdominal preadipocytes. **a** Acute glycerol release after 5 h isoproterenol exposure (2 μM). Data is represented as mean ± SEM. Abdominal cells, $n = 5$ donors and facial cells, $n = 7$ donors. *Student *t* test $P < 0.05$. **b** Facial and abdominal adipocytes images before and after 1 week of isoproterenol treatment. **c** After 1 week of isoproterenol exposure, the expression of lipid processing genes was evaluated by qPCR analysis and was normalized by expression of the *RPL13A* gene. Abdominal cells, $n = 6$ donors and facial cells, $n = 4$ donors. *CD* control differentiation, *CDR* CD + 10 μM rosiglitazone, *ADRB2* β_2-adrenergic receptor

upregulated in abdominal fat as compared to facial fat, with fold change not less than 2 and the overall FDR less than 0.05. Using the same criteria, 157 probe sets were significantly downregulated in abdominal fat as compared to facial fat (Fig. 14.6b). The detail expression profiles of these 283 probe sets are also illustrated in this heat map Fig. 14.6b. The lists of the top ten genes regulated differentially were summarized in Table 14.1. Interestingly, the majority of upregulated genes in abdominal compared to facial adipocytes, in other words, significantly downregulated in facial

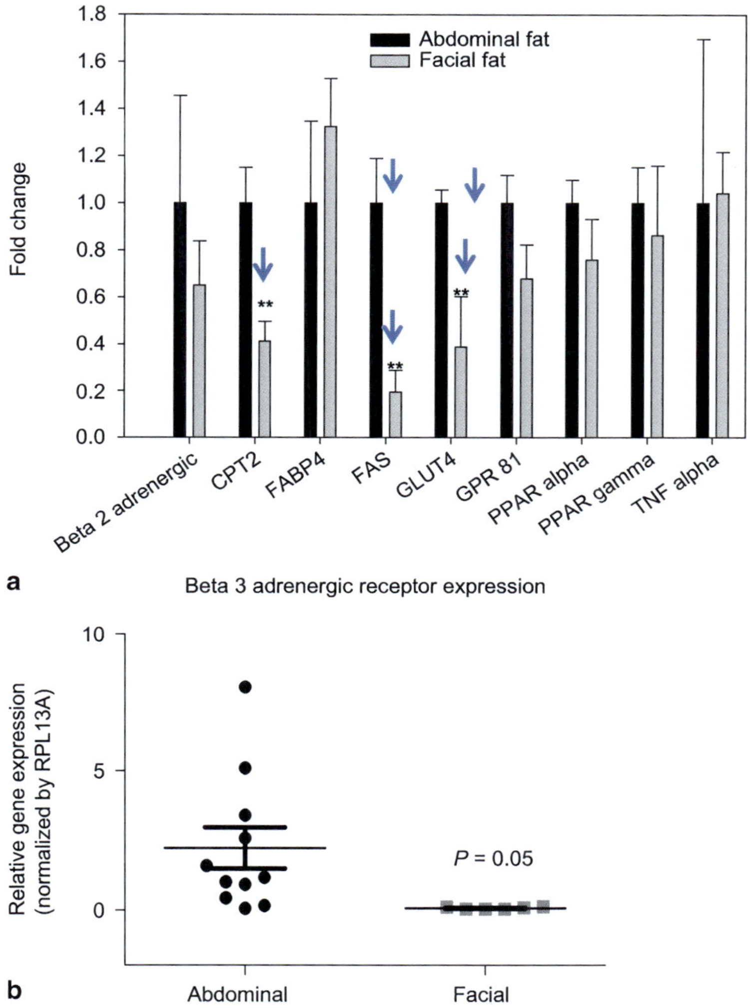

Fig. 14.5 Expression of adipocyte markers in facial and abdominal subcutaneous fat tissues. mRNA was isolated immediately following tissue harvest from females 50–60 years of age. Gene expression was evaluated by qPCR analysis; expression of fat cell markers was normalized by expression of the *RPL13A* gene (mean ± SEM). **a** Lipid metabolic gene expression in subcutaneous fat tissues ($n = 4$ donors per group), **b** β_3-adrenergic receptor expression in facial (n = 5 donors) and abdominal (n = 10 donors) tissues. $^*P < 0.05$ versus abdominal cells. *BLD* below limit of detection, *CPT2* carnitine palmitoyltransferase 2, *FABP4* fatty acid binding protein 4, *FAS* fatty acid synthase, *GLUT4* glucose transporter 4, *GPR81* G-protein coupled receptor 8, *PPAR-α* peroxisome proliferator-activated receptor alpha, *PPAR-γ* peroxisome proliferator-activated receptor gamma, *TNF-α* tumor necrosis factor alpha

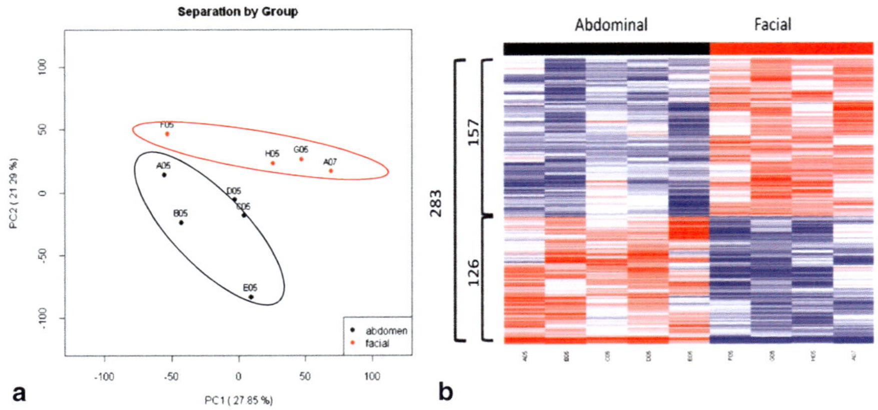

Comparison Facial Fat cells versus Abdominal Fat cells	Direction	Fold Change Cutoff	FDR Cutoff	Number of Selected Probe Sets
Facial fat Abdominal Fat	Down	>=2	<=0.05	126
Facial fat vs Abdominal Fat	Up	>=2	<=0.05	157

Fig. 14.6 Depot specific gene expression profile and comparison in human facial versus abdominal fat cells by Microarray analysis. **a** PCA analysis demonstrated a distinct gene expression pattern between two depots (Abdominal fat cells; $n = 5$ donors, Facial fat cells; $n = 4$ donors). **b** Heat-map illustration showing identified 283 probes differentially regulated between the two depots. Around 126 probe sets were significantly downregulated in facial fat cells as compared to abdominal fat cells. Additionally, 157 probe sets were significantly upregulated in facial fat cells as compared to abdominal fat cells

were the HOX transcription factors, which have also been identified in other depot comparisons; for example, abdominal versus gluteal regions (Gehrke et al. 2013).

Conclusion

A higher proportion of facial cells differentiate, as determined by morphologic examination and lipid droplet quantization. In addition, facial cells retain their ability to differentiate through more subpassages than abdominally derived cells. This is of prime interest since abdominal preadipocytes have not been shown to differentiate in culture beyond five passages (Wall et al. 2007; Zenbio INC. 2012).The clearest difference between abdominal and facial cells or freshly isolated adipose tissues was in the expression of β-adrenergic receptor. Functionally, this was observed as a greater lipolytic response to isoproterenol in abdominal adipocytes. This increased response was consistent with the observation of higher expression of β_2- and β_3-adrenergic receptors in the abdominal cells when compared with facial cells.

Table 14.1 Genes differentially regulated in abdominal adipocytes compared to facial adipocytes

Fold change	Gene symbol	Description
Upregulated in abdominal versus facial		
55.12	*HOXC10*	Homeobox C10
33.78	*HOXA10*	Homeobox A10
31.85	*HOXA9*	Homeobox A9
25.78	*HOXA7*	Homeobox A7
10.27	*GPRC5A*	G–protein coupled receptor, family C, group 5, member A
9.93	*HOXA10*	Homeobox A10
9.68	*TMEM132C*	Transmembrane protein 132C
9.36	*HOXA9*	Homeobox A9
9.01	*HOXA3*	Homeobox A3
7.84	*PARP8*	Poly (ADP-ribose) polymerase family, member 8
Downregulated in abdominal versus facial		
-16.7	*SLITRK1*	SLIT and NTRK-like family, member 1
-9.4	*EGFL6*	EGF-like-domain, multiple 6
-8.9	*DIO2*	Deiodinase, iodothyronine, type II
-8.6	*MOXD1*	Monooxygenase, DBH-like 1
-8.3	*NPNT*	Nephronectin
-7.4	*GPM6B*	Glycoprotein M6B
-6.2	*THSD4*	Thrombospondin, type I
-6.2	*PAX3*	Paired box 3
-6.0	*PLCXD3*	Phosphatidylinositol-specific phospholipase C, X domain containing 3
-5.9	*EVI1*	Ecotropic viral integration site 1

The robust adipogenic capacity of facial preadipocytes was not expected and yet very intriguing. It will be interesting to see if the adipogenic capacity from facial preadipocytes is altered during the aging process or not. Unlike abdominal fat, facial preadipocytes are more often exposed to the cold or UV radiation, which may certainly have an indirect effect on the volume loss, as cold exposure could potentially induce some of the metabolic aspects of beige adipose tissue, such as accumulation mitochondria and production of mitochondrial uncoupling protein (Seale et al. 2009). The UV effect was further researched and it was reported that UV exposure can also lead to a cascade of reactions responsible for facial fat loss (Li et al. 2013).

This work demonstrates that the greater responsiveness of facial preadipocytes to rosiglitazone could potentially be explained either by the absence of the natural PPAR-γ agonist in facial subcutaneous tissue or by the presence of an antagonist or inhibitor of lipogenesis. Since addition of rosiglitazone can fully restore the differentiation on the facial preadipocytes strategies to compensate for facial fat loss by subcutaneous injection of such PPAR-γ agonist, could be less painful and more efficient than a fat transplant.

Distinct regulation of *HOX* genes expression in facial adipocytes is intriguing observation. The HOX network are transcriptional factors that comprise of 39 genes regulating the developmental process during embryogenesis, but also plays an important role in metabolic, somatic, and congenital functions in adults (Cantile et al. 2003). Potential interaction of this molecular network and human adipogenesis has been suggested (Cantile et al. 2003). Recently, Gehrke et al. suggested epigenetic regulation, in particular, DNA methylation of *HOX* genes as a mechanism how this transcriptional factors exhibit such a distinct depot specific expression pattern (Gehrke et al. 2013). A strong correlation between *HOX* gene expression and body fat mass has been observed (Dankel et al. 2010; Gesta et al. 2006). Dankel et al. found the significant upregulation of *HOX* gene expression in the abdominal subcutaneous adipose from the obese patient after profound fat loss via bariatric surgery (Dankel et al. 2010). Nevertheless, how *HOX* genes are participating in fat mass control is still unknown.

Fat and volume loss in the aging face can induce gravimetrical sagging, less firm dermis, and possibly wrinkled epidermis. The induction of subcutaneous facial adipocyte differentiation could potentially result in an improvement of facial contour, as well as a reduction of the appearance of skin defect around the same area. Therefore, characterization and better understanding of facial adipocytes could be useful in developing treatments for facial fat loss during aging.

References

Adams M, Montague CT, Prins JB, Holder JC, Smith SA, Sanders L, Digby JE, Sewter CP, Lazar MA, Chatterjee VK, O'Rahilly S. Activators of peroxisome proliferator-activated receptor gamma have depot-specific effects on human preadipocyte differentiation. J Clin Invest. 1997;100(12):3149–53.

Arner P. Differences in lipolysis between human subcutaneous and omental adipose tissues. Ann Med. 1995;27(4):435–8.

Beeson W, Woods E, Agha R. Tissue engineering, regenerative medicine, and rejuvenation in 2010: the role of adipose-derived stem cells. Facial Plast Surg. 2011;27(4):378–87.

Bolinder J, Kager L, Ostman J, Arner P. Differences at the receptor and postreceptor levels between human omental and subcutaneous adipose tissue in the action of insulin on lipolysis. Diabetes. 1983;32(2):117–23.

Bucky LP, Kanchwala SK. The role of autologous fat and alternative fillers in the aging face. Plast Reconstr Surg. 2007;120(6 Suppl):89S–97S.

Cantile M, Procino A, D'Armiento M, Cindolo L, Cillo C. *HOX* gene network is involved in the transcriptional regulation of in vivo human adipogenesis. J Cell Physiol. 2003;194(2):225–36.

Cartwright MJ, Tchkonia T, Kirkland JL. Aging in adipocytes: potential impact of inherent, depot-specific mechanisms. Exp Gerontol. 2007;42(6):463–71.

Cohen SR, Mailey B. Adipocyte-derived stem and regenerative cells in facial rejuvenation. Clin Plast Surg. 2012;39(4):453–64.

Coleman S, Saboeiro A, Sengelmann R. A comparison of lipoatrophy and aging: volume deficits in the face. Aesthet Plast Surg. 2009;33(1):14–21.

Dankel SN, Fadnes DJ, Stavrum AK, Stansberg C, Holdhus R, Hoang T, Veum VL, Christensen BJ, Vage V, Sagen JV, Steen VM, Mellgren G. Switch from stress response to homeobox transcription factors in adipose tissue after profound fat loss. PLoS ONE. 2010;5(6):e11033.

Gehrke S, Brueckner B, Schepky A, Klein J, Iwen A, Bosch TC, Wenck H, Winnefeld M, Hagemann S. Epigenetic regulation of depot-specific gene expression in adipose tissue. PLoS ONE. 2013;8(12):e82516.

Gesta S, Bluher M, Yamamoto Y, Norris AW, Berndt J, Kralisch S, Boucher J, Lewis C, Kahn CR. Evidence for a role of developmental genes in the origin of obesity and body fat distribution. Proc Natl Acad Sci U S A. 2006;103(17):6676–81.

Karagiannides I, Tchkonia T, Dobson DE, Steppan CM, Cummins P, Chan G, Salvatori K, Hadzopoulou-Cladaras M, Kirkland JL. Altered expression of C/EBP family members results in decreased adipogenesis with aging. Am J Physiol Regul Integr Comp Physiol. 2001;280(6):R1772–80.

Li WH, Pappas A, Zhang L, Ruvolo E, Cavender D. IL-11, IL-1alpha, IL-6, and TNF-alpha are induced by solar radiation in vitro and may be involved in facial subcutaneous fat loss in vivo. J Dermatol Sci. 2013;71(1):58–66.

Mauriege P, Galitzky J, Berlan M, Lafontan M. Heterogeneous distribution of beta and alpha-2 adrenoceptor binding sites in human fat cells from various fat deposits: functional consequences. Eur J Clin Invest. 1987;17(2):156–65.

McQuaid SE, Humphreys SM, Hodson L, Fielding BA, Karpe F, Frayn KN. Femoral adipose tissue may accumulate the fat that has been recycled as VLDL and nonesterified fatty acids. Diabetes. 2010;59(10):2465–73.

Seale P, Kajimura S, Spiegelman BM. Transcriptional control of brown adipocyte development and physiological function-of mice and men. Gene Dev. 2009;23(7):788–97.

Sepe A, Tchkonia T, Thomou T, Zamboni M, Kirkland JL. Aging and regional differences in fat cell progenitors-a mini-review. Gerontology. 2011;57(1):66–75.

Tang W, Zeve D, Suh JM, Bosnakovski D, Kyba M, Hammer RE, Tallquist MD, Graff JM. White fat progenitor cells reside in the adipose vasculature. Science. 2008;322(5901):583–6.

Tchkonia T, Morbeck DE, Von ZT, Van DJ, Lustgarten J, Scrable H, Khosla S, Jensen MD, Kirkland JL. Fat tissue, aging, and cellular senescence. Aging Cell. 2010;9(5):667–84.

Wall ME, Bernacki SH, Loboa EG. Effects of serial passaging on the adipogenic and osteogenic differentiation potential of adipose-derived human mesenchymal stem cells. Tissue Eng. 2007;13(6):1291–98.

Zenbio INC. Subcutaneous human adipocyte instruction manual. ZBM0001.03. 2012.

Part VI
Preclinical Models and Skin Phenotype (Lipid Gene Knockouts that Result in Hair Loss or Dry Skin and Impaired Barrier)

Chapter 15
Stearoyl-CoA Desaturases are Regulators of Lipid Metabolism in Skin

James M. Ntambi

Core Messages

An important class of lipid metabolism enzymes expressed in skin is the delta-9 desaturases, which catalyze the synthesis of delta-9 monounsaturated lipids, primarily oleoyl-(18:1n9) and palmitoyl-CoA (16:1n7), the major monounsaturated fatty acids (MUFA) of cutaneous lipids.

Mice with a deletion of the delta-9 desaturase-1 isoform (SCD1) either globally or specifically in the skin exhibit increased whole-body energy expenditure, protection against dietary-induced adiposity, hepatic steatosis, and glucose intolerance.

The increased energy expenditure in these mice does not result simply from heat loss through the skin. Rather, thermogenesis appears to be constitutively activated in these mice regardless of changes in the ambient temperature.

Structure and Endocrine Nature of Skin

The skin is the largest single organ in humans, serving as a major barrier to infection, water loss, and abrasion (Madison 2003). Skin is a stratified tissue composed of epidermis, dermis, and subcutaneous fat layers. The epidermis is the thinnest of the three layers but is the mitotically most active stratum due to the continued differentiation of keratinocytes into the cornified epithelium exposed to the environment. The dermis is thicker than the epidermis and contains a number of specialized structures such as sweat glands, hair follicles, and sebaceous glands that impart functional diversity to the organ. This layer functions as an insulator, shock absorber, and energy storage depot.

J. M. Ntambi (✉)
Department of Biochemistry and of Nutritional Sciences,
University of Wisconsin-Madison, WI 53706, USA
e-mail: ntambi@biochem.wisc.edu

A. Pappas (ed.), *Lipids and Skin Health,* DOI 10.1007/978-3-319-09943-9_15

The diverse roles of the skin require the synthesis of large amounts of lipids, such as triglycerides, wax esters, ceramides, cholesterol esters, cholesterol, and retinyl esters. Some of these lipids are used as cell membrane components, signaling molecules, and as a source of energy but some others are excreted out of the cells. The turnover of these lipids is high due to the continued shedding of dead keratinocytes from the epidermis.

In rodents, a competent skin barrier to water loss is formed 2 or 3 days prior to birth. Acquisition of barrier function during gestation correlates with the formation of a stratum corneum. The main lipid classes in the stratum corneum are free fatty acids, cholesterol, and ceramides (Doering et al. 2002). Stratum corneum ceramides consist of at least 12 ceramide species that differ from each other in head group architecture and in the main fatty acid length. The fatty acid esterified to the amide of the phytosphingosine head group can be either an α-hydroxy or non-hydroxy fatty acid (Doering et al. 2002). The fatty acid length varies between C16–33 with C24 and C26 chain lengths being predominant.

Information on the relationship between lipid organization and composition is of great importance to unravel the mechanisms controlling skin barrier function. This is particularly evident in some skin diseases in which impaired barrier function can be ascribed to altered lipid composition and organization; however, very little information is available about the relation between stratum corneum lipid composition and organization.

With a few exceptions, all hair-bearing mammals possess sebaceous glands embedded in the skin over most of the body and are in higher concentrations in the scalp, face, forehead, and eyes (Ge et al. 2003). In the absence of sebaceous glands, mice tend to have patchy, abnormal skin. It is known that the major function of sebaceous glands is to secrete lipid complex-lubricants, termed sebum (Miyazaki et al. 2001c; Zheng et al. 1999; Sundberg et al. 2000). These lubricants contain the neutral lipids of wax esters, triglycerides and cholesterol esters. Wax ester synthesis appears to be common to all sebaceous glands and is a marker of sebaceous cell function and differentiation (Pappas et al. 2002). Therefore, understanding of the role of fatty acid metabolism in wax ester synthesis is critical in understanding sebogenesis. Wax esters in sebum may lubricate the skin and also impart unique physical (e.g., melting temperature, viscosity) or antibacterial properties to these secretions (Downing et al. 1989). The studies on isolated sheep and human hair follicles indicate that an attached sebaceous gland is required for inner root sheath breakdown. There is strong evidence for interdependence of the sebaceous gland and hair follicle: in situations in which one organ collapses the other is often also lost. In addition, the sebaceous gland can be regenerated by the reservoir of stem cells in the hair follicle bulge. Nevertheless, the interdependence is not obligatory (Coulombe et al. 1989; Hughes et al. 1996). Sebaceous glands are present in some mouse mutants that lack hair follicles, and can be induced in footpad epidermis, an anatomic region normally devoid of hair follicles and sebocytes (Coulombe et al. 1989; Hughes et al. 1996). Retroviral lineage marking has provided strong evidence that the sebaceous gland can arise and be maintained independently of the hair follicle bulge. While much is known concerning the lipid content of sebum (Downing and Strauss 1974; Nikkari 1974; Lampe et al. 1983), the functions of the individual lipid components within them remain unknown.

Stearoyl-CoA Desaturase: What Is This Enzyme and What Does It Do?

Stearoyl-CoA desaturase (SCD) is an endoplasmic reticulum enzyme that catalyzes the biosynthesis of MUFA from saturated fatty acids that are either synthesized de novo or derived from the diet. SCD in conjunction with nicotinamide adenine dinucleotide hydrogen (NADH), the flavoprotein cytochrome b_5 reductase, the electron acceptor cytochrome b_5, and molecular oxygen introduces a single double bond in a spectrum of methylene-interrupted fatty acyl-CoA substrates. The preferred substrates are palmitoyl-(16:0) and stearoyl-CoA (18:0), which get converted into palmitoleoyl-(16:1n7) and oleoyl-CoA (18:1n9), respectively (Ntambi 1999; Sessler and Ntambi 1998; Hellerstein 1999). These products are the most abundant MUFA in various kinds of tissue lipids including phospholipids, triglycerides, cholesterol esters, wax esters, and alkyldiacylglycerols (Miyazaki et al. 2000, 2001a, 2002; Zheng et al. 2001; Miyazaki et al. 2001b). Apart from being components of lipids, MUFA also serve as mediators of signal transduction and cellular differentiation (Miyazaki and Ntambi 2003; Ntambi and Miyazaki 2003, 2004. MUFA also influence apoptosis and may have some role in mutagenesis in some tumors (Miyazaki and Ntambi 2003; Ntambi and Miyazaki 2003, 2004). Therefore, given the multiple roles of MUFA, variation in SCD activity in mammals would be expected to have an effect on a variety of key physiological variables, which include skin development, differentiation, and metabolism.

The MUFA are produced by four SCD gene isoforms (SCD1–4) in mouse and two gene isoforms (hSCD1 and hSCD5) in humans. SCD1, 2, and 3 are expressed in mouse skin (Miyazaki et al. 2001a, c, 2002; Zheng et al. 1999; Sundberg et al. 2000; Zheng et al. 2001; Miyazaki et al. 2001). Although expression of SCD1 and SCD3 are limited to the sebaceous glands, the distribution within the gland is different. In situ hybridization shows that the transcript of SCD1 is located within the undifferentiated sebocytes, while SCD3 mRNA resides in the differentiated sebocytes (Zheng et al. 2001). SCD2 is expressed outside the sebaceous gland mainly in the epidermis. The fact that three genes of the SCD family are found in mouse skin raises the question of the specific function of each gene in this tissue.

Stearoyl-CoA Desaturase-2 Gene Expression is Required for Lipid Synthesis During Early Skin Development

To study the role of SCD2 isoform in lipid synthesis and development, SCD2 deficient (*Scd2* $^{-/-}$) mice were generated (Miyazaki et al. 2005). Although Scd2 $^{-/-}$ mice were born in the expected Mendelian manner, most of the mice died within 24 h after birth, most likely as a result of severe dehydration due to skin permeability barrier dysfunction. The mechanism by which SCD2 deficiency leads to defective permeability barrier function is most probably due to alterations in epidermal lipid metabolism. Electron microscopic analysis of the epidermal stratum corneum of the

Scd2 $^{-/-}$ mice revealed immature lamellar membranes and reductions in the internal contents of epidermal lamellar bodies, suggesting impaired delivery of lipid to lamellar granules, leading to insufficient deposition of lipids in the stratum corneum interstices. It has been known that the precursors of the stratum corneum lipids in lamellar granules are mainly cholesterol, phospholipids, and glucosylceramides (Wertz 2000; Madison 2003). In agreement with electromagnetic (EM) analysis, lipid analysis showed that the content of glucosylceramide was significantly reduced in Scd2 $^{-/-}$ mice despite no alterations in the content of either phospholipids or cholesterol. Further conversion of epidermal glycosylceramides and acylglycosylceramides to ceramides and acylceramides, respectively, by β-cerebrosidase (Gba) is required for skin permeability barrier homeostasis (Doering et al. 1999). The amount of acylceramide, a unique and essential epidermal lipid (Wertz 2000; Madison 2003), was reduced by 30 % in the epidermis of Scd2 $^{-/-}$ mice, despite no alteration in Gba expression. Recently, Stone et al. (2004; Cases et al. 2001) reported that mice lacking diacylglycerol acyltransferase-2, a rate-limiting enzyme of triglycerides (TG) synthesis, exhibited defective skin permeability barrier with lethality similar to Scd2 $^{-/-}$ mice. We also found a drastic reduction of TG and cell envelope (CE) in the epidermis of Scd2 $^{-/-}$ neonatal mice, although the expression of epidermal diacylglycerol acyl transferase-2 was not altered. Thus, our results suggest that the reduction in lipid contents, including acylceramide, TG, and CE, in the epidermis of Scd2 $^{-/-}$ mice is due to lack of sufficient levels of endogenous MUFA.

Some aspects of the skin phenotype of Scd2 $^{-/-}$ mice resemble an essential fatty acid deficiency that has been known to exist in mouse skin for decades. In this model, it has been postulated that linoleic acid [18:2 n-6] is replaced by the accumulation of 20:3 n-9, which is an elongation and desaturation product of 18:1 n-9. Essential fatty acids are required for normal growth and skin function, and a deficiency in these fatty acids is characterized by growth retardation, skin abnormalities, and increased transepidermal water loss. In addition, linoleic acid is the major component of acylceramide, the major lipid involved in epidermal permeability barrier function (Hansen and Jensen 1985; Melton et al. 1987; Wertz and Downing 1990). Analysis of fatty acid composition of epidermal lipids showed that, although the total amounts of linoleic acid in the lipid fractions and FFA pool were not different between Scd2 $^{+/+}$and Scd2 $^{-/-}$ mice, the content of linoleic acid in acylceramide was reduced by >80 %, whereas that in the phospholipid fraction was increased by >30 %. In addition, very little linoleic acid was detected in TG and CE fractions of epidermis of either Scd2 $^{-/-}$ or wild-type mice. The mechanism consistent with these observations that in order for the Scd2 $^{-/-}$ mice to compensate for the reduction in the levels of MUFA required for the synthesis of phospholipids and to maintain membrane fluidity and cell structure, linoleic acid has to come into the embryo from maternal plasma therefore is partitioned away from acylceramide synthesis into phospholipid synthesis. The reduced levels of acylceramide and the deficiency of linoleic acid in the acylceramide fraction would be consistent with a skin permeability barrier defect in the Scd2 $^{-/-}$ mice. The mechanism of how the decision is made to partition linoleic acid is not known but is probably at the level of enzymes that activate linoleic acid before channeling into the phospholipid fraction and other metabolic pathways.

The expression of SCD1 was induced in the epidermis of Scd2 $^{-/-}$ neonates, and it is possible those mice that had the highest induction produced adequate amounts of MUFA that overcame the linoleic acid deficiency in the acylceramide fraction and had a functional skin permeability barrier. The degree of SCD1 induction in skin of Scd2 $^{-/-}$ mice was varied, and Scd2 $^{-/-}$/Scd1 $^{+/-}$ double mutant mice on 129 background showed 100% lethality, suggesting that adequate levels of SCD1 expression in the skin were required for survival of the 30% of the Scd2 $^{-/-}$ mice that made it to adulthood. The epidermis of Scd2 $^{-/-}$ mice exhibited a reduction in the expression of involucrin and transglutaminase-1, makers of mature keratinocytes (Schmuth et al. 2004; Tu et al. 2001), suggesting that SCD2 deficiency attenuates keratinocyte differentiation. Thus SCD2, as well as its products MUFA, may play a crucial role in keratinocyte differentiation and skin homeostasis in neonatal mice. Further experiments are required to address this point.

Involvement of SCD-1 in Skin and Sebaceous Gland Lipid Production and Retinoid Metabolism

Studies on the role of lipids in cutaneous biology have mainly focused on polyunsaturated fatty acids mainly alpha-linoleic acid and their role in skin barrier function (Marcelo et al. 1992). The role of MUFA in skin homeostasis is unknown. However, it is possible that these fatty acids, which are the products of SCD, play an important role in differentiation of the sebaceous gland and a parallel can be drawn to the activation of SCD and production of MUFA during 3T3-L1 preadipocyte differentiation (Casimir and Ntambi 1996; Christianson et al. 2008). Mice with a global disruption in the SCD1 gene (SCD1 $^{-/-}$ or GKO mice) show cutaneous abnormal lipid levels with atrophic sebaceous glands, suggesting an important role of MUFA in skin lipid homeostasis (Miyazaki et al. 2001c; Zheng et al. 1999; Sundberg et al. 2000; Miyazaki et al. 2001; Miyazaki and Ntambi 2003; Ntambi and Miyazaki 2003, 2004). Our previous studies showed that the skin of SCD1 $^{-/-}$ mice are deficient in triglycerides, cholesterol esters, and wax esters that have reduced content in C16:1n7 and C18:1n9 fatty acids (Miyazaki et al. 2001c). Recent studies have attributed the failure of SCD1 $^{-/-}$ mice to regulate adaptive thermogenesis and to protect against diet-induced obesity to the defective skin lipid barrier (Binczek et al. 2007). However, earlier studies had showed that the SCD1 deficient mice have a normal skin barrier but had reduced hydration due to a deficiency in glycerol (Sundberg et al. 2003). In our hands the defective skin lipid barrier is due to the deficiency of SCD2 isoform in the epidermis of neonates (Miyazaki et al. 2005). Further, there are studies showing that obese humans have increased TEWL (Loffler et al. 2002) supporting the contention that the hypermetabolism in SCD1 deficient mice may not be due to TEWL and an impaired epidermal lipid barrier.

It was unexpectedly found that skin retinol and retinoic acid (RA) levels as well as a marked retinoic-acid responsive genes due to skin *Scd1* deficiency were elevated (Flowers et al. 2011). This is reminiscent of *Dgat1*-deficient mice, which have been shown to have a reduced capacity for skin retinol esterification leading to elevated levels of skin retinol and RA, and resulting in cyclical hair loss (Shih et al. 2009). Severe alterations in retinol availability or RA levels can affect the regulation of adaptive thermogenesis and adiposity, two processes that are fundamental to energy balance (Bonet et al. 2003). For example, retinaldehyde inhibits adipogenesis and blocking retinaldehyde metabolism to RA reduces diet-induced obesity (Ziouzenkova et al. 2007). The synthesis of RA from its precursor retinol, or vitamin A, is a major mode in the regulation of RA levels in cells and tissues (Napoli 1999). To generate RA, retinol is oxidized in two sequential reactions, that are catalyzed by retinol and retinal dehydrogenases (Napoli 1999), whose activities regulate RA homeostasis. The balance between retinol and retinyl esters may also regulate the availability of retinol for these reactions. Retinol esterification is carried out by two distinct enzymatic activities. One is mediated by lecithin retinol acyltransferase (LRAT), which catalyzes the covalent joining of a fatty acyl moiety from lecithin (phosphatidylcholine) to retinol that is bound to cellular retinol-binding protein-1 (CRBP-1; MacDonald and Ong 1988). Unbound retinol is esterified by acyl-CoA: retinol acyltransferase (ARAT), an activity that has been very difficult to purify. However, DGAT-1 has ARAT activity and it has recently been shown that DGAT-1 functions as the major ARAT in murine skin (Yen et al. 2005) where it acts to maintain retinoid homeostasis and prevent retinoid toxicity. When dietary retinol is abundant, DGAT-1 deficiency results in elevated levels of RA in skin and cyclical hair loss; both are prevented by dietary retinol deprivation (Shih et al. 2009). Since the skin and metabolic phenotypes of DGAT-1 and SCD1 knockout mice are similar (Smith et al. 2000), it is very likely that the endogenous MUFA produced by SCD1 are not available for esterification of excess unbound retinol to retinyl esters by DGAT-1 leading to accumulation of retinol. Both DGAT-1 and SCD-1 are therefore important for handling free retinol that exceeds the capacity of LRAT. Some of the excess retinol is subsequently converted to RA, which initiates the signaling that leads to the phenotype observed in both the SCD1 and DGAT-1 knockout mice. Interestingly, SCD1-gene expression is activated by retinol in liver (Miller et al. 1997) and muscle, possibly to enhance esterification of excess retinol. In addition, we demonstrated previously that SCD1 and DGAT-2 colocalize in ER of liver (Man et al. 2006). It is very possible SCD1 and DGAT-1 colocalize in ER of the sebaceous glands.

Despite reducing liver retinol stores by more than 400-fold with RD-intervention in both Lox and SKO mice, skin RA levels and normal hair growth were not normalized in SKO mice subjected to a retinol-deficient diet intervention (Flowers et al. 2011). In contrast, studies by Shih et al. in *Dgat1*-deficient mice were able to normalize skin RA levels and hair growth with a similar intervention (Shih et al. 2009). This may indicate that lack of *Scd1* causes a more pronounced alteration in RA metabolism. Consistent with this notion, the skin phenotype elicited by *Scd1* deficiency is more severe than that observed due to *Dgat1* deficiency

(Shih et al. 2009). The persistent elevation in skin RA levels may also be due to decreased RA catabolism. We observed robust suppression of *Cyp2e1* and *Cyp1a1*, which encode two key cytochrome P450 enzymes that catalyze RA 4-hydroxylase activity in murine skin (Roos et al. 1998; Jugert et al. 1994). Alternatively, non-retinoid pathways may also be contributing to the skin phenotype. MUFAs are also important for cellular cholesterol ester synthesis (Zheng et al. 2001) and SKO mice have elevated levels of skin free cholesterol (Miyazaki et al. 2001c). Thus, a decrease in MUFA availability may affect both cholesterol and retinol homeostasis, in addition to triglyceride synthesis.

The therapeutic and pathological effects of retinoids on the skin have been known for many years. In vivo, both topical and orally administered retinoids stimulate keratinocyte proliferation and differentiation, resulting in increased number of epidermal cell layers and epidermal thickness, widening of spaces between keratinocytes and changes in the thickness and organization of the stratum corneum (Fisher and Voorhees 1996; Elias 1987). Furthermore, retinoid treatment causes a dose-dependent increase in TEWL, potentially due to loosening and fragility of the stratum corneum (Reichrath et al. 2007; Elias et al. 1981). The high responsiveness of various skin cells, such as epidermal keratinocytes, follicular keratinocytes, and sebocytes, to retinoid treatment is explained by their prominent expression of enzymes, binding proteins, and nuclear receptors involved in RA synthesis and signaling (Everts et al. 2007).

One of the most well-studied retinoid drugs is isotretinoin (13-*cis* RA), which likely acts as a prodrug that becomes selectively activated in the sebocyte after isomerization to tretinoin (all-*trans* RA; Tsukada et al. 2000). Isotretinoin treatment causes apoptosis and reduced size of the sebaceous gland, and the sebocytes appear undifferentiated and have decreased lipid accumulation (Nelson et al. 2006, 2009). Skin biopsies taken from patients treated with isotretinoin also showed high expression of *LCN2* (lipocalin-2; Flo et al. 2004). In addition, isotretinoin treatment of SEB-1 human sebocyte cultures causes cell cycle arrest and apoptosis, concomitant with increased expression of *LCN2* (Flo et al. 2004; Nelson et al. 2008). The *LCN2* promoter has binding sites for both RAR and RXR, suggesting that the toxic effect of RA on sebocytes is occurring via a lipocalin-2-mediated mechanism (Nelson et al. 2008). Since we observed a robust 27.7-fold elevation of *Lcn2* in the skin of SKO mice, it is possible that the primary disturbance in the skin of SKO mice is RA-induced LCN2 leading to sebocyte dysfunction and sebaceous gland atrophy.

In summary, our knowledge of the types of lipids present in skin, and in some cases, of their biosynthetic enzymes, is considerable. However, the individual roles of major lipid species such as ceramides, cholesterol esters, and wax esters in the tissue remain unknown, and in many cases, the fatty acids components, sub cellular locations, and regulation of the metabolic pathways that produce these molecules are unfamiliar. To gain insight into these areas, it is proposed that mice that lack SCD1, SCD2, and SCD3 should be characterized and studied with respect to skin barrier function, sebaceous gland function and lipid metabolism.

Acknowledgements Our research has been supported by NIH grant NIDDK-RO162388, AHA Grant-in Aid, Wisconsin Affiliate, USDA (Hatch) grant # 142-4306.

References

Binczek E, Jenke B, Holz B, Gunter RH, Thevis M, Stoffel W. Obesity resistance of stearoyl-CoA desaturase-deficient (SCD1-/-) mouse results from disruption of epidermal lipid barrier and adaptive thermoregulation. J Biol Chem. 2007;388:405–18.

Bonet ML, Ribot J, Felipe F, Palou A. Vitamin A and the regulation of fat reserves. Cell Mol Life Sci. 2003;7:1311–21.

Cases S, Stone SJ, Zhou P, Yen E, Tow B, Lardizabal KD, Voelker T, Farese RV. Cloning of DGAT2, a second mammalian diacylglycerol acyltransferase, and related family members. J Biol Chem. 2001;276:38870–76.

Casimir DA, Ntambi JM. cAMP activates the expression of stearoyl-CoA desaturase gene 1 during early preadipocyte differentiation. J Biol Chem. 1996;271:29847–53.

Christianson JL, Nicoloro S, Straubhaar J, Czech MP. Stearoyl-CoA desaturase 2 is required for Peroxisome proliferator-activated receptor gamma expression and adipogenesis in cultured 3T3-L1 cells. J Biol Chem. 2008;283:2906–16.

Coulombe PA, Kopan R, Fuchs E. Expression of keratin K14 in the epidermis and hair follicle: insights into complex programs of differentiation. J Cell Biol. 1989;109:2295–312.

Doering T, Holleran WM, Potratz A, Vielhaber G, Elias PM, Suzuki K, Sandhoff K. Sphingolipid activator proteins are required for epidermal permeability barrier formation. J Biol Chem. 1999;274:11038–45.

Doering T, Brade H, Sandhoff K. Sphingolipid metabolism during epidermal barrier development in mice. J Lipid Res. 2002;43:1727–33.

Downing DT, Strauss JS. Synthesis and composition of surface lipids of human skin. J Invest Dermatol. 1974;162:228–44.

Downing DT, Stewart ME, Strauss JS. Changes in sebum secretions and the sebaceous gland. Clin Geriatr Med. 1989;5:109–14.

Elias PM. Retinoid effects on the epidermis. Dermatologica. 1987;175(Suppl 1):28–36.

Elias PM, Fritsch PO, Lampe M, Williams ML, Brown BE, et al. Retinoid effects on epidermal structure, differentiation, and permeability. Lab Invest. 1981;44:531–40.

Everts HB, Sundberg JP, King LE Jr, Ong DE. Immunolocalization of enzymes, binding proteins, and receptors sufficient for retinoic acid synthesis and signaling during the hair cycle. J Invest Dermatol. 2007;127:1593–604.

Fisher GJ, Voorhees JJ. Molecular mechanisms of retinoid actions in skin. FASEB J. 1996;10:1002–13.

Flo TH, Smith KD, Sato S, Rodriguez DJ, Holmes MA, et al. Lipocalin 2 mediates an innate immune response to bacterial infection by sequestrating iron. Nature. 2004;432:917–21.

Flowers MT, Paton CM, O'Byrne SM, Scheisser K, Dawson J, Blaner WS, Kendziorski C, Ntambi JM. Metabolic changes in skin caused by Scd1 deficiency: a focus on retinol metabolism. PLoS ONE. 2011;5(5):e19734.

Ge L, Gordon JS, Hsuan C, Prouty SM. Identification of the delta-6 desaturase of human sebaceous glands: expression and enzyme activity. J Invest Dermatol. 2003;120:707–14.

Hansen HS, Jensen B. Essential function of linoleic acid esterified in acylglucosylceramide and acylceramide in maintaining the epidermal water permeability barrier. Evidence from feeding studies with oleate, linoleate, arachidonate, columbinate and alpha-linolenate. Biochim Biophys Acta. 1985;834:357–63.

Hellerstein MK. De novo lipogenesis in humans: metabolic and regulatory aspects. Eur J Clin Nutr. 1999;53:S53–65.

Hughes BR, Morris C, Cunliffe WJ, Leigh IM. Keratin expression in pilosebaceous epithelia in truncal skin of acne patients. Br J Dermatol. 1996;134:247–56.

Jugert FK, Agarwal R, Kuhn A, Bickers DR, Merk HF, et al. Multiple cytochrome P450 isozymes in murine skin: induction of P450 1A, 2B, 2E, and 3A by dexamethasone. J Invest Dermatol. 1994;102:970–75.

Lampe MA, Williams ML, Elias PM. Human epidermal lipids: characterization and modulations during differentiation. J Lipid Res. 1983;24:131–40.

Loffler H, Aramaki JU, Effendy I. The influence of body mass index on skin susceptibility to sodium lauryl sulphate. Skin Res Technol. 2002;8:19–22.

MacDonald PN, Ong DE. Acyl-CoA-independent esterification of retinol bound to cellular retinol-binding protein (type II) by microsomes from rat small intestine. J Biol Chem. 1988;263:12478–82.

Madison KC. Barrier function of the skin: "La Raison d'Etre" of epidermis. J Invest Dermatol. 2003;121:231–41.

Man WC, Miyazaki M, Chu K, Ntambi JM. Co-localization of SCD1 and DGAT2: implying preference for endogenous monounsaturated fatty acids in triglyceride synthesis. J Lipid Res. 2006;47:1928–39.

Marcelo CL, Duell EA, Rhodes LM, Dunham WR. In vitro model of essential fatty acid deficiency. J Invest Dermatol. 1992;99:703–8.

Melton JL, Wertz PW, Swartzendruber DC, Downing DT. Effects of essential fatty acid deficiency on epidermal O-acylsphingolipids and transepidermal water loss in young pigs. Biochim Biophys Acta. 1987;921:191–7.

Miller CW, Waters KM, Ntambi JM. Regulation of hepatic stearoyl-CoA desaturase gene 1 by vitamin A. Biochem Biophys Res Commun. 1997;231:206–10.

Miyazaki M, Ntambi JM. Physiological role of stearoyl-CoA desaturase. Prostaglandins Leukot Essent Fatty Acids. 2003;68:113–21.

Miyazaki M, Kim YC, Gray-Keller MP, Attie AD, Ntambi JM. The biosynthesis of hepatic cholesterol esters and triglycerides is impaired in mice with a disruption of the gene for stearoyl-CoA desaturase 1. J Biol Chem. 2000;275:30132–8.

Miyazaki M, Kim YC, Ntambi JM. A lipogenic diet in mice with a disruption of the stearoyl-CoA desaturase 1 gene reveals a stringent requirement of endogenous monounsaturated fatty acids for triglyceride synthesis. J Lipid Res. 2001a;42:1018–24.

Miyazaki M, Kim HJ, Man WC, Ntambi JM. Oleoyl-CoA is the major de novo product of stearoyl-CoA desaturase 1 gene isoform and substrate for the biosynthesis of the Harderian gland 1-alkyl-2,3-diacylglycerol. J Biol Chem. 2001b;276:39455–61.

Miyazaki M, Man WC, Ntambi JM. Targeted disruption of stearoyl-CoA desaturase1 gene in mice causes atrophy of sebaceous and meibomian glands and depletion of wax esters in the eyelid. J Nutr. 2001c;131:2260–8.

Miyazaki M, Gomez FE, Ntambi JM. Lack of stearoyl-CoA desaturase-1 function induces a palmitoyl-CoA delta-6 desaturase and represses the stearoyl-coA desaturase-3 gene in the preputial glands of the mouse. J Lipid Res. 2002;43:2146–54.

Miyazaki M, Dobrzyn A, Elias PM, Ntambi JM. Stearoyl-CoA desaturase-2 gene expression is required for lipid synthesis during early skin and liver development. Proc Natl Acad Sci U S A. 2005;102:12501–6.

Napoli JL. Interactions of retinoid binding proteins and enzymes in retinoid metabolism. Biochim Biophys Acta. 1999;1440:139–62.

Nelson AM, Gilliland KL, Cong Z, Thiboutot DM. 13-cis Retinoic acid induces apoptosis and cell cycle arrest in human SEB-1 sebocytes. J Invest Dermatol. 2006;126:2178–89.

Nelson AM, Zhao W, Gilliland KL, Zaenglein AL, Liu W, et al. Neutrophil gelatinase-associated lipocalin mediates 13-cis retinoic acid-induced apoptosis of human sebaceous gland cells. J Clin Invest. 2008;118:1468–78.

Nelson AM, Zhao W, Gilliland KL, Zaenglein AL, Liu W, et al. Temporal changes in gene expression in the skin of patients treated with isotretinoin provide insight into its mechanism of action. Dermatoendocrinology. 2009;1:177–87.

Nikkari N. Comparative chemistry of sebum. J Invest Dermatol. 1974;62:257–67.

Ntambi JM. Regulation of stearoyl-CoA desaturase by polyunsaturated fatty acid and cholesterol. J Lipid Res. 1999;40:1549–58.

Ntambi JM, Miyazaki M. Recent insights into stearoyl-CoA desaturase 1. Curr Opin Lipidol. 2003;14:255–61.

Ntambi JM, Miyazaki M. Regulation of stearoyl-CoA desaturases and role in metabolism. Prog Lipid Res. 2004;43:91–104.

Pappas A, Anthonavage M, Gordon JS. Metabolic fate and selective utilization of major fatty acids in human sebaceous gland. J Invest Dermatol. 2002;118:164–71.

Reichrath J, Lehmann B, Carlberg C, Varani J, Zouboulis CC. Vitamins as hormones. Horm Metab Res. 2007;39:71–84.

Roos TC, Jugert FK, Merk HF, Bickers DR. Retinoid metabolism in the skin. Pharmacol Rev. 1998;50:315–33.

Schmuth M, Elias PM, Hanley K, Lau P, Moser A, Willson TM, Bikle DD, Feingold KR. The effect of LXR activators on AP-1 proteins in keratinocytes. J Invest Dermatol. 2004;123:41–8.

Sessler AM, Ntambi JM. Polyunsaturated fatty acid regulation of gene expression. J Nutr. 1998;128:923–26.

Shih MY, Kane MA, Zhou P, Yen CL, Streeper RS, et al. Retinol esterification by DGAT1 is essential for retinoid homeostasis in murine skin. J Biol Chem. 2009;284:4292–9.

Smith SJ, Cases S, Jensen DR, Sande E, Tow B, Sanan D, Raber J, Eckel RH, Farese RV Jr. Obesity resistance and multiple mechanisms of triglyceride synthesis in mice lacking DGAT. Nat Genet. 2000;25:87–90.

Stone SJ, Myers HM, Watkins SM, Brown BE, Feingold KR, Elias PM, Farese RV. Jr. Lipopenia and skin barrier abnormalities in DGAT2-deficient mice. J Biol Chem. 2004;279:11767–76.

Sundberg JP, Boggess D, Sundberg BA, Eilertsen K, Parimoo S, Filippi M, Stenn K. Asebia-2J (Scd1(ab2J)): a new allele and a model for scarring alopecia. Am J Pathol. 2000;156:2067–75.

Sundberg JP, Feingold K, Elias PA. Glycerol regulates stratum corneum hydration in sebaceous gland deficient (asebia) mice. J Invest Dermatol. 2003;120:728–73.

Tsukada M, Schroder M, Roos TC, Chandraratna RA, Reichert U, et al. 13-cis retinoic acid exerts its specific activity on human sebocytes through selective intracellular isomerization to all-trans retinoic acid and binding to retinoid acid receptors. J Invest Dermatol. 2000;115:321–7.

Tu CL, Chang W, Bikle DD. The extracellular calcium-sensing receptor is required for calcium-induced differentiation in human keratinocytes. J Biol Chem. 2001;276:41079–85. A

Wertz PW. Lipids and barrier function of the skin. Acta Dermatol Venereol. 2000;208:7–11.

Wertz PW, Downing DT. Metabolism of linoleic acid in porcine epidermis. J Lipid Res. 1990;31:1839–44.

Yen C-LE, Monetti M, Burri BJ, Farese RV Jr. The triacylglycerol synthesis enzyme DGAT1 also catalyzes the synthesis of diacylglycerols, wax esters, and retinyl esters. J Lipid Res. 2005;46(7):1502–11.

Zheng Y, Eilertsen KJ, Ge L, Zhang L, Sundberg JP, Prouty SM, Stenn KS, Parimoo S. Scd1 is expressed in sebaceous glands and is disrupted in the asebia mouse. Nat Genet. 1999;23:268–70.

Zheng Y, Prouty SM, Harmon A, Sundberg JP, Stenn KS, Parimoo S. Scd3–a novel gene of the stearoyl-CoA desaturase family with restricted expression in skin. Genomics. 2001;71:182–91.

Ziouzenkova O, Orasanu G, Sharlach M, Akiyama TE, Berger JP, Viereck J, Hamilton JA, Tang G, Dolnikowski GG, Vogel S, et al. Retinaldehyde represses adipogenesis and diet-induced obesity. Nat Med. 2007;6:695–702.

Chapter 16
Triglyceride Synthesis Enzymes in Skin/Fur

Carrie Grueter

Core Messages

- The final reaction of triglyceride synthesis is catalyzed by acyl-CoA:diacylglycerol acyltransferase (DGAT) enzymes, DGAT1 and DGAT2.
- Despite their ability to similarly catalyze TG synthesis, DGAT1 and DGAT2 belong to two separate gene families, share neither DNA nor protein sequence homology and differ in their biochemical, cellular, and physiological functions.
- DGAT2 only catalyzes the synthesis of TG whereas DGAT1 also catalyzes the synthesis of diacyalglcyerols, retinyl esters, and wax esters.
- DGAT1 functions primarily as a wax synthase and an acyl-CoA:retinol acyltransferase (ARAT) to regulate normal fur lipid composition and retinoid homeostasis, respectively.
- DGAT2 functions as a TG-synthesizing enzyme in murine skin to regulate acylaceramides, which are required for normal barrier function.

Introduction

In this chapter, we will discuss the two enzymes known to synthesize triacylglycerol (or triglyceride, TG) and their role in murine skin and fur physiology. However, it is imperative to note that these enzymes were named based on the initial discovery that they possessed the ability to synthesize TG. Moreover, it is likely that the additional enzymatic activities of one of the enzymes play more important roles in regulating lipid homeostasis in the skin than its ability to synthesize TG.

C. Grueter (✉)
Vanderbilt University Medical Center, Nashville, USA
e-mail: carrie.grueter@Vanderbilt.Edu

A. Pappas (ed.), *Lipids and Skin Health,* DOI 10.1007/978-3-319-09943-9_16

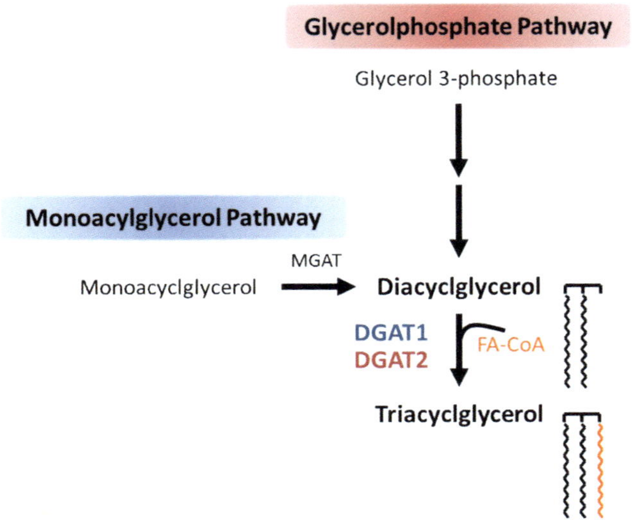

Fig. 16.1 Acyl-CoA:diacylglycerol acyltransferase (*DGAT*) enzymes and triacylglycerol (or triglyceride) biosynthesis. There are two major pathways for triacylglycerol synthesis: the glycerol phosphate pathway (or Kennedy pathway) and the monoacylglycerol pathway. DGAT enzymes are involved the final reaction of both pathways by catalyzing an ester linkage between the free hydroxyl group of diacylglycerol and a fatty acyl-CoA (*FA-CoA*; the active form of fatty acids) to form triacylglycerol. *MGAT* acyl CoA:monoacylglycerol acyltransferase

Triglyceride Synthesis

Two major pathways for TG biosynthesis, elucidated in the 1950s and 1960s are known as the monoacylglycerol pathway (Yen et al. 2008) and the glycerol phosphate pathway, or the Kennedy pathway (Kennedy 1957; Fig.16.1). The monoacylglycerol pathway is the dominant pathway in cell types such as enterocytes, hepatocytes, and adipocytes that participate in the reesterification of hydrolyzed dietary fats, whereas the glycerol phosphate pathway is present in most cells types (Kayden et al. 1967). In the final reaction of both pathways, diacylglycerol and a fatty acyl-CoA, the "active" form of fatty acids (Coleman et al. 2002), are covalently bound to form TG. This reaction is catalyzed by the acyl-CoA:diacylglycerol acyltransferase (DGAT, E.C.2.3.1.20) enyzmes DGAT1 and DGAT2.

Both DGAT1 and DGAT2 reside in the endoplasmic reticulum, which is believed to be the main site of newly synthesized TG (Weiss et al. 1960). TGs synthesized by DGAT enzymes are then either stored in cytosolic lipid droplets or, in some organs such as the liver and small intestine, secreted as components of lipoproteins. Although several models have emerged (Guo et al. 2009; Walther and Farese 2009) the exact mechanism(s) by which TGs are deposited into lipid droplets remains unclear.

DGAT Enzymes

DGAT1

Despite their ability to similarly catalyze TG synthesis, DGAT1 and DGAT2 belong to two separate gene families, share neither DNA nor protein sequence homology and differ in their biochemical, cellular, and physiological functions (Yen et al. 2008). DGAT1 belongs to a large family of membrane-bound *O*-acyltransferases known as the MBOAT family. The genes encoding human and murine DGAT1 were identified by their sequence homology to acyl-CoA:cholesterol acyltransferase enzymes but were shown to encode a protein having DGAT activity (Cases et al. 1998). In humans, the DGAT1 gene is located on chromosome 8 and comprises 17 exons spanning 10.62 kb. In most species, the encoded protein has a predicated molecular weight of ~55 kDa and consists of about 500 amino acids. In humans, DGAT1 is ubiquitously expressed, with the highest mRNA expression in tissues that make considerable amounts of TG, such as the small intestine, liver, and adipose tissue (Cases et al. 1998). In mice, the expression pattern is similar to that of humans, except in that mRNA levels are relatively low in liver (Cases et al. 1998). As for the mRNA expression of DGAT specifically in the skin, it is scarcely detectable in the epidermis or dermis of neonatal mice (Stone et al. 2004). Although it is expressed at very low levels in the epidermis of adult mice, it is highly expressed in sebaceous glands (Chen et al. 2002a).

Regarding biochemical function, DGAT1 not only catalyzes the synthesis of TGs, but also the synthesis of diacyalglcyerols, retinyl esters, and wax esters via its acyl-CoA:monoacylglycerol acyltransferase (MGAT), acyl-CoA:retinol acyltransferase (ARAT) and wax synthase activities, respectively (Yen et al. 2005a). In effect, studies in vivo show that its ARAT and wax synthase activities play an important role in skin and hair physiology (Chen et al. 2002a; Shih et al. 2009). Furthermore, DGAT1 functions as such a potent ARAT that it may have been named differently had it been initially identified to possess this activity (Yen et al. 2005a).

DGAT2

DGAT2 belongs to the DGAT2/MGAT gene family (Cases et al. 2001). In addition to DGAT2 and MGATs 1, 2, and 3 (Yen et al. 2002; Yen and Farese 2003; Cheng et al. 2003), this family also consists of wax monoester synthases (Cheng and Russell 2004; Turkish et al. 2005) and a multifunctional *O*-acyltransferase (Yen et al. 2005b). Interestingly, this multifunctional enzyme catalyzes the synthesis of diacyalglcyerols, retinyl esters, and wax esters (Yen et al. 2005b) similarly to the multiple enzymatic activities of DGAT1.The fact that members of the DGAT2/MGAT gene family comprise enzymes that possess wax synthase activity highlights the overlapping similarities between the gene families of DGAT1 and DGAT2.

In humans, the DGAT2 gene is located on chromosome 11 and comprises eight exons spanning 42.03 kb. In most species, the encoded protein has a predicated molecular weight ranging from 40 to 44 kDa and consists of 350–400 amino acids. Like DGAT1, DGAT2 is expressed in most mammalian tissues, with the highest mRNA levels in organs that make considerable amounts of TG (small intestine, liver, and adipose tissue; Cases et al. 2001). In mice, DGAT2 mRNA is more widely expressed than humans with less variation in levels across tissues (Cases et al. 2001). As for the mRNA expression of DGAT2 specifically in the skin, it is expressed at higher levels in the epidermis than dermis of neonatal mice (Stone et al. 2004). In the epidermis of adult mice, DGAT2 mRNA expression is much more robust than DGAT1 and the majority of its expression is accounted for in the basal layer (Stone et al. 2004).

DGAT-Deficient Mice

Understanding of the physiologic function(s) of DGAT enzymes chiefly arose from genetically modified mice. The generation of mice lacking *Dgat1* (*Dgat1*$^{-/-}$) and *Dgat2* (*Dgat2*$^{-/-}$) provided the opportunity to study the importance of TG synthesis in systemic- and tissue-specific metabolic homeostasis. Importantly, observations from these mice demonstrate the distinct biochemical functions of DGAT1 and DGAT2, which are mirrored in their divergent physiological roles. Moreover, mRNA expression is differentially regulated under various metabolic conditions and the absence of one DGAT enzyme is not compensated by the other in either DGAT-deficient mouse model.

Dgat1$^{-/-}$ mice are viable and have been studied extensively. These mice exhibit multiple phenotypes that include extended longevity (Streeper 2012), resistance to diet-induced obesity (due to increased energy expenditure) and glucose intolerance (Smith et al. 2000), enhanced insulin and leptin sensitivity (Chen et al. 2002b), impaired mammary gland development (Cases et al. 2004), and skin and fur abnormalities (Chen et al. 2002a). DGAT2-deficient mice, on the-other-hand, have been studied significantly less as homozygous DGAT2-deficient mice survive only a few hours after birth (Stone et al. 2004) and heterozygous DGAT2-deficient mice, although viable, exhibit few detectable phenotypes (unpublished). The early mortality of *Dgat2*$^{-/-}$ mice is contributed to by insufficient substrates required for energy metabolism and impaired permeability barrier function of the skin. Together, these findings suggest that DGAT2 is the dominant DGAT enzyme in regulating TG homeostasis.

Skin and Fur Abnormalities in Dgat1-Deficient Mice

Although *Dgat1*$^{-/-}$ mice have been primarily studied in regards to whole-body energy metabolism, important observations have been made in respect to DGAT1 function in skin and fur (Chen et al. 2002a; Shih et al. 2009). Adult *Dgat1*$^{-/-}$ mice develop dry fur and hair loss, which are associated with atrophic sebaceous glands and fur lipid abnormalities. Young *Dgat1*$^{-/-}$ mice exhibit normal fur appearance at weaning. However, after puberty (6–8 weeks of age) the fur of *Dgat1*$^{-/-}$ mice appears dry and less sheen compared to that of wild-type mice and subsequent hair loss is observed (Chen et al. 2002a). The hair loss phenotype is more prominent in young males than in females (Chen et al. 2002a) but is similar regardless of gender as it worsens with age (unpublished).

In a wild-type mouse, the sheen appearance of the fur is due to lipids (sebum) secreted from the sebaceous glands associated with hair follicles. In 6-week-old *Dgat1*$^{-/-}$ mice, the sebaceous glands are normal in appearance and located in a typical association with normal hair follicles. However, in 3-month-old *Dgat1*$^{-/-}$ mice, sebaceous glands are atrophied along the ventral and dorsal surfaces and few are associated with hair follicles. Although fur lipids from both wild-type and *Dgat1*$^{-/-}$ mice are composed of similar levels of sterol esters, free cholesterol, and surprisingly, triglycerides, *Dgat1*$^{-/-}$ mice lack several fur lipids, with the most prominent of these being type II wax diesters. Like the hair loss phenotype, the difference in fur lipid composition is also age-dependent where the differences are less striking in young mice but more pronounced in adults. Of note, the fur of heterozygous (*Dgat1*$^{+/-}$) mice is normal in appearance throughout their lifespan.

As a result of atrophied sebaceous glands and decreased fur lipids, *Dgat1*$^{-/-}$ mice exhibit impaired water repulsion and hypothermia following water immersion. The fur of *Dgat1*$^{-/-}$ mice retains significantly more water compared to wild-type controls, which are almost completely dry, 5 min after water immersion (Chen et al. 2002a). Consequently, *Dgat1*$^{-/-}$ mice become hypothermic and this persists for an hour following immersion.

Farese and colleagues found the skin and fur phenotypes of *Dgat1*$^{-/-}$ mice to be unexpectedly modulated by the adipocyte-secreted hormone, leptin (Chen et al. 2002a). Leptin is a peptide hormone known to reduce food intake and enhance energy expenditure through its receptors in the central nervous system (specifically in the hypothalamus; Morton and Schwartz 2011) but also functions to regulate energy metabolism directly through its receptors in peripheral tissues, including the skin (Frank et al. 2000; Poeggeler et al. 2010). When DGAT1 deficiency is introduced into mice with a spontaneous mutant in the leptin gene (*Dgat1*$^{-/-}$*ob*/*ob*), the hair loss, water repulsion, and sebaceous gland abnormalities are almost absent. However, it is important to note that the sebaceous gland size and fur lipid levels are slightly decreased in *Dgat1*$^{-/-}$*ob*/*ob* mice compared to controls (*Dgat1*$^{+/+}$ *ob*/*ob*). Following 2 weeks of peripheral *or* central leptin infusion, the size of sebaceous glands and fur lipid levels are slightly reduced in controls but both are markedly reduced in *Dgat1*$^{-/-}$*ob*/*ob* mice. Furthermore, these changes are reverted to pretreatment states 2 weeks after leptin administration is withdrawn. Since both peripheral and central administrations of leptin have similar effects on sebaceous glands and fur lipid composition, it is reasonable that these observations can be mediated entirely through the hypothalamus. One plausible mechanism by which leptin centrally modulates the skin and fur phenotypes of *Dgat1*$^{-/-}$ mice is by activation of the sympathetic nervous system to suppress the expression of DGAT2. Indeed, mice deficient in DGAT1 are known to exhibit enhanced leptin sensitivity (Chen et al. 2002b) and DGAT2 mRNA is significantly decreased in the skin of these mice (Chen et al. 2002a). Moreover, compared to wild-type controls, DGAT2 mRNA expression is increased at similar levels in the skin of leptin-deficient mice despite the presence or absence of DGAT1.

Upon further investigation of the hair loss observed in *Dgat1*$^{-/-}$ mice, Farese and colleagues discovered this phenotype to also be, in part, due to impaired retinoid (retinol and its derivatives) homeostasis and that the ARAT activity of DGAT1 plays an important role in murine skin (Shih et al. 2009). As mentioned above, DGAT1 not only catalyzes the synthesis of TGs, but also functions as a potent ARAT synthesizing retinyl esters through the esterification of retinol (vitamin A) with fatty acyl-CoA substrates (Yen et al. 2005a). Retinol is obtained through the diet and mostly stored in the form of retinyl esters in the cytosolic lipid droplets of cells providing then a local source of retinol. Retinol is the precursor for retinoic acids (all-*trans* and 9-*cis*-retinoic acid), which are ligands for nuclear hormone receptors (retinoic acid receptor and retinoid-x receptor) and are known to affect sebaceous gland function (Zouboulis et al. 1991; Strauss et al. 1987) and hair growth (Bazzano et al. 1993). In whole skin of *Dgat1*$^{-/-}$ mice compared to controls, ARAT activity is significantly reduced by ~90% and levels of unesterified retinol and all-*trans* retinoic acid are increased by ~22 and 40%, respectively (Shih et al. 2009). Studies from *Dgat1*$^{-/-}$ mice suggest that when dietary retinol is sufficient, DGAT1 deficiency results in retinoic acid toxicity in the skin, which subsequently leads to cyclical hair loss. Consistent with retinoic acid toxicity, genes regulated by the retinoic acid receptor are also elevated in the skin of these mice. When dietary retinol is deficient however, neither retinoic acid toxicity nor hair loss is observed in *Dgat1*$^{-/-}$ mice. Moreover, mice deficient of DGAT1 specifically in the skin exhibit cyclical hair loss and increased expression of retinoic acid receptor target genes indicating that dysregulation of retinoid homeostasis is due loss of the enzyme in the skin rather than systemic changes in metabolism (Shih et al. 2009). Collectively, these findings reveal DGAT1 to be a major ARAT in murine skin and therefore essential for maintaining retinoid homeostasis and preventing retinoic acid toxicity-induced hair loss. Of note, it remains to be determined if the sebaceous gland and fur lipid phenotypes of *Dgat1*$^{-/-}$ mice are also due to the dysregulation of retinoid homeostasis.

Similarities Between Dgat1- and Scd1-Deficient Mice

The pleiotropic phenotype of *Dgat1*$^{-/-}$ mice is remarkably similar to mice that lack stearoyl-CoA desaturase1 (*Scd1*$^{-/-}$), an enzyme that catalyzes the synthesis of the delta-9-monounsaturated fatty acids (described in Chapter "Stearoyl-CoA Desaturases are Regulators of Lipid Metabolism in Skin" this volume). Like *Dgat1*$^{-/-}$ mice, *Scd1*$^{-/-}$ mice exhibit increased energy expenditure, resistance to diet-induced obesity and glucose intolerance, enhanced insulin and leptin sensitivity, sebaceous gland atrophy, and hair loss (Ntambi and Miyazaki 2003). Interestingly, mice deficient of SCD1 specifically in the skin (SKO) show increased energy expenditure and are protected from diet-induced obesity, recapitulating the global SCD1 deficiency phenotype. Furthermore, thorough analysis of the skin of SKO mice show reduced mRNA expression of enzymes involved in TG synthesis, including Dgat1, increased mRNA expression of retinoic acid regulated genes (Chapter "Stearoyl-CoA Desaturases are Regulators of Lipid Metabolism in Skin" this volume), and a robust increase in retinol and retinoic acid (Flowers et al. 2011). Given the similarities of altered epidermal retinoid metabolism and energy balance observed in both the global DGAT1- and skin-specific SCD1-deficent mice, it is possible that the skin-specific deletion of *Dgat1* could also result in increased energy expenditure and protection from diet-induced obesity. In effect, unpublished data suggest that the deletion of *Dgat1* specifically in the skin results in an increase in heat loss, energy expenditure and decreased adiposity on a regular chow diet. Collectively, one can speculate that a reduction of ARAT activity in the skin and subsequent retinoic acid toxicity results in a loss of insulating factors (fur and skin/fur lipids) and consequently increases thermogenesis in order to maintain normal core body temperature. This in-turn results in an increase in energy expenditure and therefore a reduction in adiposity. Further experimentation is required to determine if skin-specific Dgat1-deficient mice are protected from diet-induced obesity and if this effect is lost by a high-fat, retinol-deficient diet.

Skin Abnormalities in Dgat2-Deficent Mice

Studies from *Dgat2*$^{-/-}$ mice show that DGAT2 also plays an important role in the skin. As mentioned previously, *Dgat2*$^{-/-}$ mice survive only a few hours after birth which is contributed to, in part, by severe skin abnormalities (Stone et al. 2004). The skin of newborn *Dgat2*$^{-/-}$ mice is shiny, lacks elasticity, and exhibits impaired permeability barrier function leading to increased transepidermal water loss and rapid dehydration. While there is no evidence of abnormal epidermal differentiation, the structure of the skin from neonatal *Dgat2*$^{-/-}$ mice exhibits thinning of the epidermis, compact hyperkeratosis of affected stratum corneum and effacement of the epidermal rete ridges/papillary projections which leads to a flattened dermal–epidermial interface. In normal skin barrier function, lamellar bodies are secreted from keratinocytes in the stratum spinosum/granulosum into the stratum corneum, resulting in the formation of an impermeable, lipid-containing membrane that serves as a water barrier. In the epidermis of *Dgat2*$^{-/-}$ mice, examination by electron microscopy revealed a typical number of lamellar bodies but a reduction in the lamellar body content and the number of lamellar membranes in the stratum corneum extracellular space. Further analysis of the skin of *Dgat2*$^{-/-}$ mice showed a 96 % decrease in TG content while other lipids were similar to those of control mice. Assessment of the composition of these lipids revealed a significant reduction (>90 %) in linoleic acid containing TGs and free fatty acids. Also observed was a 60 % decrease in skin levels of acyl-ceramide, a skin lipid also composed of linoleic acid, which is thought to be required for the maintenance of the skin permeability barrier. Skin grafting experiments also suggest that the skin defects in *Dgat2*$^{-/-}$ mice are partially due to a lack of the enzyme systemically in addition to the loss specifically in the skin. Skin from *Dgat2*$^{-/-}$ and control mice grafted onto athymic nude mice showed similar fur development and transepidermal water loss 3 weeks following transplantation. Importantly, plasma levels of TG and free fatty acids were reduced by 64 and 80 %, respectively. Taken together, decreased triglyceride synthesis, via DGAT2 deficiency, leads to a dysregulation of systemic and epidermal lipid metabolism resulting in abnormal lamellar body secretory system, impaired skin permeability barrier function, and ultimately rapid dehydration.

Similarities Between Dgat2- and Scd2-Deficient Mice

Interestingly, the skin phenotype of *Dgat2*$^{-/-}$ mice is similar to mice that lack SCD2, an isoform of SCD that is important in lipid synthesis in early development and is required for survival (Miyazaki et al. 2005). *Scd2*$^{-/-}$ neonatal mice exhibit increased transepidermal water loss that associated with a reduction of linoleic acid incorporation into acyl-aceramides. Futhermore, only ~30 % of *Scd2*$^{-/-}$ mice survive to adulthood. One can speculate the possibility that SCD2 provides DGAT2 with monounsaturated fatty acids required for TG biosynthesis during early skin development and the maintenance of skin permeability barrier function.

In summary, studies from DGAT-deficient mice reveal that the enzymatic activities of DGAT1 and DGAT2 play important roles in regulating murine skin and fur physiology. Although DGAT1 and DGAT2 were initially discovered to possess diacylglycerol acyltransferase activity, it is likely that the additional enzymatic activities of DGAT1 may be dominant. At least in murine skin, it seems as though DGAT1 primarily functions as a wax synthase and an ARAT to regulate normal fur lipid composition and retinoid homeostasis whereas DGAT2 functions as a TG synthesizing enzyme to regulate acylaceramides, which are required for normal barrier function. Since the expression profile of DGAT1 and DGAT2 are similar to that of humans, it is likely that DGATs also play an important role in regulating lipid metabolism in the skin of humans.

References

Bazzano G, et al. Effect of retinoids on follicular cells. J Invest Dermatol. 1993;101:138–42.

Cases S, et al. Identification of a gene encoding an acyl CoA: diacylglycerol acyltransferase, a key enzyme in triacylglycerol synthesis. Proc Natl Acad Sci U S A. 1998;95:13018–23.

Cases S, et al. Cloning of DGAT2, a second mammalian diacylglycerol acyltransferase, and related family members. J Biol Chem. 2001;276:38870–6.

Cases S, et al. Development of the mammary gland requires DGAT1 expression in stromal and epithelial tissues. Development. 2004;131:3047–55.

Cheng JB, Russell DW. Mammalian wax biosynthesis. II. Expression cloning of wax synthase cDNAs encoding a member of the acyltransferase enzyme family. J Biol Chem. 2004;279:37798–807.

Chen HC, Smith SJ, Tow B, Elias PM, Farese RV Jr. Leptin modulates the effects of acyl CoA:diacylglycerol acyltransferase deficiency on murine fur and sebaceous glands. J Clin Invest. 2002a;109:175–81.

Chen HC, et al. Increased insulin and leptin sensitivity in mice lacking acyl CoA:diacylglycerol acyltransferase 1. J Clin Invest. 2002b;109:1049–55.

Cheng D, et al. Identification of acyl coenzyme A: monoacylglycerol acyltransferase 3, an intestinal specific enzyme implicated in dietary fat absorption. J Biol Chem. 2003;278:13611–4.

Coleman RA, Lewin TM, van Horn CG, Gonzalez-Baró MR. Do long-chain acyl-CoA synthetases regulate fatty acid entry into synthetic versus degradative pathways?. J Nutr. 2002;132:2123–6.

Flowers MT, et al. Metabolic changes in skin caused by Scd1 deficiency: a focus on retinol metabolism. PloS ONE. 2011;6:e19734.

Frank S, Stallmeyer B, Kämpfer H, Kolb N, Pfeilschifter J. Leptin enhances wound re-epithelialization and constitutes a direct function of leptin in skin repair. J Clin Invest. 2000;106:501–9.

Guo Y, Cordes KR, Farese RV Jr, Walther TC. Lipid droplets at a glance. J Cell Sci. 2009;122:749–52.

Kayden HJ, Senior JR, Mattson FH. The monoglyceride pathway of fat absorption in man. J Clin Invest. 1967;46:1695–703.

Kennedy EP. Metabolism of lipides. Ann Rev Biochem. 1957;26:119–48.

Miyazaki M, Dobrzyn A, Elias PM, Ntambi JM. Stearoyl-CoA desaturase-2 gene expression is required for lipid synthesis during early skin and liver development. Proc Natl Acad Sci U S A. 2005;102:12501–6.

Morton GJ, Schwartz MW. Leptin and the central nervous system control of glucose metabolism. Physiol Rev. 2011;91:389–411.

Ntambi JM, Miyazaki M. Recent insights into stearoyl-CoA desaturase-1. Curr Opin Lipidol. 2003;14:255–61.

Poeggeler B, et al. Leptin and the skin: a new frontier. Exp Dermatol. 2010;19:12–8.

Shih MY, et al. Retinol esterification by DGAT1 is eessential for retinoid homeostasis in murine skin. J Biol Chem. 2009;284:4292–9.

Smith S, et al. Obesity resistance and multiple mechanisms of triglyceride synthesis in mice lacking Dgat. Nat Genet. 2000;25:87–90.

Stone SJ, et al. Lipopenia and skin barrier abnormalities in DGAT2-deficient mice. J Biol chem. 2004;279:11767–76.

Strauss JS, Stewart ME, Downing DT. The effect of 13-cis-retinoic acid on sebaceous glands. Arch Dermatol. 1987;123:1538a-41.

Streeper R, Salomonis N, Cases S, Grueter C. Deficiency of the lipid synthesis enzyme, DGAT1, extends longevity in mice. Aging (Albany). 2012. http://www.ncbi.nlm.nih.gov/pmc/articles/PMC3292902/.

Turkish AR, et al. Identification of two novel human acyl-CoA wax alcohol acyltransferases: members of the diacylglycerol acyltransferase 2 (DGAT2) gene superfamily. J Biol Chem. 2005;280:14755–64.

Walther TC, Farese RV Jr. The life of lipid droplets. Biochim Biophys Acta. 2009;1791:459–66.

Weiss SB, Kennedy EP Kiyasu JY. The enzymatic synthesis of triglycerides. J Biol Chem. 1960;235:40–4.

Yen C, Farese R. MGAT2, a monoacylglycerol acyltransferase expressed in the small intestine. J Biol Chem. 2003;278;18532–7.

Yen CE, Stone SJ, Cases S, Zhou P, Farese RV Jr. Identification of a gene encoding MGAT1, a monoacylglycerol acyltransferase. Proc Natl Acad Sci U S A. 2002;99:8512–7.

Yen CE, Monetti M, Burri BJ, Farese RV Jr. The triacylglycerol synthesis enzyme DGAT1 also catalyzes the synthesis of diacylglycerols, waxes, and retinyl esters. J Lipid Res. 2005a;46:1502–11.

Yen CE 4th, Brown CH, Monetti M, Farese RV Jr. A human skin multifunctional O-acyltransferase that catalyzes the synthesis of acylglycerols, waxes, and retinyl esters. J Lipid Res. 2005b;46:2388–97.

Yen CE, Stone SJ, Koliwad S, Harris C, Farese RV Jr. Thematic review series: glycerolipids. DGAT enzymes and triacylglycerol biosynthesis. J Lipid Res. 2008;49:2283–301.

Zouboulis CC, et al. Effects of 13-cis-retinoic acid, all-trans-retinoic acid, and acitretin on the proliferation, lipid synthesis and keratin expression of cultured human sebocytes in vitro. J Invest Dermatol. 1991;96:792–7.

Chapter 17
Metabolic and Structural Functions of Lipoxygenases in Skin

Sabine Rosenberger, Gerhard Fürstenberger and Peter Krieg

Core Messages

- Lipoxygenases (LOX) play important roles as key enzymes in the biosynthesis of signaling molecules involved in the regulation of skin proliferation, differentiation, and cancer.
- The LOX isoenzymes 12R-LOX and epidermis-type lipoxygenase-3 (eLOX-3) are essential for the epidermal barrier function.
- Loss of 12R-LOX or eLOX-3 function causes ichthyosis in humans and results in neonatal death in mice due to severe barrier defects.
- 12R-LOX and eLOX-3 act in tandem to oxidize the linoleate moiety of ω-hydroxyacyl ceramide.
- This is an essential step in the formation of the cornified lipid envelope, an indispensable structural component of the permeability barrier function.

Abbreviations

ARCI	Autosomal recessive congenital ichthyoses
CE	Cornified cell envelope
CLE	Corneocyte lipid envelope
e12-LOX	Epidermis-type 12-lipoxygenase
EFA	Essential fatty acid(s)
eLOX-3	epidermis-type lipoxygenase-3

P. Krieg (✉) · S. Rosenberger · G. Fürstenberger
Genome Modifications and Carcinogenesis, Deutsches Krebsforschungszentrum, Im Neuenheimer Feld 280, 69120 Heidelberg, Germany
e-mail: p.krieg@dkfz.de

S. Rosenberger
e-mail: s.rosenberger@dkfz.de

G. Fürstenberger
e-mail: g.fuerstenberger@gmx.de

A. Pappas (ed.), *Lipids and Skin Health,* DOI 10.1007/978-3-319-09943-9_17

EOS	Esterified omega-hydroxyacyl-sphingosine(s)
(F)FA	(Free) fatty acid(s)
GlcCer	Glucosylceramide(s)
H(P)ETE	Hydro(pero)xyeicosatetraenoic acid
H(P)ODE	Hydro(pero)xyoctadecadienoic acid
HxA, HxB	Hepoxilin A or B
l12-LOX	Leukocyte-type 12-lipoxygenase
LB	Lamellar body
LOX	Lipoxygenases(s)
OS	Omega-hydroxyacyl-sphingosine(s)
p12-LOX	Platelet-type 12-lipoxygenase
PUFA	Polyunsaturated fatty acids
TEWL	Transepidermal water loss
VLC-FA	Very long chain fatty acid(s)

Introduction

The chief function of the mammalian skin is to provide a barrier between the external environment and the internal milieu of the host. The skin barrier protects from environmental insults including mechanical and chemical impacts, UV light, and pathogenic microorganisms and prevents the loss of water and electrolytes. The function of this epidermal permeability barrier mainly resides in the stratum corneum and is mediated by specialized structures referred to as bricks and mortar (Candi et al. 2005; Feingold 2009; Proksch et al. 2008). The stratum corneum consists of flat, dead cells, called corneocytes (the "bricks"), which are surrounded by an inner rigid protein envelope, the cornified cell envelope (CE), and an outer corneocyte lipid envelope (CLE), a lipid monolayer covalently bound to the CE. The corneocytes are embedded in an intercellular lipid matrix (the "mortar") which forms lamellar membranes with an unique composition of ceramides, cholesterol and free fatty acids (FFA).

From the early observation of Burr, that dietary deficiency of essential fatty acids (EFA) results in a characteristic scaly skin disorder and excessive water loss (Burr and Burr 1930), it became clear that fatty acid metabolism is critically involved in the normal process of forming and maintaining the water-impermeable skin barrier. The addition of certain fatty acids such as linoleic acid to the diet could reverse some of the cutaneous symptoms of EFA deficiency and several lines of evidence suggested that the curative effects in the EFA-deficient animal involve the lipoxygenase (LOX)-catalyzed conversion of polyunsaturated fatty acids (PUFA) to oxygenated products (Nugteren et al. 1985; Nugteren and Kivits 1987).

LOX are a class of nonheme iron containing dioxygenases that catalyze the stereo- and regio-specific incorporation of molecular oxygen into PUFA. The primary reaction products are hydroperoxides which are rapidly reduced into their hydroxyl analogs or are substrates for subsequent enzymatic transformations (for reviews see Brash 1999; Haeggstrom and Funk 2011). No fewer than six functional LOX genes are identified in humans and seven in mice. Classification of LOX is carried out according to their positional specificity of oxygen insertion into arachidonic acid, e.g., 5-, 8-, 12-, and 15-LOX and to the stereoselectivity of oxygen insertion, e.g., 12S-, 12R-LOX. An additional classification considers the prototypical tissue of their expression, e.g., platelet-type 12-LOX, leukocyte-type 12-LOX, and epidermis-type 12-LOX. Based on the phylogenetic relationship mammalian LOX are divided into four subfamilies: 5-LOX, platelet-type 12-LOX, 12/15-LOX including human reticulocyte-type 15-LOX-1 and mouse leukocyte-type 12-LOX, and epidermis-type LOX which compass epidermis-type 12-LOX, 15-LOX-2 and its mouse ortholog 8-LOX, 12R-LOX and epidermis-type LOX-3.

The Role of LOX in Epidermal Proliferation, Differentiation and Carcinogenesis

The discovery of the first mammalian LOX (Hamberg and Samuelsson 1974) initiated a more detailed analysis of LOX activities and their products in human and mouse skin (Hammarström et al. 1975). These studies revealed skin as tissue with abundant and diverse LOX metabolism as evidenced by the identification of a multiplicity of arachidonic acid metabolites such as 12-hydroxyeicosatetraenoic acid (12-HETE), 15-hydroxyeicosatetraenoic acid (15-HETE) hepoxilins, trioxilines, leukotrienes, and linoleic acid derivatives including 13-hydroxyoctadecadienoic acid (13-HODE) and 9-hydroxyoctadecadienoic acid (9-HODE) and corresponding metabolites of docosahexaenoic acid and higher PUFA. Moreover, LOX are known to oxygenate esterified structural lipids such as ceramides. The latter exhibits an essential role for the proper function of the skin barrier.

All known LOX genes are expressed in skin of humans and mice, as detected by polymerase chain reaction of the mRNA, immunohistochemistry, enzyme activity, and determination of products. Induction of LOX expression and activity has been associated with inflammatory processes, modulation of epithelial proliferation, and differentiation and cancer development in human and mouse skin (reviewed in Fürstenberger et al. 2006; Wang and Dubois 2010).

5-LOX-produced leukotrienes act as proinflammatory agents in both irritant-induced acute and chronic skin inflammation such as atopic dermatitis and psoriasis, while lipoxins have been shown to be involved in the resolution of skin inflammation (for review see Kantarci and Van Dyke 2003).

The p12-LOX and 12/15-LOX metabolites 12-HETE and 13-HODE, respectively, have been reported to modulate both keratinocyte proliferation and differentiation, in that 12-HETE stimulates proliferation and inhibits terminal differentiation of keratinocytes (Hagerman et al. 1997). In addition, selectively expressed in the germinal layer of human skin, p12-LOX was found to be overexpressed in hyperproliferative psoriatic skin (Hussain et al. 1994). On the other hand, 13-HODE, the linoleic acid-derived product of l12-LOX counteracts the effects of 12-HETE by downregulation of p12-LOX expression (Fischer et al. 1996; Yoo et al. 2008).

Modulation of differentiation and suppression of growth in keratinocytes and other epithelial cells is also brought about by the orthologous 8-LOX and 15-LOX-2 and their products (Muga et al. 2000). Accordingly, transgenic mice that constitutively express 8-LOX under the control of the loricrin promoter showed a more differentiated epidermal phenotype (Kim et al. 2005). Similarly, 15-HETE exhibited a differentiation-modulating activity in prostate epithelial cells through activation of nuclear receptor PPARγ (Bhatia et al. 2005; Tang et al. 2007). Induction of both 15-LOX-2 and 8-LOX expression in the mouse papilloma cell line 308 caused suppression of cell growth that was associated with inhibition of DNA synthesis along a p38 mitogen-activated protein kinase dependent pathway (Schweiger et al. 2007).

Emerging evidence suggests that LOX pathways are involved in carcinogenesis in skin and other epithelia. 5-LOX and p12-LOX have potential pro-carcinogenic activities, whereas 8-LOX/15-LOX-2 and l12-LOX are thought to exhibit anti-carcinogenic effects. In line with this is the observation that inhibition of 5-LOX delayed tumor development in various animal models and oral carcinogenesis in humans (Wang and Dubois 2010). Aberrant expression and activity of p12-LOX is widely found in human skin cancer and a consistent feature of experimental mouse skin tumors (Fürstenberger et al. 2006). Genetic ablation of p12-LOX led to a significant reduction of carcinoma formation suggesting a protumorigenic activity of p12-LOX (Virmani et al. 2001). On the other hand, permanent elevation of l12-LOX expression that goes along with aberrant 13-HODE formation brings about an inhibition of tumor development in the skin of mice. Epidermis-type 12-lipoxygenase (e12-LOX) expression was found to be downregulated in the course of mouse skin carcinogenesis suggesting an anti-tumorigenic effect of the latter (Müller et al. 2002). However, the precise physiological function of e12-LOX in normal mouse epidermis, especially considering its nonfunctioning in humans, remains to be determined. Similarly, conversion of benign papillomas to squamous cell carcinomas in mouse skin carcinogenesis was accompanied by a complete loss of 8-LOX expression suggesting downregulation of 8-LOX to be a prerequisite for malignant transformation (Bürger et al. 1999). In addition, mice that constitutively express 8-LOX under control of the loricin promoter showed a reduced multiplicity of benign papillomas (Muga et al. 2000). Similarly, the expression of the human orthologous 15-LOX-2 has been shown to be lost in the course of malignant progression of oral keratinocytes (Wang et al. 2006) and in neoplasia of several other epithelia including prostate suggesting that the enzyme may exhibit tumor suppressor function (Shappell et al. 1999). Thus, distinct LOX may exhibit antipodal functions in carcinogenesis (for review see Fürstenberger et al. 2006).

The Role of 12R-LOX/eLOX-3 in Epidermal Barrier Function

The discovery of two previously unknown LOX isoforms in skin that exhibit unconventional enzymatic features has paved the way to new insights into the role of LOX metabolism in epidermal barrier function.

12R-LOX and eLOX-3, Epidermal LOX with Unique Enzymatic Features

The identification of the unusual arachidonic acid metabolite 12R-HETE in psoriatic scales indicated the presence of a 12R-oxygenase in skin, which initially was suspected to belong to the cytochrome P450 family (Hammarström et al. 1975; Woollard 1986). In 1998, Brash and colleagues provided evidence that a 12R-LOX can account for the selective formation of 12R-HETE in psoriatic skin and cloned the cDNA for the human 12R-LOX. The homologous mouse cDNA was cloned by our group by using a PCR strategy with degenerate primers corresponding to highly conservative LOX sequences (Boeglin et al. 1998; Krieg et al. 1998).

Up to now, 12R-LOX is the only mammalian LOX that forms R-enantiomeric products. Human 12R-LOX oxygenates arachidonic acid with low catalytic activity exclusively to 12R-HPETE and few other PUFA at the ω-9 position. The mouse enzyme, in contrast, prefers arachidonic acid methyl ester but not the free acid as a substrate. Acidic pH, elevated calcium concentrations—conditions resembling the situation in suprabasal differentiated keratinocytes—and DMSO increase the catalytic activity of 12R-LOX, and under these conditions mouse 12R-LOX also oxygenates few other substrates (Boeglin et al. 1998; Siebert et al. 2001). 12R-LOX expression and activity is low in normal human skin but strongly elevated in the inflammatory and hyperproliferative conditions of psoriasis, in particular in psoriatic scales (Baer et al. 1995; Sun et al. 1998; Woollard 1987). Since the proinflammatory activity of 12R-HETE is weak, it has been suggested that its increased synthesis is the result of keratinocyte hyperproliferation (Brash et al. 2007). Recent data indicate that a linoleic acid ester rather than free arachidonic acid comes into consideration as 12R-LOX substrate. In fact, O-linoleoyl-ω-hydroxylacyl-sphingosine (esterified omega-hydroxyacyl-sphingosine (EOS) alias ceramide 1) was identified as natural substrate for both mouse and human12R-LOX (Zheng et al. 2011).

The other key LOX involved in the epidermal barrier function, eLOX-3, was cloned in our lab from mouse epidermis and thereafter from human keratinocytes (Kinzig et al. 1999; Krieg et al. 2001). The identification as a member of the mammalian LOX family was based on the primary sequence of the mouse protein displaying the closest similarity to 12R-LOX (54% identity) and 15-LOX-2 (51% identity). However, no catalytic activity of eLOX-3 could be detected when incubated under standard conditions with prototypical LOX substrates including esterified PUFA. In skin, eLOX-3 and 12R-LOX are coexpressed with the highest expression found in differentiated layers of epidermis indicating functional linkage and a role in terminal differentiation. By immunofluorescence analyses 12R-LOX and eLOX-3 proteins were found colocalized at the surface of the keratinocytes in the stratum granulosum of mouse skin (Epp et al. 2007). In situ hybridization showed induction of 12R-LOX expression in embryonic skin at embryonic day 15.5, the onset of skin development (Sun et al. 1998).

In 2003, the catalytic activity of eLOX-3 was unveiled by Brash and colleagues with the finding that eLOX-3 functions as a hydroperoxide isomerase that converts fatty acid hydroperoxides to specific hepoxilin-type derivatives, i.e., the

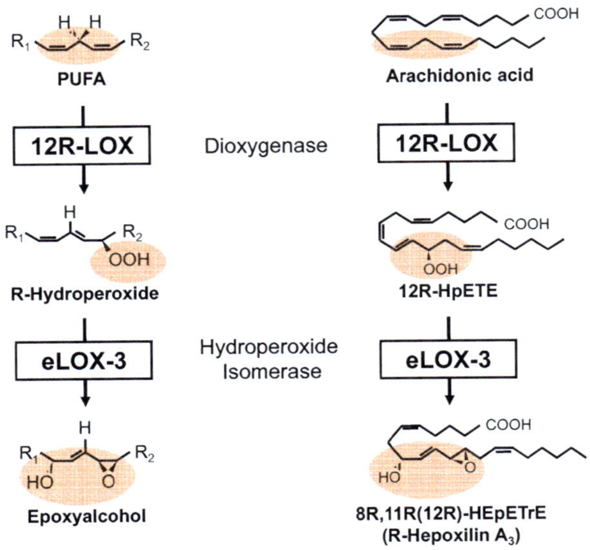

Fig. 17.1 12R-LOX and eLOX-3 act in sequence to form R-enantiomeric epoxyalcohols. The dioxygenase 12R-LOX converts PUFA containing a cis-cis-pentadiene structure to the corresponding R-hydroperoxide which in turn is converted by the hydoperoxide isomerase activity of eLOX-3 to the corresponding R-epoxyalcohol derivative. The prototypical LOX substrate arachidonic acid is metabolized via 12R-HPETE to a R-epoxyalcohol of the hepoxilin A-type

corresponding epoxyalcohol and keto-derivatives (Yu et al. 2003; Zheng and Brash 2010; Fig. 17.1). Human and mouse eLOX-3 transform a variety of LOX-derived hydroperoxides with somewhat different substrate specificities (Yu et al. 2006). Most importantly, eLOX-3 converts the natural 12R-LOX generated hydroperoxide of ceramide 1 to the corresponding epoxyalcohol-and keto ester of ceramide 1 (Zheng et al. 2011).

Loss-of-Function Mutations in the Genes of 12R-LOX and eLOX-3 are the Second Most Common Cause of ARCI

The discovery of a genetic connection between 12R-LOX and eLOX-3 and an inherited skin disorder points to an essential function of these LOX in the proper functioning of the epidermal barrier. In fact, a major breakthrough in understanding the physiological function of these LOX in skin was provided by a genetic study of Fischer and colleagues in 2002 (Jobard et al. 2002). They showed that mutations in *ALOX12B* and *ALOXE3*, the genes encoding 12R-LOX and eLOX-3, are linked to the development of autosomal recessive congenital ichthyosis (ARCI) supporting the long suspected role of LOX in the maintenance of skin permeability barrier.

ARCI refers to nonsyndromic congenital ichthyoses including harlequin ichthyosis, lamellar ichthyosis, and congenital ichthyosiform erythroderma (Oji et al. 2010). This clinically and genetically heterogeneous group of cornification disorders is characterized by a generalized scaling of the skin and has a prevalence of approximately 1 in 200,000 persons in the European and northern American populations (Oji and Traupe 2006). Affected newborns are often encased in a collodion membrane. After loss of this encasement in the first weeks of life, patients exhibit a generalized scaling, which varies in extent, color, and degree of adherence. An underlying mostly mild erythema is often seen. The typical histological phenotype of skin from ichthyosis patients is characterized by variable epidermal hyperplasia and marked hyperkeratosis. It is generally acknowledged that the ARCI phenotype results from a physical compensation for the defective permeability barrier that underlies all ichthyosis disorders.

To date, apart from the two LOX genes, seven other genes have been identified that underlie the ARCI phenotype, including *TGM1* encoding transglutaminase 1, *ABCA12* encoding a lipid transporter, a cytochrome P450 gene *CYP4F22*, a putative receptor gene *NIPAL4/ ichthyin*, two lipase genes *LIPN* and *PNPLA1* and *CERS3* encoding ceramide synthase 3 (Eckl et al. 2013; Fischer 2009; Grall et al. 2012; Israeli et al. 2011). Genetic studies indicated that mutations in *ALOX12B* and *ALOXE3* are the second most common cause of ARCI found in about 10% of the cases (Eckl et al. 2005, 2009; Vahlquist et al. 2010). To the present, 40 mutations have been identified in *ALOX12B* and 13 in *ALOXE3* (Akiyama et al. 2010; Eckl et al. 2005, 2009; Harting et al. 2008; Israeli et al. 2013; Jobard et al. 2002; Lesueur et al. 2007; Rodriguez-Pazos et al. 2011; Vahlquist et al. 2010). Most of the mutations are single nucleotide exchanges and only few are made of small deletions or small inserts. About a quarter of them are nonsense mutations resulting in a premature stop codon while the majority represent missense mutations. Examination of the molecular data revealed allelic heterogeneity for *ALOX12B,* with only some mutations located in the N-terminal beta barrel domain but most distributed within the catalytic domain with a minor focus on exons 9 and 12. In *ALOXE3* two recurrent mutations (p.Arg234X found in 13 and p.Pro630Leu found in 15 alleles) have been identified as mutational hotspots (Eckl et al. 2009; Fischer 2009).

While the nonsense mutations identified in the LOX genes definitely eliminate the catalytic activities the functional impact of the missense mutations on enzyme activities has been evaluated in detail. Although most of the mutations apparently did not interfere with substrate binding or catalysis directly, all of them led to complete loss of the catalytic activity of the LOX enzymes (Eckl et al. 2005, 2009; Yu et al. 2005) supporting the concept that the loss of function of 12R-LOX or eLOX-3 is fundamental to the pathogenesis of the LOX-dependent form of ARCI.

Animal Models to Study LOX Function in Skin

To investigate the physiological role of the 12R-LOX/eLOX-3 pathway and the molecular mechanisms of its action in the formation and maintenance of the epidermal barrier mutant mice for targeted inactivation of the 12R-LOX and eLOX-3 genes were established in our group.

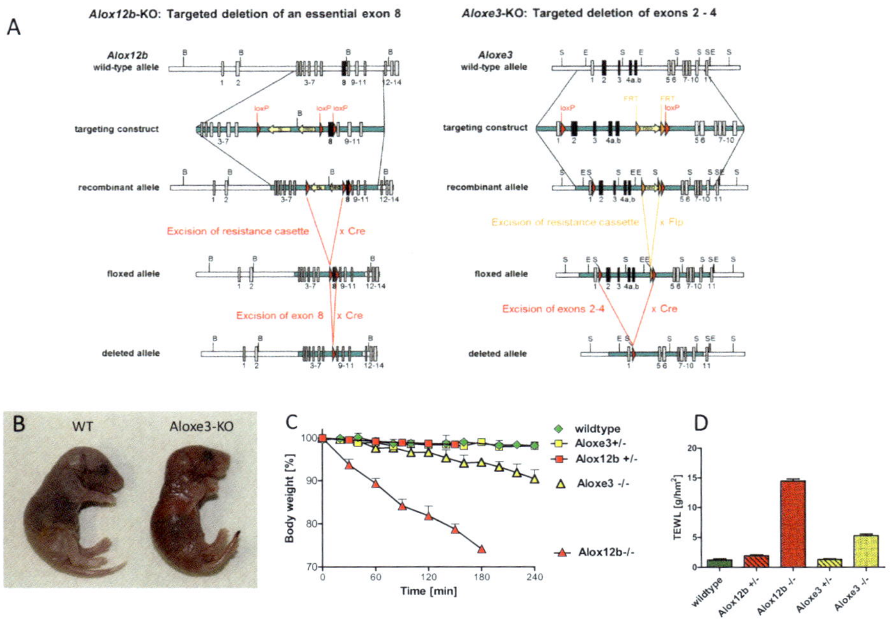

Fig. 17.2 12R-LOX and eLOX-3 knockout models. Targeted disruption of *Alox12b* and *Aloxe3*, the genes encoding 12R-LOX and eLOX-3, results in early neonatal death due to permeability barrier defects. **a** Strategy of LOX gene targeting using Cre-LoxP and Flp-FRT technology for the generation of inactivated LOX alleles. **b** Cross morphology of eLOX-3 deficient neonates. Homozygous knockout mice are born with red, shiny skin that rapidly desiccates. **c** Dehydration over time as indicated by loss of body weight. **d** Transepidermal water loss (*TEWL*) is increased up to eightfold in 12R-LOX deficient mice and about four-fold in eLOX-3 knockouts

General 12R-LOX and eLOX-3 Gene Knockout Results in Perinatal Death due to Severe Barrier Defects

A conditional null allele of the 12R-LOX gene *Alox12b* was generated by targeting exon 8 for Cre-mediated deletion and mice with general 12R-LOX gene knockout were obtained by mating mice harboring this mutant allele with transgenic CMV-Cre deleter mice (Fig. 17.2) (Epp et al. 2007). Exon 8 encodes a highly conserved region with two of the iron-binding histidines that are absolutely required for catalytic activity. Thus, deletion was expected to produce a truncated non-functional protein but actually resulted in the complete loss of 12R-LOX protein expression.

While heterozygous mice with disrupted *Alox12b* allele did not show any phenotype and reproduced normally homozygous mutant mice died within few hours after birth due to severely impaired barrier function. The mutant mice were born with red, shiny skin that rapidly desiccated. The neonates did not feed and lost 10 % of their weight per hour owing to an eight-fold increased transepidermal water loss

(TEWL) as compared to wild-type and heterozygous littermates. Dye penetration assays indicated that both the inside-out and the outside-in water barrier function were severely affected. The same lethal phenotype associated with 12R-LOX deficiency was reported by another group for an ethylnitrosurea-induced mouse mutant harboring a loss-of function mutation in *Alox12b* (Moran et al. 2007).

With the exception of a more tightly packed stratum corneum the neonatal 12R-LOX-deficient skin did not show gross abnormalities in the structural organisation of the epidermis (Epp et al. 2007). Only at the ultrastructural level, anomalies were observed including vesicular structures in the upper granular layers of the epidermis pointing to perturbance of the assembly/extrusion of lamellar bodies (LB) that play a crucial role in formation of the water impermeable barrier. Studies on the molecular mechanisms involved in the disruption of the barrier function showed that 12R-LOX-deficiency severely affected lipid components of the barrier but also proteins of the CE. Complete lack of filaggrin monomers, crucial components of the CE, associated with enhanced levels of high molecular intermediate species indicated impairment of the proteolytic processing of profilaggrin to filaggrin in 12R-LOX knockout epidermis. The loss of monomeric filaggrin was associated with increased fragility of mutant corneocytes supposedly contributing to the early neonatal death of the mutant mice. Moreover, a disordered composition of ceramides, in particular a decrease of protein-bound ceramide species could be identified as the decisive defect compromising barrier function in 12R-LOX-deficient skin. Ester-bound ceramides are the major constituents of the CLE. The CLE is formed by transglutaminase 1-catalyzed covalent attachment of the ceramides ω-hydroxyacyl-sphingosine (OS) and their sphingosine-free derivatives to involucrin and other CE peptides. The OS species especially those containing very long chain fatty acids (VLC-FA>C32) were severely reduced to only a few percent of the wild-type level. In line with these findings, Brash and coworkers showed that the CLE was largely absent in the epidermis of 12R-LOX deficient mice (Zheng et al. 2011).

For inactivating the eLOX-3 gene *Aloxe3* the Cre-loxP and flp-FRT systems were used to generate a conditional null allele with two loxP sites flanking exons 2–5 (Krieg et al. 2013). Cre-mediated deletion of the floxed segment results in an out of frame-fusion of the remaining exons thus causing a premature stop codon in exon 6 (Fig. 17.2). In the constitutive knockout model, homozygous mutant mice developed a similar but somewhat less severe skin phenotype as observed in the 12R-LOX knockout animals. Whereas 12R-LOX knockout mice all died rapidly within 3 h after birth, the eLOX-3 knockouts survived up to 12 h exhibiting less severe water loss as indicated by four-fold elevated TEWL. Accordingly, eLOX-3-deficient skin showed less pronounced structural and biochemical abnormalities, notably lack of abnormal vesicular structures and unimpaired profilaggrin processing. Of importance, lipid analysis demonstrated that the severity of barrier failure was related to the level of reduction in covalently bound ceramides, which was in eLOX-3 knockout skin to about half of the normal level as compared to only a few percent in 12R-LOX-deficient mice.

Strongly reduced levels of hepoxilin metabolites in eLOX-3-deficient epidermis furthermore indicated a function of eLOX-3 in mammalian skin as hepoxilin synthase linked to the 12S-LOX pathway. Hepoxilins and their triol derivatives have been found in normal human skin and in elevated levels in human psoriatic scales (Anton et al. 1998; Anton and Vila 2000). These metabolites have been shown to play a role in a variety of physiological processes including inflammation and neurotransmission. The majority of activities appear to be mediated by the stimulation of intracellular calcium or increased calcium transport across the membrane (Pace-Asciak 2009). In skin various potential mechanisms of action for these metabolites have been discussed including an action via a putative membrane receptor NIPAL/ichthyin or through activation of PPAR receptors (Lefevre et al. 2004; Yu et al. 2007). The exact mode of action and physiological role of hepoxilin-type metabolites in epidermis, however, remains to be established.

Adult 12R-LOX-Deficient Skin Develops an Ichthyosiform Phenotype

As a compensatory response to the excessive water loss impaired barrier function typically results in the development of a skin phenotype characterized by thickening of the epidermis (epidermal hyperplasia) and abnormal formation and desquamation of the SC (hyperkeratosis). Although some morphological and biochemical anomalies indicated the onset of hyperkeratosis and defects in or delay of the desquamation process, an obvious cutaneous phenotype was not observed in the neonatal skin of LOX-deficient mice, supposedly due to their early neonatal death. By utilizing neonatal skin grafting and the conditional knockout model with inducible gene inactivation in the skin the adult skin phenotype of 12R-LOX-deficient skin could be analyzed. *Alox12b*-deficient skin that matured after transplantation developed an ichthyosiform appearance with thickening of the epidermis, hyperproliferation, hypergranulosis, parakeratosis, and severe hyperkeratosis recapitulating the phenotype observed in skin from ichthyosis patients (De Juanes et al. 2009). Biochemically, the grafted skin showed restoration of profilaggrin processing and overexpression of CE components such as filaggrin, involucrin, and repetin. As a result, the TEWL was remarkably decreased as compared to neonatal knockout skin but still higher as compared to normal skin indicating improvement of the barrier defect by compensatory hyperkeratosis.

For inducible LOX ablation in the skin mice harboring the floxed LOX gene alleles were mated with a transgenic mouse line expressing a tamoxifen-dependent Cre recombinase under the control of the Keratin 14 promoter. Tamoxifen induced ablation of 12R-LOX, and similarly of eLOX-3, resulted in the development of a severe phenotype associated with growth retardation, dramatic loss of body weight and premature death (unpublished data). Morphological changes observed in mutant mice included focal alopecia, scaling of the skin and palmoplantar keratoderma

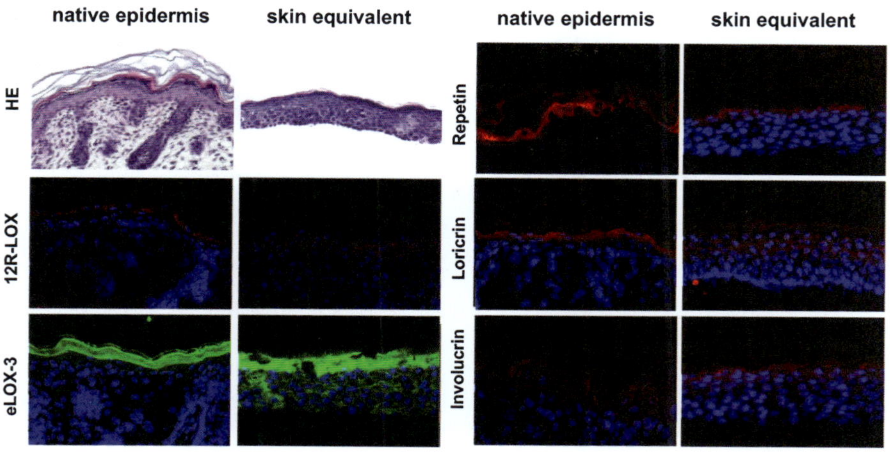

Fig. 17.3 Organotypic mouse skin equivalents. Skin equivalents established from primary keratinocytes represent suitable in vitro models resembling the natural tissue architecture and expression pattern of LOX and epidermal differentiation markers as found in native mouse skin

while histological analyses showed hyperplasia and strong hyperkeratosis resembling the ichthyosis phenotype.

Organotypic Mouse Skin Equivalents

With the aim to obtain an experimental in vitro model to study molecular mechanisms of LOX action more closely organotypic mouse skin models were established (Rosenberger et al. 2014). These in vitro systems are three-dimensional two-layered configurations consisting of epidermal keratinocytes growing on top of a dermal equivalent. The keratinocytes are freshly isolated from skins of neonatal mice and cocultivated with dermal mouse fibroblasts which are embedded in a scaffold of native collagen type I obtained from rat tail tendon. Closely resembling the in vivo situation these skin equivalents grow air-exposed with nutrients being supplied from underneath by diffusion of medium of a defined composition.

With this setup the keratinocytes are able to undergo proper epidermal morphogenesis. They form a multilayered well organized stratified epithelium consisting of basal, spinous, granular and cornified strata closely resembling the natural tissue architecture found in native mouse skin. The keratinocytes of the different layers of the epidermal equivalent show the respective characteristic morphological features and properly express various differentiation markers, such as keratin 1 and 10, filaggrin, involucrin, loricrin, and repetin (Fig. 17.3).

The major components of tight junctions are present as well as epidermal barrier lipids. Furthermore ultrastructural analyses indicate that these components are not only present but that they are processed and assembled correctly forming all structures of the epidermal barrier. These results indicate that our organotypic mouse skin models are valuable tools to study the molecular basis for epidermal barrier defects due to LOX deficiency and to test potential therapeutic approaches.

Our Current View on the Mechanism of Action of 12R-LOX/eLOX-3 in Barrier Function: Induction of Structural Changes Rather than Eicosanoid Signaling

Spatial and temporal colocalisation and a differentiation-dependent expression pattern in epithelial tissues early on suggested a common physiological role of 12R-LOX and eLOX-3 (Krieg et al. 2001). Their pivotal role in epidermal barrier function was unveiled by the finding that inactivating mutations in the genes of 12R-LOX and eLOX-3 are linked to the development of ARCI (Jobard et al. 2002). The fact that the same skin phenotype is caused by mutation of either gene suggested that both LOX may function in the same metabolic pathway and that the product of one of these enzymes may be the substrate of the other. Biochemical studies indeed showed that the two enzymes act in tandem to convert fatty acid substrates via R-hydroperoxides to specific epoxyalcohol derivatives (Yu et al. 2003). Thus, the common LOX substrate arachidonic acid is metabolized via 12R-HPETE to a R-epoxyalcohol of the hepoxilin family (Fig. 17.1). As 12R-HPETE turned out to be a preferred substrate of eLOX-3 in vitro it appeared that this specific hepoxilin pathway is the mechanistic link between 12R-LOX/eLOX-3 and their action in barrier function (Yu et al. 2006).

This view, however, is compromised by the following facts: (i) In contrast to the human enzyme the mouse 12R-LOX does not metabolize free arachidonic acid but only esterified substrates raising the question as to the functional homology of the mouse and human enzyme and to the nature of the natural substrate for this enzyme (Siebert et al. 2001). (ii) The data from our 12R-LOX-deficient mouse models clearly documented a functional homology of the mouse and human enzymes in skin. (iii) Several lines of evidence from fatty acid analyses and from feeding studies in animals indicate that not arachidonic acid but linoleic acid is the physiologically relevant EFA in relation to mammalian epidermal barrier function. (iv) Our studies furthermore revealed that the loss of covalently bound ceramide species as the major constituents of the CLE may be the decisive defect compromising barrier function in LOX-deficient skin.

Putting together these findings Brash and colleagues then recently suggested a new model of the action of LOX in barrier function that links the 12R-LOX/eLOX-3 pathway to oxidation of linoleate-containing ceramides and consequently to the formation of CLE (Zheng et al. 2011). The CLE is a hydrophobic envelope surrounding the corneocytes that is assumed to play an important role as a scaffold for the extracellular lipid lamellae or in corneocyte cohesion (Elias et al. 2000). This structure is enriched in ω-hydroxy ceramides (ω-hydroxyacyl-sphingosine, OS) and very long chain fatty acids (VLC-FA) linked covalently by ester bonds to glutamine/

glutamate residues within the outer CE. The CLE is presumed to originate from LB by fusion of their limiting membrane with the apical plasma membrane of the outermost stratum granulosum cells (Fig. 17.4). The LB membrane is enriched in glycosylated linoleate-containing esterified ω-hydroxyacyl-sphingosine ceramides (glucosyl-EOS), the precursor molecules of the CLE constituents. Upon fusion and secretion a multistep processing of glucosyl-EOS has been proposed to occur including deglucosylation by glucocerebrosidase, hydrolysis of the linoleate to form ω-hydroxy ceramide and covalent attachment to the CE by transglutaminase. The fatty acid moiety of the CLE originates from ceramidase-catalyzed hydrolysis of OS which may occur before or after the attachment step.

The novel theory on LOX action in barrier formation places the 12R-LOX/eLOX-3 sequence as another pivotal step among this ceramide processing cascade (Fig. 17.4). It was suggested that the hydrolysis of the linoleate moiety from ceramide EOS requires an "oxidation signal" which is provided by the action of 12R-LOX and eLOX-3. The LOX-catalyzed oxidation of the linoleate moiety of EOS may provoke structural changes and clearance of the oxidized lipids as the more polar carbon chain no longer strongly favors the lipid environment and this subsequently may activate a yet undefined downstream esterase/hydrolase which cleaves the oxidized linoleate to form OS with a free ω-hydroxyl. Supporting this theory Brash and colleagues could show that EOS ceramide is a natural substrate for the 12R-LOX/eLOX-3 pathway being oxygenated in the linoleate moiety by the consecutive actions of 12R-LOX and eLOX-3 to a 9R-hydroperoxide- and a corresponding 9R,10R-epoxy-13R-hydroxy-derivative (Zheng et al. 2011). Furthermore, they were able to demonstrate low levels of those oxidized species of linoleate in the EOS ceramides in normal pig and murine epidermis. These oxidized ceramides, and consequently the CLE, were almost absent in the epidermis of 12R-LOX knockout mice. The model is further supported by observations from our knockout models showing that the gradual reduction of covalently bound ceramides in the epidermis of eLOX-3 and 12R-LOX knockout mice is accompanied by barrier defects of increasing severity (Krieg et al. 2013). In the eLOX-3 knockout, a 12R-LOX mediated partial processing of the linoleate moiety to the 9R-hydroperoxide derivative may permit a limited hydrolysis of the linoleate (Fig. 17.4) and subsequent covalent linkage of resultant ω-hydroxy ceramides, whereas interruption of the first oxidation step by 12R-LOX deficiency completely prevents hydrolysis of the EOS ceramides and thus blocks the ceramide processing cascade and the formation of CLE.

Taken together, from our current perspective the action of 12R-LOX/eLOX-3 in barrier formation can now be mainly attributed to induction of structural changes in the ceramides EOS as a prerequisite for their downstream hydrolysis. It remains to be established whether in addition the free oxygenated linoleate cleavage products and their downstream metabolites may have signaling functions in barrier formation or other aspects of skin physiology. It has been suggested that these metabolites may be activators of PPARs or ligands for the plasma membrane receptor NIPAL4/ichthyin, another gene implicated in the pathogenesis of ARCI.

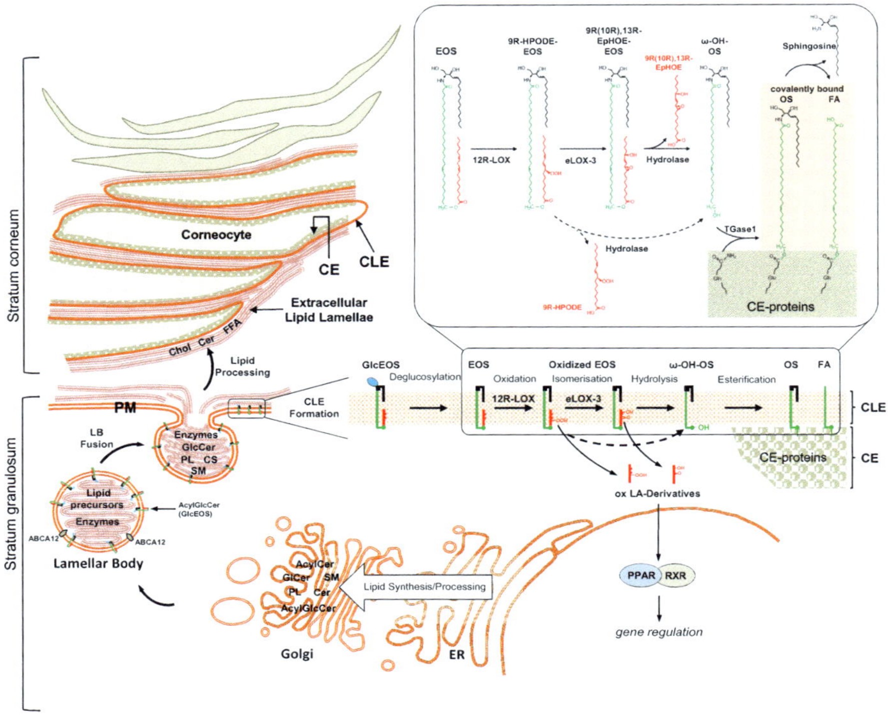

Fig. 17.4 Schema illustrating the formation of lipid structures essential for the epidermal barrier function and the role of 12R-LOX/eLOX-3 in these processes. The permeability barrier of the skin is localized in the stratum corneum. This function mainly resides in the extracellular lipid domains (extracellular lipid lamellae) between the outermost enucleated cells (corneocytes). The corneocyte is surrounded by an inner protein envelope (CE) and an outer corneocyte lipid envelope (CLE), a lipid monolayer covalently bound to the CE. The CLE is thought to be indispensable as a scaffold for the lamellar organization of the extracellular lipids. The lipid components of these structures originate from precursor lipids synthesized and processed mainly in the endoplasmic reticulum (ER) and Golgi apparatus. The precursor lipids, including acylceramide (AcylCer), glucosylceramide (GlcCer), sphingomyelin (SM), glycerophospholipids (PL), and cholesterol sulfate (CS), are packaged together with processing enzymes into lamellar bodies (LB) within the upper cell layers of the epidermis. Lipid packaging into the LB requires function of ABC transporter protein, ABCA12, localized to the limiting membranes of the LB. This LB membrane furthermore, is enriched in acylglucosylceramides (AcylGlcCer, including GlcEOS), the precursor molecules of the CLE constituents. Fusion of the LB with the apical plasma membrane (PM) in the uppermost nucleated cell layer of the epidermis (stratum granulosum), allows extrusion of lipid precursors into the extracellular domains. Subsequent enzymatic processing of precursor lipids generates the major lipid classes forming the extracellular lipid lamellae, i.e. ceramides (Cer), cholesterol (Chol), and free fatty acids (FFA). Formation of the CLE requires several processing steps of acyl-GlcCer including an essential 12R-LOX/eLOX-3 catalysed step. The successive oxygenation of the linoleate moiety of EOS by 12R-LOX and eLOX-3 generates an "oxygen signal" that permits esterase-catalyzed hydrolysis of the oxidized linoleate to release the ω-hydroxyl function on the ceramides. The ω-OH group of these Cer is required for CLE formation, as this moiety represents the attachment point between the Cer and specific amino acids of cornified cell envelope (CE) to be crosslinked by transglutaminase 1. The released oxidized linoleic acid derivatives (ox LA-Derivatives), 9R-HPODE and the hepoxilin-like metabolite (9R(10R), 13R-EpHOE), are hypothesized to act as ligands for nuclear hormone receptors (PPAR/RXR) that regulate genes involved in epidermal differentiation

The concept of LOX action in ceramide processing finally also provides an explanation for the long known effects of EFA in barrier formation and the manifestation of an EFA-deficiency phenotype. Linoleic acid-deficiency prohibits the formation of linoleate-containing EOS. LOX-catalyzed oxygenation of the linoleoyl residue is a prerequisite for ester hydrolysis and subsequent binding of the OS to CLE proteins. Thus, either lack of linoleic acid or deficiency of 12R-LOX/eLOX-3 compromise CLE formation leading to severe barrier defects.

Understanding the mechanism of LOX action in barrier function and the availability of suitable mouse knockout models and in vitro systems thereof may help to develop novel molecular approaches towards a therapy of ARCI associated with impaired LOX metabolism which could also be beneficial in the treatment of patients with other forms of ichthyosis.

Summary

LOX are key enzymes in the biosynthesis of a variety of oxylipins which act as signaling molecules involved in the modulation of keratinocyte proliferation and differentiation, in acute and chronic inflammatory processes, in epithelial regeneration and cancer development in the skin. Two recently discovered LOX, 12R-LOX and eLOX-3, have been presumed to play a key role in epidermal barrier function and loss of function mutations in the corresponding genes *AlOX12B* and *AlOXE3* have been found to be the second most cause in autosomal recessive congenital ichthyosis.

Emerging evidence indicates that the consecutive actions of both LOX oxidize linoleic acid esterified in ω-hydroxy-ceramide, the natural substrate in the skin, to a hepoxilin-related derivatives. Upon hydrolysis of the linoleate moiety the free ω-hydroxyl of the ω-hydroxy-ceramide is bound to proteins of the CE, constituting an important step in the formation of the CLE. This view is supported from mouse gene knockout studies. Targeted disruption of the corresponding LOX genes results in neonatal death due to a severely impaired permeability barrier function. Transplantation of *Alox12b*-deficient skin that matured after transplantation developed an ichthyosiform appearance with thickening of the epidermis, hyperproliferation, hypergranulosis, focal parakeratosis and marked hyperkeratosis recapitulating the phenotype observed in skin of ichthyosis patients. In the conditional knockout model, a similar ichthyosiform phenotype is observed upon tamoxifen-induced LOX ablation in the epidermis. In addition, an organotypic mouse skin model was developed that turned out to be a valuable tool to study the molecular basis for epidermal barrier defects due to LOX-deficiency and to test potential therapeutic approaches.

Acknowledgments This work was supported by grants from the Deutsche Forschungsgemeinschaft (KR 905/6-2, KR 905/7-1).

References

Akiyama M, Sakai K, Yanagi T, et al. Partially disturbed lamellar granule secretion in mild congenital ichthyosiform erythroderma with ALOX12B mutations. Brit J Dermatol. 2010;163:201–4.
Anton R, Puig L, Esgleyes T, et al. Occurrence of hepoxilins and trioxilins in psoriatic lesions. J Invest Dermatol. 1998;110:303–10.
Anton R, Vila L. Stereoselective biosynthesis of hepoxilin B3 in human epidermis. J Invest Dermatol. 2000;114:554–59.
Baer AN, Klaus MV, Green FA. Epidermal fatty acid oxygenases are activated in non-psoriatic dermatoses. J Invest Dermatol. 1995;104:251–5.
Bhatia B, Tang S, Yang P, et al. Cell-autonomous induction of functional tumor suppressor 15-lipoxygenase 2 (15-LOX2) contributes to replicative senescence of human prostate progenitor cells. Oncogene. 2005;24:3583–95.
Boeglin WE, Kim RB, Brash AR. A 12R-lipoxygenase in human skin: mechanistic evidence, molecular cloning, and expression. Proc Natl Acad Sci U S A. 1998;95:6744–49.
Brash AR. Lipoxygenases: occurrence, functions, catalysis, and acquisition of substrate. J Biol Chem. 1999;274:23679–82.
Brash AR, Yu Z, Boeglin WE, et al. The hepoxilin connection in the epidermis. FEBS J. 2007;274:3494–502.
Bürger F, Krieg P, Kinzig A, et al. Constitutive expression of 8-lipoxygenase in papillomas and clastogenic effects of lipoxygenase-derived arachidonic acid metabolites in keratinocytes. Mol Carcinog. 1999;24:108–17.
Burr GO, Burr MM. On the nature and role of the fatty acids essential in nutrition. J Biol Chem. 1930;86:587–621.
Candi E, Schmidt R, Melino G. The cornified envelope: a model of cell death in the skin. Nat Rev Mol Cell Biol. 2005;6:328–40.
De Juanes S, Epp N, Latzko S, et al. Development of an ichthyosiform phenotype in Alox12b-deficient mouse skin transplants. J Invest Dermatol. 2009;129:1429–36.
Eckl KM, Krieg P, Kuster W, et al. Mutation spectrum and functional analysis of epidermis-type lipoxygenases in patients with autosomal recessive congenital ichthyosis. Hum Mutat. 2005;26:351–61.
Eckl KM, De Juanes S, Kurtenbach J, et al. Molecular analysis of 250 patients with autosomal recessive congenital ichthyosis: evidence for mutation hotspots in ALOXE3 and allelic heterogeneity in ALOX12B. J Invest Dermatol. 2009;129:1421–8.
Eckl KM, Tidhar R, Thiele H, et al. Impaired epidermal ceramide synthesis causes autosomal recessive congenital ichthyosis and reveals the importance of ceramide acyl chain length. J Invest Dermatol. 2013;133:2202–11.
Elias PM, Fartasch M, Crumrine D, et al. Origin of the corneocyte lipid envelope (CLE): observations in harlequin ichthyosis and cultured human keratinocytes. J Invest Dermatol. 2000;115:765–9.
Epp N, Fürstenberger G, Müller K, et al. 12R-lipoxygenase deficiency disrupts epidermal barrier function. J Cell Biol. 2007;177:173–82.
Feingold KR. The outer frontier: the importance of lipid metabolism in the skin. J Lipid Res. 2009;50 Suppl:S417–22.
Fischer J. Autosomal recessive congenital ichthyosis. J Invest Dermatol. 2009;129:1319–21.
Fischer SM, Hagerman RA, Li-Stiles E, et al. Arachidonate has protumor-promoting action that is inhibited by linoleate in mouse skin carcinogenesis. J Nutr. 1996;126:1099S–104S.
Fürstenberger G, Krieg P, Müller-Decker K, et al. What are cyclooxygenases and lipoxygenases doing in the driver's seat of carcinogenesis? Int J Cancer. 2006;119:2247–54.
Grall A, Guaguere E, Planchais S, et al. PNPLA1 mutations cause autosomal recessive congenital ichthyosis in golden retriever dogs and humans. Nat Genet. 2012;44:140–7.
Haeggstrom JZ, Funk CD. Lipoxygenase and leukotriene pathways: biochemistry, biology, and roles in disease. Chem Rev. 2011;111:5866–98.

Hagerman RA, Fischer SM, Locniskar MF. Effect of 12-O-tetradecanoylphorbol-13-acetate on inhibition of expression of keratin 1 mRNA in mouse keratinocytes mimicked by 12(S)-hydroxyeicosatetraenoic acid. Mol Carcinog. 1997;19:157–64.

Hamberg M, Samuelsson B. Prostaglandin endoperoxides. Novel transformations of arachidonic acid in human platelets. Proc Natl Acad Sci U S A. 1974;71:3400–4.

Hammarström S, Hamberg M, Samuelsson B, et al. Increased concentrations of nonesterified arachidonic acid, 12L-hydroxy-5,8,14-eicosatetraenoic acid, prostaglandin E2, and prostaglandin F2alpha in epidermis of psoriasis. Proc Natl Acad Sci U S A. 1975;72:5130–4.

Harting M, Brunetti-Pierri N, Chan CS, et al. Self-healing collodion membrane and mild nonbullous congenital ichthyosiform erythroderma due to 2 novel mutations in the ALOX12B gene. Arch Dermatol. 2008;144:351–6.

Hussain H, Shornick LP, Shannon VR, et al. Epidermis contains platelet-type 12-lipoxygenase that is overexpressed in germinal layer keratinocytes in psoriasis. Am J Physiol. 1994;266:C243–53.

Israeli S, Khamaysi Z, Fuchs-Telem D, et al. A mutation in LIPN, encoding epidermal lipase N, causes a late-onset form of autosomal-recessive congenital ichthyosis. Am J Hum Genet. 2011;88:482–7.

Israeli S, Goldberg I, Fuchs-Telem D, et al. Non-syndromic autosomal recessive congenital ichthyosis in the Israeli population. Clin Exp Dermatol. 2013;38(8):911–6.

Jobard F, Lefevre C, Karaduman A, et al. Lipoxygenase-3 (ALOXE3) and 12(R)-lipoxygenase (ALOX12B) are mutated in non-bullous congenital ichthyosiform erythroderma (NCIE) linked to chromosome 17p13.1. Hum Mol Genet. 2002;11:107–13.

Kantarci A, Van Dyke TE. Lipoxins in chronic inflammation. Crit Rev Oral Biol Med. 2003;14:4–12.

Kim E, Rundhaug JE, Benavides F, et al. An antitumorigenic role for murine 8S-lipoxygenase in skin carcinogenesis. Oncogene. 2005;24:1174–87.

Kinzig A, Heidt M, Fürstenberger G, et al. cDNA cloning, genomic structure, and chromosomal localization of a novel murine epidermis-type lipoxygenase. Genomics. 1999;58:158–64.

Krieg P, Kinzig A, Heidt M, et al. cDNA cloning of a 8-lipoxygenase and a novel epidermis-type lipoxygenase from phorbol ester-treated mouse skin. Biochim Biophys Acta. 1998;1391:7–12.

Krieg P, Marks F, Fürstenberger G. A gene cluster encoding human epidermis-type lipoxygenases at chromosome 17p13.1: cloning, physical mapping, and expression. Genomics. 2001;73:323–30.

Krieg P, Rosenberger S, De Juanes S, et al. Aloxe3 knockout mice reveal a function of epidermal lipoxygenase-3 as hepoxilin synthase and its pivotal role in barrier formation. J Invest Dermatol. 2013;133:172–80.

Lesueur F, Bouadjar B, Lefevre C, et al. Novel mutations in ALOX12B in patients with autosomal recessive congenital ichthyosis and evidence for genetic heterogeneity on chromosome 17p13. J Invest Dermatol. 2007;127:829–34.

Lefevre C, Bouadjar B, Karaduman A, et al. Mutations in ichthyin a new gene on chromosome 5q33 in a new form of autosomal recessive congenital ichthyosis. Hum Mol Genet. 2004;13:2473–82.

Moran JL, Qiu H, Turbe-Doan A, et al. A mouse mutation in the 12R-lipoxygenase, Alox12b, disrupts formation of the epidermal permeability barrier. J Invest Dermatol. 2007;127:1893–97.

Muga SJ, Thuillier P, Pavone A, et al. 8S-lipoxygenase products activate peroxisome proliferator-activated receptor alpha and induce differentiation in murine keratinocytes. Cell Growth Differ. 2000;11:447–54.

Müller K, Siebert M, Heidt M, et al. Modulation of epidermal tumor development caused by targeted overexpression of epidermis-type 12S-lipoxygenase. Cancer Res. 2002;62:4610–16.

Nugteren DH, Kivits GA. Conversion of linoleic acid and arachidonic acid by skin epidermal lipoxygenases. Biochim Biophys Acta. 1987;921:135–41.

Nugteren DH, Christ Hazelhof E, Van Der Beek A, et al. Metabolism of linoleic acid and other essential fatty acids in the epidermis of the rat. Biochim Biophys Acta. 1985;834:429–36.

Oji V, Traupe H. Ichthyoses: differential diagnosis and molecular genetics. Eur J Dermatol. 2006;16:349–59.

Oji V, Tadini G, Akiyama M, et al. Revised nomenclature and classification of inherited ichthyoses: results of the first ichthyosis consensus conference in Soreze 2009. J Am Acad Dermatol. 2010;63:607–41.

Pace-Asciak CR. The hepoxilins and some analogues: a review of their biology. Br J Pharmacol. 2009;158:972–81.
Proksch E, Brandner JM, Jensen JM. The skin: an indispensable barrier. Exp Dermatol. 2008;17:1063–72.
Rodriguez-Pazos L, Ginarte M, Fachal L, et al. Analysis of TGM1, ALOX12B, ALOXE3, NIPAL4 and CYP4F22 in autosomal recessive congenital ichthyosis from Galicia (NW Spain): evidence of founder effects. Br J Dermatol. 2011;165:906–11.
Rosenberger S, Dick A, Latzko S, et al. A mouse organotypic tissue culture model for autosomal recessive congenital ichthyosis. Br J Dermatol. 2014. doi: 10.1111/bjd.13308.
Schweiger D, Fürstenberger G, Krieg P. Inducible expression of 15-lipoxygenase-2 and 8-lipoxygenase inhibits cell growth via common signaling pathways. J Lipid Res. 2007;48:553–64.
Shappell SB, Boeglin WE, Olson SJ, et al. 15-lipoxygenase-2 (15-LOX-2) is expressed in benign prostatic epithelium and reduced in prostate adenocarcinoma. Am J Pathol. 1999;155:235–45.
Siebert M, Krieg P, Lehmann WD, et al. Enzymic characterization of epidermis-derived 12-lipoxygenase isoenzymes. Biochem J. 2001;355:97–104.
Sun D, Mcdonnell M, Chen XS, et al. Human 12(R)-lipoxygenase and the mouse ortholog: molecular cloning, expression, and gene chromosomal assignment. J Biol Chem. 1998;50:33540–47.
Tang DG, Bhatia B, Tang S, et al. 15-lipoxygenase 2 (15-LOX2) is a functional tumor suppressor that regulates human prostate epithelial cell differentiation, senescence, and growth (size). Prostaglandins Other Lipid Mediat. 2007;82:135–46.
Vahlquist A, Bygum A, Ganemo A, et al. Genotypic and clinical spectrum of self-improving collodion ichthyosis: ALOX12B, ALOXE3, and TGM1 mutations in Scandinavian patients. J Invest Dermatol. 2010;130:438–43.
Virmani J, Johnson EN, Klein-Szanto AJ, et al. Role of 'platelet-type' 12-lipoxygenase in skin carcinogenesis. Cancer Lett. 2001;162:161–5.
Wang D, Dubois RN. Eicosanoids and cancer. Nat Rev Cancer. 2010;10:181–93.
Wang D, Chen S, Feng Y, et al. Reduced expression of 15-lipoxygenase 2 in human head and neck carcinomas. Tumour Biol. 2006;27:261–73.
Woollard PM. Stereochemical difference between 12-hydroxy-5,8,10,14-eicosatetraenoic acid in platelets and psoriatic lesions. Biochem Biophys Res Commun. 1986;136:169–76.
Woollard PM. Novel stereoisomer of 12-hydroxy-5,8,10,14-eicosatetraenoic acid in psoriasis. Adv Prostaglandin Thromboxane Leukot Res. 1987;17B:627–31.
Yoo H, Jeon B, Jeon MS, et al. Reciprocal regulation of 12- and 15-lipoxygenases by UV-irradiation in human keratinocytes. FEBS Lett. 2008;582:3249–53.
Yu Z, Schneider C, Boeglin WE, et al. The lipoxygenase gene ALOXE3 implicated in skin differentiation encodes a hydroperoxide isomerase. Proc Natl Acad Sci U S A. 2003;100:9162–7.
Yu Z, Schneider C, Boeglin WE, et al. Mutations associated with a congenital form of ichthyosis (NCIE) inactivate the epidermal lipoxygenases 12R-LOX and eLOX3. Biochim Biophys Acta. 2005;1686(3):238–47.
Yu Z, Schneider C, Boeglin WE, et al. Human and mouse eLOX3 have distinct substrate specificities: implications for their linkage with lipoxygenases in skin. Arch Biochem Biophys. 2006;455:188–96.
Yu Z, Schneider C, Boeglin WE, et al. Epidermal lipoxygenase products of the hepoxilin pathway selectively activate the nuclear receptor PPARalpha. Lipids. 2007;42:491–97.
Zheng Y, Brash AR. Dioxygenase activity of epidermal lipoxygenase-3 unveiled: typical and atypical features of its catalytic activity with natural and synthetic polyunsaturated fatty acids. J Biol Chem. 2010;285:39866–75.
Zheng Y, Yin H, Boeglin WE, et al. Lipoxygenases mediate the effect of essential fatty acid in skin barrier formation: a proposed role in releasing omega-hydroxyceramide for construction of the corneocyte lipid envelope. J Biol Chem. 2011;286:24046–56.

Chapter 18
PPAR Gamma Receptor, Skin Lipids and Hair

Barbara Toffoli and Béatrice Desvergne

Core Messages

- Peroxisome proliferator-activated receptors (PPARs) are ligand activated transcription factors belonging to the nuclear hormone receptor superfamily.
- PPARγ is involved in many different activities in the epidermis, such as keratinocyte differentiation, permeability barrier recovery, dermal wound closure, sebaceous gland formation, sebocyte differentiation, and melanogenesis.
- Preclinical studies with PPARγ ligands on various skin diseases have been performed and they could represent a new strategy in the treatment of scarring alopecia.
- PPARγ deserves further studies as therapeutic target, likely not with the current drugs, but with future new classes of safer molecules and in combined therapies.

Introduction

PPARs are ligand activated transcription factors belonging to the nuclear hormone receptor superfamily. Three related PPAR isotypes have been identified in vertebrates, including Xenopus, mouse, rat, hamster, and human: PPARα (NR1C1), PPARβ or δ (NR1C2), and PPARγ (NR1C3; Desvergne and Wahli 1999). PPAR genes are differentially expressed in a wide range of tissues in the adult organism (Braissant et al. 1996; Kliewer et al. 1994) and they are involved in important

B. Desvergne (✉) · Barbara Toffoli
Center for Integrative Genomics, University of Lausanne, Lausanne, Switzerland
e-mail: Beatrice.Desvergne@unil.ch

A. Pappas (ed.), *Lipids and Skin Health,* DOI 10.1007/978-3-319-09943-9_18

pathways that regulate lipid metabolism, insulin action, cell proliferation, and inflammation (Shearer and Billin 2007; Lee et al. 2003). In particular, PPARγ is considered the "master" regulator of adipogenesis; it is induced during the differentiation of preadipocytes into adipocytes and is strongly expressed in both white and brown adipose tissue (Tontonoz and Spiegelman 2008). Ectopic expression of PPARγ can also stimulate adipose differentiation from fibroblasts (Tontonoz et al. 1994) and myoblasts (Hu et al. 1995). Reciprocally, loss-of-function studies both in cell culture and in vivo provided clear evidences that PPARγ is essential for the formation of fat (Rosen et al. 1999; Barak et al. 1999). The ultimate demonstration comes with PPAR$\gamma^{-/-}$ mice, rescued from the embryonic lethality by using an epiblastic specific deletion of PPARγ that keeps the functional gene in the placenta. The PPAR$\gamma^{-/-}$ mice are totally deprived of adipose tissue and represent a clean model of total body lipodystrophy (Nadra et al. 2010).

The role of PPARγ in fat metabolism highlights only one facet of the activities of PPARγ in metabolism, which go beyond this tissue. Both in animal models and in human, PPARγ agonists, which are mainly represented by the thiazolidinediones (TZDs; e.g., rosiglitazone, ciglitazone, pioglitazone, and troglitazone) are potent insulin sensitizers acting at cellular levels in various metabolic organs, giving a systemic output with controlled glycemia in type 2 diabetes. Albeit TZDs are being retrieved from the market, due to some undesirable side-effects, they are still experimentally extremely useful. PPARγ also controls inflammation (Tontonoz and Spiegelman 2008). This is demonstrated at the cellular levels, and particularly in macrophages activity, but also in a variety of context in vivo (e.g., Desreumaux et al. 2001). Finally, PPARγ ligands are being tested for their pro-differentiation and pro-apoptotic activities, making them potential targets for drugs aimed at tumor control.

The skin is an amazing tissue that is sensitive to—and may reflect—many aspects of systemic disorders (inflammatory diseases, metabolic diseases, auto-immune diseases, etc.). It is also a complex tissue with high rate of cell renewal, high rates of fatty acid, and cholesterol metabolism mostly involved in the formation of a competent permeability barrier, presence of cells belonging to the immune-inflammation system, and finally with specific appendages such as hair follicles and sebaceous glands (Feingold and Elias 2014; Eming et al. 2007), all aspects potentially affected or controlled by PPARγ. Thus, a large interest has emerged in the study of all three PPARs, but more particularly of PPARγ in skin health, repair and disease (Di-Poi et al. 2004).

While Avci et al. (2013) have recently reviewed the numerous mouse models useful in the field of dermatology, the present chapter summarizes the observations in different preclinical models that demonstrate or suggest a role of PPARγ in various skin pathologies, pointing it as a possible therapeutic target.

PPARγ in Hyperproliferative and Inflammatory Skin Diseases

Psoriasis is a typical hyperproliferative and inflammatory skin disease, which is usually a lifelong skin pathology characterized by keratinocyte impaired differentiation, hyperproliferation, and chronic inflammation. As mentioned above, each of these features may be affected by the PPARγ activity.

First, PPARγ is expressed in keratinocytes, where it is primarily involved in the differentiation process. This was demonstrated in vivo in hairless mice treated with the highly potent PPARγ ligands troglitazone or ciglitazone. Their topical application increased the expression of important epidermal barrier proteins, like loricrin, involucrin, and filaggrin, without altering epidermal histology and thickness. Accordingly, PPARγ agonists improved homeostasis of the permeability barrier, which is the functional output of epidermal differentiation, as exemplified by accelerated recovery of the barrier acute disruption provoked by either acetone treatment or tape stripping (Mao-Qiang et al. 2004). A number of other studies further suggested the implication of PPARγ in promoting keratinocytes differentiation and in regulating skin barrier function (Mao-Qiang et al. 2004; Rivier et al. 1998; Dai et al. 2007, 2008; Jiang et al. 2008, 2011).

Second, less clear is the role of PPARγ in keratinocyte proliferation. While some authors reported that the topical treatment of normal mice with PPARγ activators promoted an increase in keratinocyte proliferation balanced by an increase in apoptosis (Mao-Qiang et al. 2004), others showed in cell culture a clear antiproliferative action of thiazolidinedione treatment (Ellis et al. 2000; Bhagavathula et al. 2004; Venkatraman et al. 2004). Conversely, mice with a specific lack of PPARγ in epidermal keratinocytes did not exhibit any gross cutaneous abnormalities or functional alterations other than a slight increase in epidermal thickness (Mao-Qiang et al. 2004). Nevertheless, when studied in the context of mouse models of hyperproliferative diseases induced by intradermal injection of IL-21 or by twice-daily tape stripping, PPARγ activation by ligands promoted an amelioration of the histological features and a reduction of the epidermal hyperplasia (Mastrofrancesco et al. 2013; Demerjian et al. 2006).

Third, many in vitro studies supported and confirmed the important role of PPARγ in modulating skin inflammatory signaling (Karnik et al. 2009; Konger et al. 2010; Chene et al. 2007), and stimulating skin innate immunity (Dai et al. 2010). This is not only relevant in psoriasis but also in other inflammatory skin diseases, such as contact dermatitis, atopic dermatitis, and rosacea. In a model of irritant contact dermatitis, troglitazone treatment inhibited epidermal hyperplasia and inflammation provoked by topical application of 12-O-tetradecanoylphorbol-13-acetate in hairless mice (Kim et al. 2006). In the same way, both topical and oral administration of BP-1017, a α-Lipoic acid-based PPARγ agonist, have been effective in a mouse model of allergic contact dermatitis (Venkatraman et al. 2004).

Thus, these observations strongly support the involvement of PPARγ as a protective factor in hyperproliferative and inflammatory skin disorders (Mastrofrancesco et al. 2013; Demerjian et al. 2006). Along this line, troglitazone treatment of lesional human psoriatic skin transplanted onto SCID mice induced a reduction in the inflammatory status and in skin thickness, accompanied by an improvement in epidermal differentiation (Ellis et al. 2000). In human, one study reported that the reduced PPARγ expression in psoriatic patient lesions correlated with the severity of psoriasis and increased incidence of metabolic complications, suggesting that PPARγ could be an important link between psoriasis and metabolic syndrome (Hegazy et al. 2012).

In general, although preclinical studies in animal models were quite promising, clinical assays in patients were disappointing. The oral treatment with troglitazone and especially pioglitazone seemed encouraging in ameliorating the psoriatic plaques, but rosiglitazone treatment reported contrasting data, suggesting that some beneficial activities might be at least in part independent of the presence of PPARγ in keratinocytes (Mao-Qiang et al. 2004; Ellis et al. 2000, 2007; Kuenzli and Saurat 2003; Bongartz et al. 2005; Shafiq et al. 2005; Malhotra et al. 2012; Pershadsingh et al. 2005; He et al. 2004). Furthermore, considering that (i) no topical treatment has ever proven efficient in patients, (ii) only systemic treatment showed positive effects, which is problematic with regard to side effects of TZDs, further studies are needed to clarify the usefulness of PPARγ agonists in psoriasis management.

PPARγ and Skin Fibrosis

Scleroderma, also known as systemic sclerosis (SSc), is a complex autoimmune disease with an early and transient inflammation and vascular injury, followed by progressive fibrosis affecting the skin and multiple internal organs (Varga and Abraham 2007). The aberrant synthesis and tissue accumulation of collagen is both the hallmark of the pathology and the leading factor to organ injury. Lesional scleroderma fibroblasts display an activated phenotype characterized by a sustained production of transforming growth factor-β (TGFβ), connective tissue growth factor, and other profibrotic cytokines and growth factors.

Subcutaneous injections of bleomycin in mice mimic many aspects of skin scleroderma. Cotreatment with rosiglitazone in vivo attenuated bleomycin-mediated skin inflammation and fibrosis as well as subcutaneous lipoatrophy and counteracted myofibroblast accumulation in damaged skin (Wu et al. 2009). Reciprocally, the generation of mice with fibroblast-specific deletion of PPARγ allowed for further in vivo studies. This deletion per se did not significantly affect skin structure and dermal homeostasis but accelerated dermal wound closure (Sha et al. 2012). In the model of bleomycin-induced fibrosis, absence of PPARγ worsened cutaneous thickening and accelerated fibrosis, inducing an increase in collagen deposition and an enhanced sensitivity to TGFβ (Kapoor et al. 2009). These results in mice clearly suggest a role for PPARγ in controlling fibrosis.

Human PPARγ activation, as assessed in human fibroblasts derived from newborn foreskins, has only a relatively modest effect on collagen levels in nonstimulated fibroblasts. However, treatment of such fibroblasts in culture with a PPARγ agonist abrogated TGFβ-induced collagen synthesis in a dose-dependent manner (Ghosh et al. 2004; Zhang et al. 2010), through the suppression of Smad3-dependent transcriptional responses (Ghosh et al. 2009). The protective antifibrotic effect of PPARγ activation was confirmed using dermal fibroblasts isolated from lesional areas of SSc patients (Shi-wen et al. 2010; Gonzalez et al. 2012), or in explanted fibroblasts and skin organ cultures concomitantly treated with TGFβ (Wu et al. 2009). Interestingly, there seem to be a mutual inhibitory cross-talk between TGFβ and PPARγ signaling in the context of skin fibrogenesis, since TGFβ treatment of human dermal fibroblasts from control individuals resulted in a time- and dose-dependent downregulation of PPARγ expression. Finally, in patients with SSc, PPARγ expression was markedly diminished in skin and lung, as well as in fibroblasts explanted from the damaged skin (Wei et al. 2010).

Recent studies proposed adiponectin, an adipokine directly regulated by PPARγ, as a possible mediator of the antifibrotic effects of PPARγ agonists, via adenosine monophosphate-activated protein kinase. Since the levels of adiponectin as well as that of its receptor are significantly impaired in scleroderma patients with progressive fibrosis, restoring adiponectin signaling in fibroblasts appears a promising therapeutic strategy to ameliorate dermal fibrosis (Fang et al. 2012; Lakota et al. 2012).

PPARγ in Scarring Alopecia and Sebaceous Gland Dysfunction

Cicatricial or scarring alopecias (CAs) consist of a permanent destruction of the pilosebaceus unit, either primarily due to a folliculocentric detrimental inflammation, or secondary to more generalized damaging events, like thermal burns or radiation, which results in follicular loss. This short definition hides a wide variety of disorders of poorly-known origins. According to the nature of the inflammatory cells that are infiltrating the pilosebaceous unit during the active phase, CAs are divided in lymphocytic, neutrophilic, and mixed CAs (Harries and Paus 2009; Harries et al. 2008).

The PPARγ connection was elegantly demonstrated by Karnik et al. who developed a mouse model in which PPARγ is specifically deleted in the stem cells of the hair follicle bulge, via a stem cell-specific promoter *Keratin 15* Cre-loxP-mediated gene targeting. The generated mice exhibited a skin and hair phenotype that emulates the human lymphocytic cicatritial alopecia, or lichen planopilaris (LPP), with dystrophic hair follicles, follicular plugging, perifollicular fibrosis and sebaceous gland atrophy. Perifollicular inflammation, characterized by a mixed mononuclear infiltrate consisting of lymphocytes, macrophages, plasma cells, and mast cells was also seen. In agreement with the microarray results obtained in LPP human biopsy, microarray data on PPARγ-deficient mice confirmed the histological data, showing a significant increase in gene expression of chemokines, extracellular matrix-associated proteins, and apoptosis related genes, suggesting a strong involvement of macrophages and T-lymphocytes in the damage. Moreover, as observed also in LPP affected patients, a decreased expression of lipid metabolism and peroxisomal genes was measured. In summary, the authors demonstrated that PPARγ is absolutely crucial for maintenance of a functional epithelial stem cell compartment in murine hair follicles (Karnik et al. 2009).

The sebaceous glands associated to the hair follicles are also fundamental for maintenance of a functional barrier and skin homeostasis (Niemann and Horsley 2012). PPARγ is not only required for the formation of sebaceous glands (Rosen et al. 1999; Fu et al. 2010), but it also regulates sebocyte functions such as cellular responses to oxidative stress (Zhang et al. 2006) and eotaxin production (Nakahigashi et al. 2012). The Gsdma3(Dfl)/+ mice are defolliculated mice, characterized by defective sebaceous glands associated with a hair loss phenotype very similar to the one observed in scarring alopecia. Since aberrant hair cycle, chronic inflammation, reduced lipogenesis in sebaceous glands, loss of hair follicles and reduced expression of PPARγ were all features present in defolliculated mice, Ruge et al. 2011 suggested the use of these mice as a good LPP model to test the effects of PPARγ agonists as well as other new therapies. Altogether, these observations plead for setting-up clinical trials assessing the efficiency of PPARγ activators in scarring alopecia. It is interesting to note that diabetic patients who were also affected by LPP experienced beneficial effects and improvements of both the diabetes and the skin phenotype under oral treatment with pioglitazone (Mirmirani and Karnik 2009; Baibergenova and Walsh 2012).

Aside from the scarring alopecia disorders, the role of PPARγ in subcutaneous fat might also affect the proper development of hair follicles. Festa et al. demonstrated the importance of the intradermal adipocyte lineage cells in mice and identified a dynamic process of adipogenesis in the skin that parallels the activation of hair follicle stem cells. Treating mice with the PPARγ antagonists (BADGE and GW9662) demonstrated that the inhibition of PPARγ before the anagen induction (from P18 to P24) caused a block of the intradermal adipose tissue regrowth, without reducing the number of adipocyte precursor cells. Furthermore, the mice were not able to enter into anagen and remained in the telogen phase of the hair cycle, clearly suggesting that preadipocytes with a functional PPARγ are important for the proper regeneration of the hair follicles (Festa et al. 2011).

Finally, acne is another disease of the pilosebaceous unit, characterized by increased and altered sebum production, inflammation, and modified keratinization (Williams et al. 2012), and the critical role of PPARγ in sebaceous gland suggested its possible involvement in the pathogenesis of acne vulgaris. However, experiments using sebocytes or sebaceous gland in culture delivered contradictory results. On the one hand, treatment of SEB-1 human sebocytes with rosiglitazone in vitro increased lipogenesis. Consistently, patients receiving thiazolidinediones for the treatment of type 2 diabetes showed a significant increase in sebum production, with no increase in the incidence of acne (Trivedi et al. 2006). On the other hand, the treatment of the whole sebaceous gland in organ cultures with the PPARγ agonists rosiglitazone and 15-deoxy-D-12,14-prostaglandin J2 revealed an opposite effect with inhibition of sebaceous lipogenesis (Downie et al. 2004). Using a different approach, Schuster et al. demonstrated that immortalized human sebocytes (SZ95) are protected from staurosporine-induced apoptosis when treated with rosiglitazone, albeit less efficiently than when treated with a PPARβ/δ agonist, suggesting that PPARγ activation might contribute to limit the release of lipids in the context of sebocyte apoptosis (Schuster et al. 2011). Altogether, due to conflicting data, further studies are necessary to assess the potential of PPARγ as a target in acne vulgaris.

PPARγ and Carcinogenesis

The possible role for PPARγ agonists to act as cancer chemopreventive agents has been regularly highlighted (Michalik et al. 2004; Berger et al. 2013; Robbins and Nie 2012). Starting from evidences showing that PPARγ signaling in the epidermis is affected by UVB irradiation (Zhang et al. 2005), Sahu et al. explored the role of PPARγ in photocarcinogenesis in hairless albino SKH-1 mice, with or without an epidermal specific PPARγ deletion obtained through a cytokeratin 14 driven Cre-LoxP strategy. Upon chronic exposure to UVB irradiation, mice lacking PPARγ in the epidermis exhibited an accelerated tumorigenesis, a stronger propensity to progress to malignancy, and a significant increase in epidermal hyperplasia as well as p53 immunopositive keratinocytes in tumor free areas. Furthermore, the mice showed a sustained increase in erythema and transepidermal water loss due to a disruption of the normal barrier function (Sahu et al. 2012). These data, coupled with previous studies conducted in PPARγ heterozygous mice exposed to a chemical mediated carcinogenesis, suggested the hypothesis that PPARγ agonists could be useful as chemopreventive agents in nonmelanoma skin cancer (Sahu et al. 2012; Nicol et al. 2004). Unfortunately, He et al. found opposite results when using two different mouse skin carcinogenesis models (He et al. 2005).

Skin cancer can also develop from melanocytes, which produce melanin, giving rise to melanoma, the most severe type of skin cancer. The antiproliferative effect of TZD treatment at high doses has been confirmed in many different in vitro studies performed on normal melanocytes and on melanoma cells, albeit the direct role of PPARγ activation remains to be characterized (Kang et al. 2004, 2006; Botton et al. 2009; Eastham et al. 2008; Grabacka et al. 2008; Placha et al. 2003; Paulitschke et al. 2012). PPARγ involvement in melanocyte apoptotic process is more controversial. On the one hand, Kang et al. demonstrated that ciglitazone promoted apoptosis in normal human melanocytes, mediated by increased caspase-3 and reduction of Bcl-2 and p-ERK (Kang et al. 2006). On the other hand, Botton et al. demonstrated that ciglitazone was not able to induce cell death in normal melanocyte, but only acted in melanoma cells. At low doses, ciglitazone could induce cell-cycle arrest in such cells, whereas apoptosis induction was caused by higher ciglitazone concentration and independently of PPARγ transcriptional activation (Botton et al. 2009).

In human clinics, a randomized multi-institutional phase II trial with a cohort of patients with advanced melanoma, refractory to at least one previous chemotherapy with maximum tolerated doses, compared the benefits of additional therapy with either metronomic chemotherapy (trofosfamide) or combined antiinflammatory/angiostatic treatments (trofosfamide plus rofecoxib and pioglitazone). The estimated progression-free survival rates at 1 year were 0 % for metronomic chemotherapy, but 9 % for additional antiinflammatory therapy (Reichle et al. 2007). This result, coupled with in vivo experiments in nude mice in which ciglitazone treatment inhibited human melanoma xenograft development (Botton et al. 2009), suggests that PPARγ ligands could be beneficial in the treatment of melanomas. Here again, further studies are needed since in general PPARγ ligands showed disappointing results when used as monotherapy in different types of cancer, and presented modest efficiency in combined therapy (Hatton and Yee 2008; Youssef and Badr 2011; Shimizu and Moriwaki 2008).

Finally, and beside these effects on proliferation/apoptosis, PPARγ activation has been shown to accelerate the pigmentation of human melanocytes and cultured skin (Lee et al. 2007; Flori et al. 2011) and to increase melanocytes migration, suggesting a possible usage of PPARγ agonists in the treatment of pigmentary disorders such as vitiligo (Lee et al. 2007).

Conclusions and Hypothesis

Altogether, the reports cited in the present review suggest that PPARγ is involved in many different activities in the epidermis, such as keratinocyte differentiation, permeability barrier recovery, dermal wound closure, sebaceous gland formation, sebocyte differentiation, and melanogenesis. While preclinical studies showed that PPARγ ligands could represent efficient strategies to treat various skin diseases involving inflammation and epidermal proliferation, such as psoriasis, rosacea, and tumors, the clinical assays in patients have been disappointing so far. In the same manner and with the same reservations, PPARγ agonists, which affect also signaling pathways involved in lipid homeostasis and in dermal fibrosis, could be extremely useful for the treatment of other common skin diseases, like acne vulgaris and scleroderma. Finally, in the light of the results obtained by Karnik et al. (2009) supporting a crucial role of PPARγ in maintaining a healthy pilosebaceous unit, the recovery of PPARγ activity may represent a new strategy in the treatment of scarring alopecia.

Unfortunately, TZDs that are effective oral insulin sensitizers for the treatment of type 2 diabetes present high risks of side effects, like fluid retention and congestive heart failure, which strongly limit their prescription to specific controlled cases. In conclusion, PPARγ deserves further studies as therapeutic target, likely not with the current drugs, but with future new classes of safer molecules and in combined therapies.

References

Avci P, et al. Animal models of skin disease for drug discovery. Expert Opin Drug Disc. 2013;8(3):331–55.

Baibergenova A, Walsh S. Use of pioglitazone in patients with lichen planopilaris. J Cutan Med Sur. 2012;16(2):97–100.

Barak Y, et al. PPAR gamma is required for placental, cardiac, and adipose tissue development. Mol Cell. 1999;4(4):585–95.

Berger H, et al. SOCS3 transactivation by PPARgamma prevents IL-17-driven cancer growth. Cancer Res. 2013;73(12):3578–90.

Bhagavathula N, et al. Rosiglitazone inhibits proliferation, motility, and matrix metalloproteinase production in keratinocytes. J Invest Dermatol. 2004;122(1):130–9.

Bongartz T, et al. Treatment of active psoriatic arthritis with the PPARgamma ligand pioglitazone: an open-label pilot study. Rheumatology. 2005;44(1):126–9.

Botton T, et al. In vitro and in vivo anti-melanoma effects of ciglitazone. J Invest Dermatol. 2009;129(5):1208–18.

Braissant O, et al. Differential expression of peroxisome proliferator-activated receptors (PPARs): tissue distribution of PPAR-alpha, -beta, and -gamma in the adult rat. Endocrinology. 1996;137(1):354–66.

Chene G, et al. n-3 and n-6 polyunsaturated fatty acids induce the expression of COX-2 via PPARgamma activation in human keratinocyte HaCaT cells. Biochim Biophys Acta. 2007;1771(5):576–89.

Dai X, et al. STAT5a/PPARgamma pathway regulates involucrin expression in keratinocyte differentiation. J Invest Dermatol. 2007;127(7):1728–35.

Dai X, et al. PPAR gamma is an important transcription factor in 1 alpha, 25-dihydroxyvitamin D3-induced involucrin expression. J Dermatol Sci. 2008;50(1):53–60.

Dai X, et al. PPARgamma mediates innate immunity by regulating the 1alpha, 25-dihydroxyvitamin D3 induced hBD-3 and cathelicidin in human keratinocytes. J Dermatol Sci. 2010;60(3):179–86.

Demerjian M, et al. Topical treatment with thiazolidinediones, activators of peroxisome proliferator-activated receptor-gamma, normalizes epidermal homeostasis in a murine hyperproliferative disease model. Exp Dermatol. 2006;15(3):154–60.

Desreumaux P, et al. Attenuation of colon inflammation through activators of the retinoid X receptor (RXR)/peroxisome proliferator-activated receptor gamma (PPARgamma) heterodimer. A basis for new therapeutic strategies. J Exp Med. 2001;193(7):827–38.

Desvergne B, Wahli W. Peroxisome proliferator-activated receptors: nuclear control of metabolism. Endocr Rev. 1999;20(5):649–88.

Di-Poi N, et al. Functions of peroxisome proliferator-activated receptors (PPAR) in skin homeostasis. Lipids. 2004;39(11):1093–9.

Downie MM, et al. Peroxisome proliferator-activated receptor and farnesoid X receptor ligands differentially regulate sebaceous differentiation in human sebaceous gland organ cultures in vitro. Br J Dermatol. 2004;151(4):766–75.

Eastham LL, Mills CN, Niles RM. PPARalpha/gamma expression and activity in mouse and human melanocytes and melanoma cells. Pharm Res. 2008;25(6):1327–33.

Ellis CN, et al. Troglitazone improves psoriasis and normalizes models of proliferative skin disease: ligands for peroxisome proliferator-activated receptor-gamma inhibit keratinocyte proliferation. Arch Dermatol. 2000;136(5):609–16.

Ellis CN, et al. Placebo response in two long-term randomized psoriasis studies that were negative for rosiglitazone. Am J Clin Dermatol. 2007;8(2):93–102.

Eming SA, Krieg T, Davidson JM. Inflammation in wound repair: molecular and cellular mechanisms. J Invest Dermatol. 2007;127(3):514–25.

Fang F, et al. The adipokine adiponectin has potent anti-fibrotic effects mediated via adenosine monophosphate-activated protein kinase: novel target for fibrosis therapy. Arthritis Res Ther. 2012;14(5):R229.

Feingold KR, Elias PM. Role of lipids in the formation and maintenance of the cutaneous permeability barrier. Biochim Biophys Acta. 2014;1841(3):280–94.

Festa E, et al. Adipocyte lineage cells contribute to the skin stem cell niche to drive hair cycling. Cell. 2011;146(5):761–71.

Flori E, et al. 2,4,6-Octatrienoic acid is a novel promoter of melanogenesis and antioxidant defence in normal human melanocytes via PPAR-gamma activation. Pigment Cell Melanoma Res. 2011;24(4):618–30.

Fu G, et al. Committed differentiation of hair follicle bulge cells into sebocytes: an in vitro study. Int J Dermatol. 2010;49(2):135–40.

Ghosh AK, et al. Disruption of transforming growth factor beta signaling and profibrotic responses in normal skin fibroblasts by peroxisome proliferator-activated receptor gamma. Arthritis Rheum. 2004;50(4):1305–18.

Ghosh AK, et al. Peroxisome proliferator-activated receptor-gamma abrogates Smad-dependent collagen stimulation by targeting the p300 transcriptional coactivator. FASEB J. 2009;23(9):2968–77.

Gonzalez EG, et al. Synthetic cannabinoid ajulemic acid exerts potent antifibrotic effects in experimental models of systemic sclerosis. Ann Rheum Dis. 2012;71(9):1545–51.

Grabacka M, et al. PPAR gamma regulates MITF and beta-catenin expression and promotes a differentiated phenotype in mouse melanoma S91. Pigment Cell Melanoma Res. 2008;21(3):388–96.

Harries MJ, Paus R. Scarring alopecia and the PPAR-gamma connection. J Invest Dermatol. 2009;129(5):1066–70.

Harries MJ, et al. Management of primary cicatricial alopecias: options for treatment. Br J Dermatol. 2008;159(1):1–22.

Hatton JL, Yee LD. Clinical use of PPARgamma ligands in cancer. PPAR Res. 2008;2008:159415.

He G, Thuillier P, Fischer SM. Troglitazone inhibits cyclin D1 expression and cell cycling independently of PPARgamma in normal mouse skin keratinocytes. J Invest Dermatol. 2004;123(6):1110–9.

He G, et al. The effect of PPARgamma ligands on UV- or chemically-induced carcinogenesis in mouse skin. Mol Carcinog. 2005;43(4):198–206.

Hegazy RA, et al. Psoriasis and metabolic syndrome: is peroxisome proliferator-activated receptor-gamma part of the missing link? Eur J Dermatol. 2012;22(5):622–8.

Hu E, Tontonoz P, Spiegelman BM. Transdifferentiation of myoblasts by the adipogenic transcription factors PPAR gamma and C/EBP alpha. Proc Natl Acad Sci U S A. 1995;92(21):9856–60.

Jiang YJ, et al. PPAR and LXR activators regulate ABCA12 expression in human keratinocytes. J Invest Dermatol. 2008;128(1):104–9.

Jiang YJ, et al. PPARgamma activators stimulate aquaporin 3 expression in keratinocytes/epidermis. Exp Dermatol. 2011;20(7):595–9.

Kang HY, et al. Expression and function of peroxisome proliferator-activated receptors in human melanocytes. Br J Dermatol. 2004;150(3):462–8.

Kang HY, et al. Peroxisome proliferator-activated receptors-gamma activator, ciglitazone, inhibits human melanocyte growth through induction of apoptosis. Arch Dermatol Res. 2006;297(10):472–6.

Kapoor M, et al. Loss of peroxisome proliferator-activated receptor gamma in mouse fibroblasts results in increased susceptibility to bleomycin-induced skin fibrosis. Arthritis Rheum. 2009;60(9):2822–9.

Karnik P, et al. Hair follicle stem cell-specific PPARgamma deletion causes scarring alopecia. J Invest Dermatol. 2009;129(5):1243–57.

Kim S, et al. Phytosphingosine stimulates the differentiation of human keratinocytes and inhibits TPA-induced inflammatory epidermal hyperplasia in hairless mouse skin. Mol Med. 2006;12(1–3):17–24.

Kliewer SA, et al. Differential expression and activation of a family of murine peroxisome proliferator-activated receptors. Proc Natl Acad Sci U S A. 1994;91(15):7355–9.

Konger RL, et al. The peroxisome proliferator-activated receptor gamma system regulates ultraviolet B-induced prostaglandin e(2) production in human epidermal keratinocytes. PPAR Res. 2010;2010:467053.

Kuenzli S, Saurat JH. Effect of topical PPARbeta/delta and PPARgamma agonists on plaque psoriasis. A pilot study. Dermatology. 2003;206(3):252–6.

Lakota K, et al. Levels of adiponectin, a marker for PPAR-gamma activity, correlate with skin fibrosis in systemic sclerosis: potential utility as biomarker? Arthritis Res Ther. 2012;14(3):R102.

Lee CH, Olson P, Evans RM. Minireview: lipid metabolism, metabolic diseases, and peroxisome proliferator-activated receptors. Endocrinology. 2003;144(6):2201–7.

Lee JS, Choi YM, Kang HY. PPAR-gamma agonist, ciglitazone, increases pigmentation and migration of human melanocytes. Exp Dermatol. 2007;16(2):118–23.

Malhotra A, et al. Thiazolidinediones for plaque psoriasis: a systematic review and meta-analysis. Evid Based Med. 2012;17(6):171–6.

Mao-Qiang M, et al. Peroxisome-proliferator-activated receptor (PPAR)-gamma activation stimulates keratinocyte differentiation. J Invest Dermatol. 2004;123(2):305–12.

Mastrofrancesco A, et al. Preclinical studies of a specific PPARgamma modulator in the control of skin inflammation. J Invest Dermatol. 2013. doi:10.1038/jid.2013.448.

Michalik L, Desvergne B, Wahli W. Peroxisome-proliferator-activated receptors and cancers: complex stories. Nat Rev Cancer. 2004;4(1):61–70.

Mirmirani P, Karnik P. Lichen planopilaris treated with a peroxisome proliferator-activated receptor gamma agonist. Arch Dermatol. 2009;145(12):1363–6.

Nadra K, et al. PPARgamma in placental angiogenesis. Endocrinology. 2010;151(10):4969–81.

Nakahigashi K, et al. PGD2 induces eotaxin-3 via PPARgamma from sebocytes: a possible pathogenesis of eosinophilic pustular folliculitis. J Allergy Clin Immunol. 2012;129(2):536–43.

Nicol CJ, et al. PPARgamma influences susceptibility to DMBA-induced mammary, ovarian and skin carcinogenesis. Carcinogenesis. 2004;25(9):1747–55.

Niemann C, Horsley V. Development and homeostasis of the sebaceous gland. Semin Cell Dev Biol. 2012;23(8):928–36.

Paulitschke V, et al. Proteome analysis identified the PPARgamma ligand 15d-PGJ2 as a novel drug inhibiting melanoma progression and interfering with tumor-stroma interaction. PloS ONE. 2012;7(9):e46103.

Pershadsingh HA, Benson SC, Ellis CN. Improvement in psoriasis with rosiglitazone in a diabetic and a nondiabetic patient. Skinmed. 2005;4(6):386–90.

Placha W, et al. The effect of PPARgamma ligands on the proliferation and apoptosis of human melanoma cells. Melanoma Res. 2003;13(5):447–56.

Reichle A, et al. Targeted combined anti-inflammatory and angiostatic therapy in advanced melanoma: a randomized phase II trial. Melanoma Res. 2007;17(6):360–4.

Rivier M, et al. Differential expression of peroxisome proliferator-activated receptor subtypes during the differentiation of human keratinocytes. J Invest Dermatol. 1998;111(6):1116–21.

Robbins GT, Nie D. PPAR gamma, bioactive lipids, and cancer progression. Front Biosci. 2012;17:1816–34.

Rosen ED, et al. PPAR gamma is required for the differentiation of adipose tissue in vivo and in vitro. Mol Cell. 1999;4(4):611–7.

Ruge F, et al. Delineating immune-mediated mechanisms underlying hair follicle destruction in the mouse mutant defolliculated. J Invest Dermatol. 2011;131(3):572–9.

Sahu RP, et al. Mice lacking epidermal PPARgamma exhibit a marked augmentation in photocarcinogenesis associated with increased UVB-induced apoptosis, inflammation and barrier dysfunction. Int J Cancer. 2012;131(7):E1055–66.

Schuster M, et al. Peroxisome proliferator-activated receptor activators protect sebocytes from apoptosis: a new treatment modality for acne? Br J Dermatol. 2011;164(1):182–6.

Sha W, et al. Loss of PPARgamma expression by fibroblasts enhances dermal wound closure. Fibrogenesis Tissue Repair. 2012;5:5.

Shafiq N, et al. Pilot trial: Pioglitazone versus placebo in patients with plaque psoriasis (the P6). Int J Dermatol. 2005;44(4):328–33.

Shearer BG, Billin AN. The next generation of PPAR drugs: do we have the tools to find them? Biochim Biophys Acta. 2007;1771(8);1082–93.
Shimizu M, Moriwaki H. Synergistic effects of PPARgamma ligands and retinoids in cancer treatment. PPAR Res. 2008;2008:181047.
Shi-wen X, et al. Rosiglitazone alleviates the persistent fibrotic phenotype of lesional skin scleroderma fibroblasts. Rheumatology. 2010;49(2):259–63.
Tontonoz P, Spiegelman BM. Fat and beyond: the diverse biology of PPARgamma. Ann Rev Biochem. 2008;77:289–312.
Tontonoz P, Hu E, Spiegelman BM. Stimulation of adipogenesis in fibroblasts by PPAR gamma 2, a lipid-activated transcription factor. Cell. 1994;79(7):1147–56.
Trivedi NR, et al. Peroxisome proliferator-activated receptors increase human sebum production. J Invest Dermatol. 2006;126(9):2002–9.
Varga J, Abraham D. Systemic sclerosis: a prototypic multisystem fibrotic disorder. J Clin Invest. 2007;117(3):557–67.
Venkatraman MS, et al. Alpha-Lipoic acid-based PPARgamma agonists for treating inflammatory skin diseases. Arch Dermatol Res. 2004;296(3):97–104.
Wei J, et al. PPARgamma downregulation by TGFbeta in fibroblast and impaired expression and function in systemic sclerosis: a novel mechanism for progressive fibrogenesis. PloS ONE. 2010;5(11):e13778.
Williams HC, Dellavalle RP, Garner S. Acne vulgaris. Lancet. 2012;379(9813):361–72.
Wu M, et al. Rosiglitazone abrogates bleomycin-induced scleroderma and blocks profibrotic responses through peroxisome proliferator-activated receptor-gamma. Am J Pathol. 2009;174(2):519–33.
Youssef J, Badr M. Peroxisome proliferator-activated receptors and cancer: challenges and opportunities. Br J Pharmacol. 2011;164(1):68–82.
Zhang Q, et al. Epidermal peroxisome proliferator-activated receptor gamma as a target for ultraviolet B radiation. J Biol Chem. 2005;280(1):73–9.
Zhang Q, et al. Involvement of PPARgamma in oxidative stress-mediated prostaglandin E(2) production in SZ95 human sebaceous gland cells. J Invest Dermatol. 2006;126(1):42–8.
Zhang GY, et al. Peroxisome proliferator-activated receptor-gamma (PPAR-gamma) agonist inhibits transforming growth factor-beta1 and matrix production in human dermal fibroblasts. J Plast Reconstr Aesthet Surg. 2010;63(7):1209–16.

Part VII
Lipids in Skincare

Chapter 19
Mineral Oil in Skin Care: Safety Profile

Georgios N Stamatas

Core Messages

- Mineral oil and vegetable-derived oils are used in skin care with the aim to enhance the skin barrier function leading to moisture retention in the stratum corneum (SC). They also act as emollients ameliorating dry skin conditions and helping to reduce skin irritation.
- Pharmaceutical grade mineral oil used in skin care products is inert, chemically stable, does not clog pores and, like many vegetable-derived oils, penetrates only the outermost layers of the SC.
- Toxicological studies in animals did not identify any hazard related to topical exposure to mineral oil at any dose.
- Pharmaceutical grade mineral oil used in personal care products has a low allergic potential and has been shown to be safe also for use on infants and young children. This conclusion is supported by the long and uneventful human use of mineral oil for medical and cosmetic applications.

Introduction

Mineral oil or liquid paraffin (*paraffinum liquidum*) has been used as an ingredient in personal care products since the late 1800s (Rawlings and Lombard 2012). Together with paraffin and petrolatum, mineral oil is derived from petroleum through a series of refining steps and purification. Pharmaceutical grade mineral oil is the most refined and purified fraction of mineral oil with approved uses as ingredient in food, cosmetics, and pharmaceutical products. It is transparent, colorless, and odorless and comprises a complex mixture of highly refined saturated straight chain, branched, and cyclic hydrocarbons. Because of their complex chemical compositions, different types of mineral oils are typically characterized by their viscosities.

G. N. Stamatas (✉)
Johnson & Johnson Santé Beauté France, Issy-les-Moulineaux, France
e-mail: gstamata@its.jnj.com

A. Pappas (ed.), *Lipids and Skin Health*, DOI 10.1007/978-3-319-09943-9_19

A principal medical and cosmetic expectation of an oil-containing skin care product is moisturization, associated with an improvement of the signs of dry skin. A variety of mineral oil-containing personal care products with different concentrations of mineral oil are available, including moisturizers, cleansers, and emollients (i.e., products to soothe skin irritation). In infant skin care, a formula containing pharmaceutical grade mineral oil in high concentration (typically more than 90 %) is known as "baby oil" (Jolly and Sloughfy 1975). Its use is linked to skin emolliency and, when used in infant massage, other positive benefits both to the child (Agarwal et al. 2000; Field et al. 1996) and the parent (Glover et al. 2002) that include developmental and psychological aspects.

The safety of mineral oil has been questioned in the popular press primarily due to its petroleum origin. In this chapter the safety aspects of the use of mineral oil in cosmetic products are addressed with a special focus on pharmaceutical grade mineral oil. Its current safety classification by national and international authorities and its chemical and biochemical properties compared to those of vegetable-derived oils used in cosmetic products is presented together with the existing safety evidence from toxicological studies in animals, as well as tests assessing its allergic and photo-allergic potential in humans. Moreover, the existing literature with regards to the occlusive potential of mineral oil as compared to that of vegetable-derived oils is reviewed and a discussion of the appropriateness of using mineral oil and vegetable-derived oils in the particular case of infant skin care is included.

Safety Classification of Mineral Oil as an Ingredient in Skin Care

Petrolatum, paraffin, and mineral oil are widely used as emollients and vehicles in cosmetics because of their low volatility and smooth texture. Mineral oil has been approved as a food additive and as "skin protectant" ingredient for topical products by the Food and Drug Administration (FDA) in the USA and is listed in European and Japanese pharmacopeia (Rawlings and Lombard 2012). A skin protectant is defined as an ingredient that temporarily protects injured or exposed skin from harmful or annoying stimuli and may provide relief to such skin.

To determine carcinogenicity of products used in humans, the World Health Organization (WHO) uses a five group classification system as developed by the International Agency for Research on Cancer (IARC; 2011). The system classifies substances, mixtures, and processes as: Group 1—Carcinogenic to humans; Group 2A—Probably carcinogenic to humans; Group 2B—Possibly carcinogenic to humans; Group 3—Unclassifiable as to carcinogenicity to humans, and Group 4—Probably not carcinogenic to humans. The WHO classifies highly refined mineral oil as a Group 3, which compares to the classification of the European directive relating to cosmetic products (European Union/European Council 2003).

This means that both organizations agree that there is not enough evidence to support its carcinogenicity to humans.

Mode of Action of Topically Applied Oils

The primary role of the top layer of the epidermis, the SC, is to provide a barrier to tissue dehydration and to penetration of extrinsic aggressors to the body. A common analogy for the barrier effect of SC is the "brick-and-mortar" model, with corneocytes depicted as the bricks and the lipid lamellae as the mortar (Elias 1988). This model has been further developed to include the contributions to the barrier function by corneodesmosomes, natural moisturization factors, antimicrobial lipids, pH buffering capacity, etc. (Rawlings 2010) However, it is generally agreed that the presence of SC lipids provides the main resistance to excessive transepidermal water loss (TEWL). Therefore, it is reasonable to assume that topical application of external oils will enhance the barrier effect by adding quantitatively to the lipid phase.

However, the way in which topically applied oils interact with SC lipids needs also to be taken into consideration. The chains of saturated hydrocarbons contained in mineral oil are flexible and can incorporate themselves easily into the lipid lamellae without grossly affecting their organization. On the other hand, the effect of unsaturated hydrocarbons, such as those contained in free fatty acids of vegetable oils, varies depending on the number and position of their unsaturated bonds. It has been shown for example in experiments in vitro and in vivo that oleic acid (one unsaturated bond) has a detrimental effect in the SC lipid organization, while linoleic acid (two unsaturated bonds) may actually be stabilizing this organization (Mack Correa et al. 2014).

The case for the vegetable oils as ingredients of cosmetic products is more complex. It has been suggested that the ratio of free oleic to linoleic acid needs to be examined as an indicator of whether a specific oil is suitable or not for skin care (Danby et al. 2013). However, this ratio is not readily known and may vary considerably for any particular vegetable oil, depending on the source, the region, storage conditions, and annual variations of the quality.

In contrast to mineral oil that is chemically and biologically inert, some vegetable oils may have a biological effect on keratinocyte metabolism. How effective a specific vegetable oil is in that respect depends on its content of triglycerides, flavonoids, phytosterols, and tocopherols. It has been reported that certain vegetable oils may support skin barrier function due to their content of essential fatty acids (Rieger and Deem 1974). However, the most important mode of action that results in SC moisturization is similar for both mineral and vegetable oil, and that is retaining moisture through the formation of a thin protective layer on skin surface (Rawlings and Lombard 2012; Stamatas et al. 2008; Patzelt et al. 2012; Rieger and Deem 1974; Powers and Fox 1959; Strübmann et al. 1993; Caussin et al. 2007).

Chemical Stability of Mineral and Vegetable Oils Used in Skin Care

By definition vegetable oils are plant-derived, while mineral oil is derived from fossilized animal material (fossil plankton and algae; Rawlings and Lombard 2012). Vegetable oils are smaller and more chemically diverse (Rawlings and Lombard 2012; Patzelt et al. 2012; Stamatas et al. 2008; DiNardo 2005; Rieger and Deem 1974; Powers and Fox 1959). The components of the vegetable oils are susceptible to oxidation and photo-oxidation and chemical instability which turns them to rancid. In contrast, mineral oil is inert and resistant to oxidation due to the saturated nature of its hydrocarbons and, in contrast to vegetable oils, its lack of the terminal carboxyl group of the fatty acids (Rawlings and Lombard 2012; Patzelt et al. 2012; Stamatas et al. 2008).

Because it is consisted of saturated hydrocarbons that are biologically inert, mineral oil cannot be easily metabolized by microbes. This, together with the absence of water, leads to the fact that bacteria do not readily grow in mineral oil (Rawlings and Lombard 2012). As an extra benefit, mineral oil washes out easier from clothes than vegetable oils because it cannot form bonds through oxidation with natural fabrics (Rawlings and Lombard 2012).

Toxicology of Mineral Oil Use in Personal Skin Care

Mineral oil as an ingredient in skin care formulations is nontoxic and generally regarded as safe (Rawlings and Lombard 2012). Due to its inertness mineral oil is recommended for use on allergic and atopic skin (Rawlings and Lombard 2012; Patzelt et al. 2012).

Animal Feeding and Topical Application Tests

A toxicological review by Nash et al. collected the reports of several animal studies that involved topical exposure to white mineral oils over the lifetime of the animals (Table 19.1; Nash et al. 1996). In each study, a complete autopsy was performed and histological evaluation was conducted including examination of the liver, kidneys, spleen, and mesenteric lymph nodes. These findings establish that repeated topical exposure to white mineral oils has not been found to produce any local or systemic

Table 19.1 Summary of topical exposure animal studies to white mineral oils. (Adapted from Nash et al. 1996)

Species	No of animals	Duration	Material	Dose mg/kg/day	Results
C3H/HeJ mice	50	3 × per week over lifetime	White oil	296	No histopathological changes, no tumors
C3H/HeJ mice	50	3 × per week over lifetime	White oil	296	No abnormalities in visceral organs
C3H/HeJ mice	140	3 × per week over lifetime	White oil	296	No abnormalities in visceral organs
C3H/HeJ mice	40	2 × per week over 24 months	White oil	296	No abnormalities in visceral organs
C3H/HeJ mice	30	3 × per week over lifetime	White oil	238	No abnormalities in visceral organs
	50			296	
F344 rats	10	2 × per week over 91 days	Mineral oil	41	Increase in liver weight
C3H mice	10		Mineral oil	143	No abnormalities in visceral organs, minor dermal irritation
New Zealand white rabbits	4	20 days followed by 14 days no treatment	99 % mineral oil	2000	No dermal, hematological or histopathological changes in all groups
			16 % mineral oil	2000	

toxicity including tumor formation. In light of these findings as well as reports of negligible epidermal penetration of topically applied white mineral oils, the same authors conclude that there is no evidence of any hazard identified for topical exposure to white mineral oils at any dose in multiple species.

Human Studies

A series of human studies have been conducted by Johnson & Johnson Consumer Companies, Inc. to test the allergic potential of skin care products with high content of pharmaceutical grade mineral oil (>90 %; Johnson & Johnson Consumer Companies Inc). The tests included human repeated insult patch test (HRIPT) on a total of 1035 subjects, cumulative irritation test on 137 subjects, phototoxicity on 60 subjects, and photoallergy tests on 171 subjects. The results of these studies are summarized in Table 19.2. In all the tests there was no allergic, phototoxic, and photoallergic responses and in the case of the cumulative irritation tests the finding was negligible irritation potential.

Table 19.2 Safety data of skin care products with high content (>90 %) of pharmaceutical grade mineral oil. (Johnson & Johnson Consumer Companies Inc)

Mineral oil content (%)	HRIPT	Cumulative irritation	Phototoxicity	Photoallergy
90.58	No induced allergic responses (0/216)[a]	Not conducted	No phototoxic responses (0/10)	No photoallergic or induced allergic responses (0/29)
91.35	Not conducted	Not conducted	No phototoxic responses (0/10)	No photoallergic or induced allergic responses (0/30)
93.08	No induced allergic responses (0/198)	Negligible irritation potential (0/672)	No phototoxic responses (0/10)	No photoallergic or induced allergic responses (0/30)
98.00	No induced allergic responses (0/220)	Negligible irritation potential (0/648)	No phototoxic responses (0/10)	No photoallergic or induced allergic responses (0/24)
98.75	Not conducted	Negligible irritation potential (0.5/744)	Not conducted	Not conducted
99.65	Not conducted	Negligible irritation potential (4/528)	Not conducted	Not conducted
99.85	No induced allergic responses (0/208)	Negligible irritation potential (0.5/696)	No phototoxic responses (0/10)	No photoallergic or induced allergic responses (0/30)
99.98	No induced allergic responses (0/193)	Not conducted	No phototoxic responses (0/10)	No photoallergic or induced allergic responses (0/28)

HRIPT human repeated insult patch test

[a] Numbers in parentheses show positive responses per number of subjects, except in the cumulative irritation test where the numbers show cumulative irritation score per maximum possible irritation score

Is Mineral Oil Comedogenic?

Mineral oil has been suspected in popular press to cause blocking of skin pores, leading to acne and preventing toxins from escaping the body through the skin. However, recent studies using advanced measuring methods clearly demonstrated that the tested oils do not clog skin pores (Patzelt et al. 2012; Stamatas et al. 2008; DiNardo 2005). These studies concluded that mineral oil and vegetable oils used in skin care products form a protective layer on the skin, enhancing moisture retention, while allowing the skin to breathe without obstructing the pores. Wiechers et al. evaluated the effect of individual, blended, and formulated emollients on the mechanical properties of skin resulting in a defined tactile sensory perception (Wiechers 2008). The study recognized the existence of occlusion as a safe moisture-control mechanism, which depends principally on the choice of the emollients and by choosing the right emulsifier (Wiechers 2008; Caussin et al. 2007; Strübmann et al. 1993).

Is Mineral Oil Penetrating to the Viable Layers of the Skin?

To address the question of potential penetration of lipid components of topically applied oils through the SC and into the viable epidermis, our group used in vivo confocal Raman micro-spectroscopy (Stamatas et al. 2008). This method can detect and quantify the presence of lipids in the SC, as well as calculate the lipid concentration profile as a function of depth. The thickness of the SC can also be measured non-invasively and the occlusive effects of topically applied oils can be inferred by the increase of SC thickness due to water retention. We have therefore employed this method to test the skin penetration and the occlusive effect of mineral oil and two vegetable oils, almond and jojoba oil. It was found that the total lipids of the tested oils penetrated to the same extent up to 6–8 μm in the SC without reaching the viable epidermis. When measuring the increase in the SC thickness, the occlusive potential of mineral oil and vegetable oils were comparable. These observations were similar on adults and on infants (Stamatas et al. 2008).

In another study Patzelt et al. used in vivo laser scanning microscopy and TEWL measurements to investigate the penetration of mineral oil and vegetable oils (jojoba, soybean, avocado, and almond oil; Patzelt et al. 2012). It was shown that the tested oils penetrate only into the upper layers of the SC, enhancing moisture retention, without blocking completely the exchange of gases such as water vapor through the SC.

Mineral Oil Use in Infant Skin Care

Topical products containing mineral oil are frequently used in baby skin care (Rawlings and Lombard 2012). Evidence suggests that the skin barrier continues to develop at least up to 12 months following birth (Nikolovski et al. 2008), therefore the case of oils in infant skin care needs special attention. What has been discussed above about the inert nature and safety profile of mineral oil needs to be taken into consideration.

Field et al. reported the benefits of using mineral oil in infant massage in a study on 60 healthy newborns, randomly assigned to a massage group with oil and a massage group without oil (Field et al. 1996). The massage had a soothing calming influence on the infants, particularly when given with oil.

In some countries olive oil is often used in infant skin care, particularly for massaging but also in diaper cleansing routines. As mentioned above, oleic acid, a free fatty acid which is naturally abundant in high acidity olive oil, can be detrimental to the integrity of the skin barrier. Interestingly, oleic acid is a commonly used penetration enhancer for transcutaneous drug delivery. Skin barrier degradation may be particularly concerning in children with atopic dermatitis, as the barrier in atopic dermatitis already has increased TEWL and oleic acid can further worsen the condition. Also oils which are naturally high in erucic acid should also be avoided for use in skin care products for young children (Darmstadt et al. 2002). Examples of such oils are mustard seed oil, rapeseed oil, and oils of other members from the plant family Brassicaceae. In contrast, clinical evidence suggests the use of mineral oil containing cleansers and skin care products for babies and young children is safe and beneficial (Blume-Peytavi et al. 2009; Simpson et al. 2010; Field et al. 1996; Loden et al. 2004).

Summary

Pharmaceutical grade mineral oil is a complex mixture of highly refined saturated hydrocarbons, which are derived from petroleum through various refining steps and purification. In skin care it is used topically to cleanse, lubricate, and massage the skin. This work focuses on the safety of use in personal care products containing mineral oil. Particular emphasis is given for the safety of skin care products used on babies and young children. It has been shown that the chemically stable, inert mineral oil applied in skin care products penetrates only into the upper layers of the SC, enhancing moisture retention, without being 100 % occlusive. These skin care products do not clog the pores, they have a low allergic potential, and have been shown to be safe. In addition, mineral oil has been shown to improve skin softness and barrier function better than some other emollients. In conclusion, safety concerns about the use of topical mineral oil cannot be supported by the current clinical evidence.

Acknowledgements The author would like to thank Vincent R. Walczac of Johnson & Johnson Consumer Companies, Inc. for providing the data in file for Table 19.2, Dr. Anneke Andriessen for editorial assistance and Drs. Beate Gerstbrein and Apostolos Pappas for critically reviewing the manuscript.

References

Agarwal K, Gupta A, Pushkarna R, Bhargava S, Faridi M, Prabhu M. Effects of massage & use of oil on growth, blood flow & sleep pattern in infants. Indian J Med Res. 2000;112:212.

Blume-Peytavi U, Cork M, Faergemann J, Szczapa J, Vanaclocha F, Gelmetti C. Bathing and cleansing in newborns from day 1 to first year of life: recommendations from a European round table meeting. J Eur Acad Dermatol Venereol. 2009;23:751–9.

Caussin J, Gooris G, Groenink H, Wiechers J, Bouwstra J. Interaction of lipophilic moisturizers on stratum corneum lipid domains in vitro and in vivo. Skin Pharmacol Physiol. 2007;20:175–86.

Danby SG, Alenezi T, Sultan A, Lavender T, Chittock J, Brown K, Cork MJ. Effect of olive and sunflower seed oil on the adult skin barrier: implications for neonatal skin care. Pediatr Dermatol. 2013;30:42–50.

Darmstadt G, Mao-Qiang M, Chi E, Saha S, Ziboh V, Black R, Santosham M, Elias P. Impact of topical oils on the skin barrier: possible implications for neonatal health in developing countries. Acta Paediatr. 2002;91:546–54.

Dinardo JC. Is mineral oil comedogenic? J Cosmet Dermatol. 2005;4:2–3.
Elias PM. Structure and function of the stratum corneum permeability barrier. Drug Dev Res. 1988;13:97–105.
European Union/European Council. Directive 2003/15/EC of the European Parliament and of the Council of 27 February 2003 amending Council Directive 76/768/EEC on the approximation of the laws of the Member States relating to cosmetic products. Off J Eur Union. 2003;66:26–35.
Field T, Schanberg S, Davalos M, Malphurs J. Massage with oil has more positive effects on normal infants. Pre Perinat Psychol J. 1996;11:75–80.
Glover V, Onozawa K, Hodgkinson A. Benefits of infant massage for mothers with postnatal depression. Semin Neonatol. 2002;7:495–500, (Elsevier).
International Agency for Research on Cancer (IARC). Agents classified by the IARC Monographs, Vol. 1–100. 2011. http://monographs.iarc.fr/ENG/Classification/ClassificationsGroupOrder.pdf. Accessed on September 25, 2014.
Johnson & Johnson Consumer Companies INC. Human test reports on the use of products with high content of mineral oil—Internal unpublished reports over the years 1993–2008.
Jolly ER, Sloughfy CA. Clinical evaluation of baby oil. J Soc Cosmet Chem. 1975;26:227–34.
Loden M, Buraczewska I, Edlund F. Irritation potential of bath and shower oils before and after use: a double-blind randomized study. Br J Dermatol. 2004;150:1142–7.
Mack Correa CM, Mao G, Saad P, Flach CR, Mendelsohn R, Walters RM. Molecular interactions of plant oil components with stratum corneum lipids correlate with clinical measures of skin barrier function. Exp Dermatol. 2014;23:39–44.
Nash J, Gettings S, Diembeck W, Chudowski M, Kraus A. A toxicological review of topical exposure to white mineral oils. Food Chem Toxicol. 1996;34:213–25.
Nikolovski J, Stamatas GN, Kollias N, Wiegand BC. Barrier function and water-holding and transport properties of infant stratum corneum are different from adult and continue to develop through the first year of life. J Invest Dermatol. 2008;128:1728–36.
Patzelt A, Lademann J, Richter H, Darvin M., Schanzer S, Thiede G, Sterry W, Vergou T, Hauser M. In vivo investigations on the penetration of various oils and their influence on the skin barrier. Skin Res Technol. 2012;18:364–9.
Powers D, Fox C. The effect of cosmetic emulsions on the stratum corneum. J Soc Cosmet Chem. 1959;10:109–16.
Rawlings AV. Recent advances in skin 'barrier' research. J Pharm Pharmacol. 2010;62:671–7.
Rawlings AV, Lombard KJ. A review on the extensive skin benefits of mineral oil. Int J Cosmet Sci. 2012;34:511–8.
Rieger MM, Deem DE. Skin moisturizers. II. The effects of cosmetic ingredients on human stratum corneum. J Soc Cosmet Chem. 1974;25:253–62.
Simpson EL, Berry TM, Brown PA, Hanifin JM. A pilot study of emollient therapy for the primary prevention of atopic dermatitis. J Am Acad Dermatol. 2010;63:587–93.
Stamatas GN, De Sterke J, Hauser M, Von Stetten O, Van Der Pol A. Lipid uptake and skin occlusion following topical application of oils on adult and infant skin. J Dermatol Sci. 2008;50:135–42.
Strübmann A, Weiben H, Wirtz A. Water vapour permeability of skin care products in relation to molecular and environmental influences. Int J Cosmet Sci. 1993;15:227–33.
Wiechers JW, Taelman MC, Dederen JC, Barlow T. Relative performance testing of formulations: emulsifiers. In "Skin Barrier: Chemistry of Delivery Systems", edited by JW Wiechers, Allured Publishing Corporation, Carol Stream, IL, USA. 2008:235–248.

Chapter 20
Lipophilic Antioxidants

Vasiliki Lagouri

> **Core Messages**
> - Skin is the organ which is exposed more to the oxidative attack of free radicals and reactive oxygen species.
> - Lipophilic nutrients such as tocopherols (vit. E) and carotenoids are receiving growing attention in the skin care industry because of their antioxidant activity and skin related benefits that recent studies have demonstrated.

Introduction

Skin is the organ of the body which is exposed more to the oxidative attack of free radicals and reactive oxygen species (ROS) originated from various environmental stimuli such as ultraviolet (UV) radiation, air pollutants, or chemicals. ROS and free radicals are also produced during normal cellular metabolism. To counteract the harmful effects of ROS, the various compartments of the skin (epidermis, dermis, subcutis) are equipped with specific antioxidant systems located in the different skin layers, which help to maintain an equilibrium between ROS and antioxidants and thus prevent oxidative stress. Lipophilic antioxidants such as tocopherols and carotenoids are all highly effective antioxidants which can protect the cell membranes from lipid oxidation. These lipophilic antioxidants are also important nutrients that are receiving growing attention in the skin care industry because of their antioxidant and mostly photoprotective properties. Therefore, it is imperative to summarize the most important lipophilic antioxidants, their antioxidant mechanisms, and their effects on skin health.

V. Lagouri (✉)
National and Kapodistrian, University of Athens, Athens, Greece
e-mail: vlagouri@gmail.com

A. Pappas (ed.), *Lipids and Skin Health*, DOI 10.1007/978-3-319-09943-9_20

Tocopherols

Tocopherols-Chemistry

The name tocopherols comes from the Greek words "τόκος" [birth] and "φέρειν" [to bear or carry], meaning in sum "to carry a pregnancy," with the ending "-ol" that depicts a chemical alcohol. Tocopherols are a class of chemical compounds many of which have vitamin E (Vit. E) activity. The vitamin activity was first identified in 1936 from a dietary fertility study in rats. Vitamin E is not a single molecule but two classes of similar molecules derived from a chromanol structure, which consists of a family of eight substances (four tocopherols and four tocotrienols), many of which with antioxidant properties. These compounds are closely related homologues and isomers depending respectively on the number and position of methyl groups on the aromatic ring of the tocol backbone in tocopherols.

Natural tocochromanols comprise two homologous series: the tocopherols with a saturated-side chain, and the tocotrienols with an unsaturated-side chain. Tocopherols are the most abundant form of vitamin E in the body, consisting of four different forms (alpha-, beta-, gamma-, and delta-tocopherol). Tocotrienols, which are found in the body to a lesser extent, also exist in four different forms (alpha-, beta-, gamma-, and delta-tocotrienol). Either tocopherols or tocotrienols may differ in the methylation of the chroman head group: β- and γ- are structural isomers (5,8-dimethyltocol and 7,8-dimethyltocol), while α- and δ- (5,7,8-trimethyltocol and 8-methyltocol) differ from each other and from β- and γ- because they possess either one more or one less methyl group in the aromatic ring (Mayer et al. 1967).

Although tocopherols and tocotrienols are available from the diet, alpha-tocopherol is the primary form of vitamin E found and maintained in the body, due to the specificity of a transport protein for alpha-tocopherol. Vitamin E is synthesized only by plants and it is a very important dietary nutrient for humans and animals (Fryer 1992). Naturally occurring vitamin E is usually labeled as "natural" or "d" vitamin E, while synthetic vitamin E is a mixture of eight isomeric forms, usually labeled "all-rac" or "dl." Tocopherols and tocotrienols are also available as derivatives in the form of esters which are more stable upon exposure to heat, light, and air. Conjugated vitamin E molecules are also used in dietary supplements. The levels of vitamin E in skin depend on its oral intake or topical delivery. The main natural sources of vitamin E are fresh vegetables, vegetable oils, cereals, and nuts.

Tocopherols-Antioxidant Activity

Lipid peroxidation (or autoxidation) is a degradative, free radical reaction responsible for the development of objectionable odors and flavors in oils, fats, and foods (Frankel 1980, 1991). It is a chain reaction that proceeds in three stages: initiation, propagation, and termination (Porter 1986). In the initiation step, a carbon-centered lipid radical (an alkyl radical) is produced by the abstraction from a polyunsaturated fatty acid moiety. The initiation reaction can be catalyzed by heat, light, and transition metals (Chan 1987). In the propagation step, the alkyl radical reacts with molecular oxygen at a very high rate, giving a peroxyl radical. The peroxyl radical, a chain-carrying radical, is able to attack another polyunsaturated lipid molecule. Although the initial peroxyl radical is converted to a hydroperoxide, this process produces a new alkyl radical, which is rapidly converted into another peroxyl radical. The chain reaction does not stop until the chain-carrying peroxyl radical meets and combines with another radical to form inactive products (termination step).

Vitamin E compounds (tocopherols and tocotrienols) are well recognized for their effective inhibition of lipid peroxidation in foods and living cells (Burton and Traber 1990). Tocopherol isomers are chain-breaking antioxidants. The antioxidant activity of the tocopherols is related to scavenging the free radicals of unsaturated lipids (Burton and Ingold 1981, 1986; Kamal-Eldin and Appelqvist 1996). a-tocopherol, the most biologically active and abundant form of vitamin E in vivo, efficiently transfers a hydrogen atom to a lipid free radical, such as peroxyl, alkoxyl, and carbon-centered radicals, giving the corresponding nonradical product of the lipid and an α-tocopheroxyl radical. The α-tocopheroxyl radical, once formed, reacts with a second free radical or each other to form a nonradical product. Each molecule of α-tocopherol consumes, thus, two lipid free radicals and terminates the free-radical chain reaction. As with phospholipids, the polar chroman ring tends to stay near the edges of the membrane, whereas the hydrophobic core will be buried deep into the membrane. When a phospholipid tail becomes peroxidized by a free radical, the tail becomes more polar and migrates to the surfaces where it can meet the tocopherol chroman ring to be neutralized, while forming a tocopheroxyl radical. The tocopheroxyl radical can be reduced (restored) to tocopherol directly by ubiquinol or vitamin C—and then by glutathione or lipoic acid (via vitamin C), which are in turn reduced by NADH or NADPH (Thiele et al. 1997a; Shang 2003; Bruno et al. 2006).

The rate at which tocopherol isomers react with peroxyl radicals is a direct measure of their antioxidant efficiency (Burton and Ingold 1981; Niki et al. 1984). It has been determined that a-tocopherol is the most efficient chain-breaking antioxidant among the four isomers. a-tocopherol also can react with alkoxyl radicals or undergo self-coupling to form dimers and trimers. a-Tocopherol at high concentrations induces the formation of lipid hydroperoxides (Koskas et al. 1984; Terao and Matsushita 1986). The pro-oxidant effect of a-tocopherol was related to the reaction of a-tocopheroxyl radicals with lipids (Mukai et al. 1993a).

Where is Vitamin E Found in Skin?

An important review (Thiele et al. 2007) outlined that a-tocopherol is the predominant vitamin E homologue in murine and human skin (Shindo et al. 1994; Thiele et al. 1997b, 1998). In addition, γ-tocopherol is present in murine and human epidermis, dermis, and stratum corneum. The a-tocopherol/γ-tocopherol molar ratio in the human dermis and epidermis is 10:1. In humans, levels of vitamin E in the epidermis are higher than in the dermis (Rhie et al. 2001). Skin may also contain measurable amounts of other diet-derived tocopherols and tocotrienols (Ikeda et al. 2001).

Vitamin E first accumulates in the sebaceous glands before it is delivered to the skin surface through sebum (Ekanayake-Mudiyanselage et al. 2004; Weber et al. 1999; Thiele et al. 1999). There are no transport proteins specific for vitamin E in the skin. Due to its lipophilic nature, vitamin E can also penetrate into all underlying layers of the skin (Thiele et al. 1999). Exposures to UV light (Thiele et al. 1998; Shindo et al. 1994; Weber et al. 1997) or ozone (Thiele et al. 1997c; Valacchi et al. 2000) lower the vitamin E content in skin, primarily in the stratum corneum. Vitamin E concentrations in the human epidermis also decline with age.

Effects of Vitamin E in Skin Health

The primary photoprotective effect of vitamin E, which is to prevent damage induced by free radicals and reactive oxygen species is attributed to its role as a lipid soluble antioxidant. The use of vitamin E in the prevention of UV-induced damage has been extensively studied. According to the review of Thiele (2007), when vitamin E is applied topically on the skin prior to UV exposure this significantly reduces acute skin responses, such as erythema and edema, sunburn cell formation (Darr et al. 1996; Lin et al. 2003; Ritter et al. 1997), lipid peroxidation (Lopez-Torres et al. 1998; Yuen and Halliday 1997), DNA-adduct formation, immunosuppression (Gensler and Magdaleno 1991; Yuen and Halliday 1997), as well as UVA-induced binding of photosensitizers (Beijersbergen van Henegouwen et al. 1995). Chronic skin reactions due to prolonged UVB/UVA-exposure, such as skin wrinkling (Jurkiewicz et al. 1995), and skin tumor incidence (Bissett et al. 1992; Burke et al. 2000) were also diminished by topical vitamin E formulations. Chung (2002), demonstrated that a topical pretreatment with 5% vitamin E for 24 h protected against UV-induced upregulation of human macrophage metalloelastase in human skin in vivo (Chung et al. 2002). Together with other studies (Lopez-Torres et al. 1998), this work suggests that topically applied vitamin E has the potential to penetrate into dermal layers, where much of oxidative protein oxidation occurs (Sander et al. 2002), and thus protects against photoaging. Vitamin E esters, particularly vitamin E acetate, were also shown to be promising agents in reducing UV-induced skin damage (Beijersbergen van Henegouwen et al. 1995; Burke et al. 2000; Record et al. 1991; Trevithick and Mitton 1993). Also vitamin E has been used successfully in chronic inflammatory skin conditions, either alone (Tsoureli-Nikita et al. 2002; Keller and Fenske 1998) or in combination with vitamin C (Hayakawa et al. 1981) or vitamin D (Javanbakht et al. 2011), thus suggesting a true anti-inflammatory action.

In a recent review (Schagen et al. 2012) which summarized multiple clinical studies that tested the effects of oral tocopherol the data seemed to be controversial between animals and human studies. An early study of vitamin E supplementation in hairless mice found no effect of dietary alpha-tocopherol acetate on UV-induced carcinogenesis (Pauling et al. 1982). Another study in mice found a reduction of

UV-induced DNA damage with dietary alpha-tocopherol acetate, but no other effects on free radical damage were observed in the skin (Record et al. 1991). One human study reported that subjects taking 400 IU/day of alpha-tocopherol had reduced UV-induced lipid peroxidation in the skin but concluded there was no overall photoprotective effect (McArdle et al. 2004). But another study was found that high doses of oral vitamin E may affect the response to UVB in humans (Boelsma et al. 2001).

A long-term study observed the effects of a combination of ascorbic acid and D-α-tocopherol (vitamin E) administered orally to human volunteers on UVB-induced epidermal damage. The treatment was well-tolerated and could be used against the hazardous effects of UV irradiation and skin cancer, according to the authors (Placzek et al. 2005). Results of another study suggest that a mixture of tocopherols and tocotrienols may be superior to alpha-tocopherol alone, as the mixture showed reduced sunburn reactions and tumor incidence after UV exposure in mice (Yamada et al. 2008).

Carotenoids

Carotenoids-Chemistry

Carotenoids are vitamin A derivates like β-carotene, astaxanthin, lycopene, which are all highly effective antioxidants and have been documented to possess photoprotective properties. Carotenoids are hydrophobic molecules commonly located within cell membranes (Semba et al. 2007). They are structurally similar to vitamin A and can be divided into two groups—those which can be converted to vitamin A, such as β-carotene, α-carotene, and β-cryptoxanthin which are called provitamin A carotenoids and those which cannot. These are nonprovitamin A carotenoids and are mainly lycopene, lutein, and its isomer zeaxanthin (McGuire and Beerman 2007).

Carotenoids are isoprenoid compounds, biosynthesized by tail-to-tail linkage of two C_{20} geranylgeranyl diphosphate molecules. This produces the parent C_{40} carbon skeleton from which all the individual variations are derived. This skeleton can be modified (1) by cyclization at one end or both ends of the molecule to give different end groups, (2) by changes in hydrogenation level, and (3) by addition of oxygen-containing functional groups. Carotenoids that contain one or more oxygen atoms are known as xanthophylls, the parent hydrocarbons as carotenes. The most characteristic feature of the carotenoid structure is the long system of alternating double and single bonds that forms the central part of the molecule. This constitutes a conjugated system in which the π-electrons are effectively delocalised over the entire length of the polyene chain. This feature is responsible for the molecular shape, chemical reactivity, and light-absorbing properties, and hence colour of carotenoids (Britton 1995). In vivo, carotenoids are commonly located in membranes where they constitute an integral part of the complex membrane structure. Moreover, the physical and chemical properties of a carotenoid are influenced by interactions with other molecules, e.g., lipids and proteins in the membranes (Britton 1995).

Carotenoids Antioxidants or Prooxidants?

In addition to the provitamin A activity of some carotenoids, the possible beneficial effects of carotenoids on health were linked to their role as antioxidants. According to Omaye (Omaye et al. 1997), much of the evidence has supported the hypothesis that lipid oxidation or oxidative stress may be the underlying mechanism in chronic diseases and that β-carotene would act as an antioxidant in vivo. Furthermore, Burton and Ingold suggested that β-carotene would be an unusual type of lipid antioxidant working at low-oxygen concentrations (Burton and Ingold 1984). In all, carotenoids have been considered as antioxidants, rather than pro-oxidants based on experimental evidence in vitro (Handelman 1996; Sies 1992).

It is known that carotenoids may act as antioxidants by quenching singlet oxygen or by reacting with free radicals. Moreover, the antioxidant/pro-oxidant properties of carotenoids are affected by the concentration of carotenoid, oxygen partial pressure, and the nature of the environment. The mechanisms of reactions between carotenoids and radical species may involve radical addition, hydrogen abstraction, and electron transfer, but the precise antioxidant/pro-oxidant mechanisms remain unclear (Britton 1995).

The antioxidant/pro-oxidant action of carotenoids on lipid oxidation has been of interest in food lipids as well as in biological membrane lipids, since the major constituents of biological membranes are lipids and proteins, lipid oxidation can damage membrane lipids. Most of the previous research on the effects of carotenoids on lipid oxidation has reported antioxidant activity in in vitro membrane lipid models (Krinsky 1993; Handelman 1996). Although carotenoids have not been considered as food antioxidants, some authors have proposed antioxidant role for carotenoids in food lipids. In different experimental conditions, there is some data discrepancy as it has shown both antioxidant (Kiritsakis and Dugan 1985; Jung and Min 1991) and pro-oxidant activity in food lipids (Warner and Frankel 1987). In all, β-carotene has been the prototype for examining antioxidant action of carotenoids in different in vitro lipid models. Thus, there has been a clear need for information on antioxidant/pro-oxidant effects of carotenoids other than β-carotene as well as carotenoid-tocopherol interaction.

Carotenoids and Skin Protection

Carotenoids have demonstrated to provide photoprotection from UV-induced free radical damage which can eventually suppress the immune function (Roberts et al. 2004). Especially lutein, may have the potential to absorb damaging blue-light wavelengths (Roberts et al. 2009). Lutein and zeaxanthin has shown protection against skin damage, such as edema and hyperplasia, caused by UV and UV light induced immunosuppression in vivo. A 2006 report using topical and oral administration suggested lutein and zeaxanthin may potentially increase skin elasticity and superficial lipids, stimulate skin hydration, and reduce lipid peroxidation (Roberts et al. 2009). β-carotene may provide some protection from UV-induced erythema especially when used with other antioxidants (Sies and Stahl 2005). A randomized, double, blind, placebo- controlled study found an antioxidant combination significantly decreased UV light induced erythema (Braun and Cohen 2007).

Take Home Messages

- The primary photoprotective effect of vitamin E—topical or oral application—to prevent skin damage induced by free radicals and reactive oxygen species, is attributed to its role as a lipid soluble antioxidant.
- Since the exact antioxidant/pro-oxidant mechanism of carotenoids remain unclear, there is a need for more information on antioxidant/pro-oxidant effects of carotenoids other than β-carotene, as well as on carotenoid-tocopherol interaction.

References

Beijersbergen van Henegouwen GMJ, Junginger HE, de Vries H. Hydrolysis of RRR-a-tocopheryl acetate (vitamin E acetate) in the skin and its UV protecting activity (an in vivo study with the rat). J Photochem Photobiol B Biol. 1995;29:45–51.

Bissett DL, Chatterjee R, Hannon DP. Protective effect of a topically applied anti-oxidant plus an anti-inflammatory agent against ultraviolet radiation-induced chronic skin damage in the hairless mouse. J Soc Cosmet Chem. 1992;43:85–92.

Boelsma E, Hendriks HF, Roza L. Nutritional skin care: health effects of micronutrients and fatty acids. Am J Clin Nutr. 2001;73:853–64.

Braun L, Cohen M. Herbs and natural supplements-an evidence based guide. 2nd ed. Amsterdam: Elsevier; 2007.

Britton G. Structure and properties of carotenoids in relation to function. FASEB J. 1995;9:1551–8.

Bruno BS, Leonard SW, Atkinson L, Montine TJ, Ramakrishnan R, Bray TM, Traber MG. Faster plasma vitamin E disappearance in smokers is normalized by vitamin C supplementation. Free Radic Biol Med. 2006;40(4):689–97. Epub 2005 Nov 15. PubMed PMID: 16458200

Burke KE, Clive J, Combs Jr G.F, Commisso J, Keen CL, Nakamura RM. Effects of topical and oral vitamin E on pigmentation and skin cancer induced by ultraviolet irradiation in skin: 2 hairless mice. Nutr Cancer. 2000;38:87–97.

Burton GW, Ingold KU. Autoxidation of biological molecules, l. The antioxidant activity of vitamin E and related chain-breaking phenolic antioxidants in vitro. J Am Chem Soc. 1981;103:6472–7.

Burton GW, Ingold KU. β-Carotene: an usual type of lipid antioxidant. Science. 1984;244:569–73.

Burton GW, Ingold KU. Vitamin E: application of the principles of physical organic chemistry to the exploration of its structure and function. Acc Chem Res. 1986;19:194–201.

Burton GW, Traber MG. Vitamin E: antioxidant activity, biokinetics, and bioavailability. Annu Rev Nutr. 1990;10:357–82.

Chan HW-S. The mechanism of autoxidation. In: Chan HW-S, editor. Autoxidation of unsaturated lipids. London: Academic; 1987. pp 1–16.

Chung JH, Seo JY, Lee MK, Eun HC, Lee JH, Kang S, Fisher GJ, Voorhees JJ. Ultraviolet modulation of human macrophage metalloelastase in human skin in vivo. J Invest Dermatol. 2002;119:507–12.

Darr D, Dunston S, Faust H, Pinnell S. Effectiveness of antioxidants (vitamin C and E) with and without sunscreens as topical photoprotectants. Acta Dermatol Venereol. 1996;76:264–8.

Ekanayake-Mudiyanselage S, Kraemer K, Thiele JJ. Oral supplementation with all-Rac- and RRR-alpha-tocopherol increases vitamin E levels in human sebum after a latency period of 14–21 days. Ann N Y Acad Sci. 2004;1031:184–94.

Frankel EN. Lipid oxidation. Prog Lipid Res. 1980;19:1–22.

Frankel EN. Recent advances in lipid oxidation. J Sci Food Agric. 1991;54:495–511.

Fryer MJ. The antioxidant effects of thylakoid vitamin E (a-tocopherol). Plant Cell Environ. 1992;15:381–92.

Gensler HL, Magdaleno M. Topical vitamin E inhibition of immunosuppression and tumorigenesis induced by ultraviolet radiation. Nutr Cancer. 1991;15:97–106.

Handelman GJ. Carotenoids as scavengers of active oxygen species. In: Cadenas E, Packer L, editors. Handbook of antioxidants. New York: Marcel Dekker; 1996. pp 259–314.

Hayakawa R, Ueda H, Nozaki T, et al. Effects of combination treatment with vitamins E and C on chloasma and pigmented contact dermatitis. A double blind controlled clinical trial. Acta Vitaminol Enzymol. 1981;3(1):31–8.

Ikeda S, Toyoshima K, Yamashita K. Dietary sesame seeds elevate alpha- and gamma-tocotrienol concentrations in skin and adipose tissue of rats fed the tocotrienol-rich fraction extracted from palm oil. J Nutr. 2001;131(11):2892–7.

Javanbakht MH, Keshavarz SA, Djalali M, et al. Randomized controlled trial using vitamins E and D supplementation in atopic dermatitis. J Dermatol Treat. 2011;22(3):144–50.

Jung MY, Min DB. Effects of quenching mechanisms of carotenoids on the photosensitized oxidation of soybean oil. J Am Oil Chem Soc. 1991;68:653–8.

Jurkiewicz BA, Bissett DL, Buettner GR. Effect of topically applied tocopherol on ultraviolet radiation-mediated free radical damage in skin. J Invest Dermatol. 1995;104:484–8.

Kamal-Eldin A, Appelqvist L-A. The chemistry and antioxidant properties of tocopherols and tocotrienols. Lipids. 1996;31:671–701.

Keller KL, Fenske NA. Uses of vitamins A, C, and E and related compounds in dermatology: a review. J Am Acad Dermatol. 1998;39(4 Pt 1):611–25.

Kiritsakis A, Dugan LR. Studies in photooxidation of olive oil. J Am Oil Chem Soc. 1985;62:892–6.

Koskas JP, Cillard J, Cillard P. Autoxidation of linoleic acid and behavior of its hydroperoxides with and without tocopherols. J Am Oil Chem Soc. 1984;61:1466–9.

Krinsky NI. Actions of carotenoids in biological systems. Annu Rev Nutr. 1993;13:561–87.

Lin JY, Selim MA, Shea CR, Grichnik JM, Omar MM, Monteiro-Riviere NA, Pinnell SR. UV photoprotection by combination topical antioxidants vitamin C and vitamin E. J Am Acad Dermatol. 2003;48:866–74.

Lopez-Torres M, Thiele JJ, Shindo Y, Han D, Packer L. Topical application of a-tocopherol modulates the antioxidant network and diminishes ultraviolet-induced oxidative damage in murine skin. Brit J Dermatol. 1998;138:207–15.

Mayer H, Metzger J, Isler O. The stereochemistry of natural gamma-tocotrienol (plastochromanol-3), plastochromanol-8 and plastochromenol-8. Helv Chim Acta. 1967;50:1376–93.

McArdle F, Rhodes LE, Parslew RA, et al. Effects of oral vitamin E and beta-carotene supplementation on ultraviolet radiation-induced oxidative stress in human skin. Am J Clin Nutr. 2004;80(5):1270–5.

McGuire M, Beerman KA. Nutritional sciences: from fundamentals to foods. Canada: Thomson Wadsworth; 2007.

Mukai K, Sawada K, Kohno Y, Terao J. Kinetic study of prooxidant effect of tocopherol. Hydrogen abstraction from lipid hydroperoxides by tocopheroxyls in solution. Lipids. 1993a;28:747–52.

Niki E, Saito T, Kawakami A, Kamiya Y. Inhibition of oxidation of methyl linoleate in solution by vitamin E and vitamin C. J Biol Chem. 1984;259:4177–82.

Omaye ST, Krinsky NI, Kagan VE, Mayne ST, Liebler DC, Bidlack WR. β-Carotene: friend or foe?. Fundam Appl Toxicol. 1997;40:163–74.

Pauling L, Willoughby R, Reynolds R, Blaisdell BE, Lawson S. Incidence of squamous cell carcinoma in hairless mice irradiated with ultraviolet light in relation to intake of ascorbic acid (vitamin C) and of D, L-alpha-tocopheryl acetate (vitamin E). Int J Vitam Nutr Res Suppl. 1982;23:53–82.

Placzek M, Gaube S, Kerkmann U, Gilbertz KP, Herzinger T, Haen E, et al. Ultraviolet B-induced DNA damage in human epidermis is modified by the antioxidants ascorbic acid and D-alpha-tocopherol. J Invest Dermatol 2005;124:304–7.

Porter NA. Mechanisms for the autoxidation of polyunsaturated lipids. Acc Chem Res. 1986;19:262–8 (Acta 1212, 4349)

Record IR, Dreosti IE, Konstantinopoulos M, Buckley RA. The influence of topical and systemic vitamin E on ultraviolet light-induced skin damage in hairless mice. Nutr Cancer. 1991;16:219–26.

Rhie G, Shin MH, Seo JY, et al. Aging- and photoaging-dependent changes of enzymic and non-enzymic antioxidants in the epidermis and dermis of human skin in vivo. J Invest Dermatol. 2001;117(5):1212–7.

Ritter EF, Axelrod M, Minn KW, Eades E, Rudner AM, Sarafin D, Klitzman B. Modulation of ultraviolet light-induced epidermal damage: beneficial effects of tocopherol. Plast Reconstr Surg. 1997;100:973–80.

Roberts RL, Green J, Lewis B. Lutein and zeaxanthin in eye and skin health. Clin Dermatol. 2009 March–April;27(2):195–201.

Sander CS, Chang H, Salzmann S, Muller CS, Ekanayake-Mudiyanselage S, Elsner P, Thiele JJ. Photoaging is associated with protein oxidation in human skin in vivo. J Invest Dermatol. 2002;118:618–25.

Schagen SK, Zampeli VA, Makrantonaki E, Zouboulis CC. Discovering the link between nutrition and skin aging. Dermatoendocrinology. 2012;4(3):298–307.

Semba RD, Lauretani F, Ferrucci L. Carotenoids as protection against sarcopenia in older adults. Arch Biochem Biophys. 2007 Feb;458(2):141–5.

Shang F. Vitamin C and vitamin E restore the resistance of GSH-depleted lens cells to H2O2. Free Radic Biol Med. 2003;34(5):521–30.

Shindo Y, Witt E, Han D, Epstein W, Packer L. Enzymic and non-enzymic antioxidants in epidermis and dermis of human skin. J Invest Dermatol. 1994;102:122–4.

Sies H, Stahl W. Bioactivity and protective effects of natural carotenoids. Biochim Biophys Acta (BBA)-Mol Basis Dis. 2005 May;1740(2):101–7.

Sies H, Stahl W, Sundquist AR. Antioxidant functions of vitamins. Vitamins E and C, beta-carotene, and other carotenoids. Ann N Y Acad Sci. 1992;368:7–19.

Terao J, Matsushita S. The peroxidizing effect of a-tocopherol on autoxidation of methyl linoleate in bulk phase. Lipids. 1986;21:255–60.

Thiele JJ, Ekanayake-Mudiyanselage S. Vitamin E in human skin: organ-specific physiology and considerations for its use in dermatology. Mol Aspects Med. 2007;28:646–67.

Thiele JJ, Traber MG, Tsang K, Cross CE, Packer L. In vivo exposure to ozone depletes vitamins C and E and induces lipid peroxidation in epidermal layers of murine skin. Free Radic Biol Med. 1997a;23(3):385–91. PubMed PMID: 9214574.

Thiele JJ, Traber MG, Polefka TG, Cross CE, Packer L. Ozone-exposure depletes vitamin E and induces lipid peroxidation in murine stratum corneum. J Invest Dermatol. 1997b;108:753–7.

Thiele JJ, Traber MG, Podda M, Tsang K, Cross CE, Packer L. Ozone depletes tocopherols and tocotrienols topically applied to murine skin. FEBS Lett. 1997c;401(2/3):167–70.

Thiele JJ, Traber MG, Packer L. Depletion of human stratum corneum vitamin E: an early and sensitive in vivo marker of UV induced photo-oxidation. J Invest Dermatol. 1998;110:756–61.

Thiele JJ, Weber SU, Packer L. Sebaceous gland secretion is a major physiologic route of vitamin E delivery to skin. J Invest Dermatol. 1999;113(6):1006–101. PubMed PMID: 10594744.

Trevithick JR, Mitton KP. Topical application and uptake of vitamin E acetate by the skin conversion to free vitamin E. Biochem Mol Biol Int. 1993;31:869–78.

Tsoureli-Nikita E, Hercogova J, Lotti T, Menchini G. Evaluation of dietary intake of vitamin E in the treatment of atopic dermatitis: a study of the clinical course and evaluation of the immunoglobulin E serum levels. Int J Dermatol. 2002;41(3):146–50.

Valacchi G, Weber SU, Luu C, Cross CE, Packer L. Ozone potentiates vitamin E depletion by ultraviolet radiation in the murine stratum corneum. FEBS Lett. 2000;466(1):165–8.

Warner K, Frankel EN. Effects of β-carotene on light stability of soybean oil. J Am Oil Chem Soc. 1987;64:213–8.

Weber C, Podda M, Rallis M, Thiele JJ, Traber MG, Packer L. Efficacy of topically applied tocopherols and tocotrienols in protection of murine skin from oxidative damage induced by UV-irradiation. Free Radic Biol Med. 1997;22(5):761–9.

Weber SU, Thiele JJ, Cross CE, Packer L. Vitamin C, uric acid, and glutathione gradients in murine stratum corneum and their susceptibility to ozone exposure. J Invest Dermatol. 1999;113(6):1128–32.

Yamada Y, Obayashi M, Ishikawa T, Kiso Y, Ono Y, Yamashita K. Dietary tocotrienol reduces UVB-induced skin damage and sesamin enhances tocotrienol effects in hairless mice. J Nutr Sci Vitaminol. 2008;54(2):117–23.

Yuen KS, Halliday GM. A-Tocopherol, an inhibitor of epidermal lipid peroxidation, prevents ultraviolet radiation from suppressing the skin immune system. Photochem Photobiol. 1997;65:587–92.

Chapter 21
Fatty Acids, Fatty Alcohols, Synthetic Esters and Glycerin Applications in the Cosmetic Industry

Steve Cochran and Michael Anthonavage

> **Core Messages**
>
> - The global cosmetic industry consumes millions of metric tons of lipids and their byproducts per year. These include such lipid classes as fatty acids, fatty alcohols, esters, and lipid byproducts such as glycerin.
> - The primary applications for these lipids are for product aesthetics, therapeutic function, and penetration enhancement in a variety of formulation types.

Introduction

The art of incorporating lipids in to personal care products spans a multitude of product forms including creams, gels, lotions, and serums as examples of aqueous based formulations, while lipsticks, lip balms, and mascaras constitute the anhydrous forms. It should be noted that lipids are incredibly versatile and useful and this chapter will focus on the utilization of fatty acids, fatty alcohols, esters, and glycerin for their application in personal care ingredient formulation. This text

M. Anthonavage (✉)
52 Water Street, Lebanon, NJ 0833, USA
e-mail: manthonavage@gmail.com

S. Cochran
Senior Formulation Scientist,
635 Pierce Street, Somerset, NJ 08858, USA
e-mail: scochran@presperse.com

A. Pappas (ed.), *Lipids and Skin Health,* DOI 10.1007/978-3-319-09943-9_21

will use the international nomenclature of cosmetic ingredients (INCI) to refer to specific lipid moiety examples throughout the text.

Fatty Acids

Fatty acids and fatty alcohols are the most widely utilized and most consumed lipids. Sourcing is generally natural and mainly from plant origins such as coconut and palm. Fatty acids contain a carboxylic acid and a branched or unbranched aliphatic chain of various lengths. Due to natural sourcing, carbon chain lengths are even in accordance with understood rules of biosynthesis. Industrial production processing usually relies on some form of hydrolysis of triglycerides. Fatty acids in general have two classifications systems. Saturated fatty acids are without double bonds and are considered linear and unsaturated fatty acids are nonlinear and contain one (monounsaturated) or multiple (polyunsaturated) double bond(s). Applications include most product categories ranging from soaps and deodorants to hair and skin care including color cosmetics. Saturated fatty acids and fatty alcohols function mainly as emulsifiers, stabilizers, and tactile sensory modifiers in these product categories. It is not surprising that with the exception of soaps and deodorants the largest application of fatty acids and alcohols are in emulsions systems. Chain lengths from 12 to 20 are typical for cosmetic application in emulsions, soaps, and deodorants as a general rule of thumb (Richter and Knaut 1984, 1985; Reusch 1977; Lehninger 1982; Condea 2000; Voeste and Buchold 1984; Anneken et al. 2006).

The linear nature of saturated fatty acids is known to support spherical interface structure in emulsification applications. Fatty acids are referred to as emulsifiers while the fatty alcohols are referred to as emulsion stabilizers or modifiers when used in this application. Both contribute significantly to tactile sensory properties. The common tactile sensory term to describe fatty acids and fatty alcohols is "waxy." Fatty alcohol chain lengths are almost always limited to C16–C18 or some blend of the two for cosmetic applications (Anneken et al. 2006; Harry and Rieger 2000; Schlossman 2009). Branched chain fatty alcohols are not common for cosmetic applications. However, cetyl alcohol (1-hexadecanol) is one example of a commonly employed fatty alcohol for cosmetic application with a low-melting point (Windholz 1983). Usage levels for emulsion application are typically 1.0–5.0 % when in the salt form. Fatty acids have low-water solubility. The salt formation with a base, usually sodium hydroxide, will result in a water dispensable structure for emulsion applications. The ideal carbon chain length for fatty acid emulsion application is C16 (palmitic) and C18 (stearic). Usage level is similar at 1.0–5.0 %. Occasionally C14 (myristic) can be found in emulsions for shaving application. The laws of thermodynamic equilibrium indicate that the lowest free energy states are achieved when considering carbon chain lengths of C16 and C18 in spherical structures between 0.1 and 1 μm diameters. It is also no surprise that the typical or common chain length of the fatty acids found in human cell membranes is 16–18. This similar molecular weight range is ideal for spherical structures when designing emulsions (Adkins 1968; Gray et al. 2002; Alberts et al. 2002; Budin and Devaraj 2011). Cetyl and stearyl alcohol are frequently paired with salts of stearic acid and a secondary emulsifier in modern oil-in-water cosmetic emulsions (Harry and Rieger 2000; Schlossman 2009). Cis unsaturated fatty acids are sometimes employed to reduce the spherical nature and increase the fluidity of the interface. Oleic acid, the major fatty acid found in olive oil, is an example of a common unsaturated fatty acid for this purpose. Oleic acid is also found in human cell membranes. The reason is speculated to be similar (Gray et al. 2002; Alberts et al. 2002; Kiritsakis 1998). The essential fatty acids, which are also cis unsaturated, have suggested application as biologically functional ingredients in skin care products. This suggestion is due to their role in mitigating inflammation. They are called essential as they must be ingested as they are not synthesized by the body. They include the omega-6 and omega-3 family of fatty acids and are naturally found in small amounts in the triglycerides of vegetable oils. Linoleic acid and longer chain gamma-linolenic acid and arachidonic acid are part of the omega-6 family. Alpha-linolenic acid and longer chain eicosapentaenoic acid and docosahexaenoic acid are part of the omega-3 family and are considered antiinflammatory (Brenner 2004; Kiritsakis 1998; Williams et al. 2013). It is noteworthy to apply a greater ratio of omega 3 fatty acids over omega 6 essential fatty acids as part of a therapeutic regime for sensitive skin formulations, as omega 6 essential fatty acids are used as a substrate for the creation of arachidonic acid, a pro inflammatory lipid mediator. When formulating products for sensitive skin, this becomes important factor to consider as some traditionally used shorter chain fatty acids such as those found in coconut oil can be irritating and comedogenic to consumers with sensitive skin (National Toxicology Program 2001). In addition, there is no specificity for either omega 3 or omega 6 fatty acids for the desaturase enzyme pool found in skin to process the essential fatty acids so favoring the omega 3's reduces the potential for arachidonic acid loading in the cell membranes. Shorter chain fatty acids applied at very small usage levels can have therapeutic effect in terms of skin penetration to enhance the delivery of other actives in the formulation. Lauric acid is an example of a shorter chain fatty acid. Higher usage levels above 1.0% can be irritating to the skin and caution should be considered when formulating (Pornpattananangkul et al. 2010).

Fatty Alcohols

The fatty alcohols are speculated to increase emulsion stability through transition temperature (T_g) modulation of the microsphere film. This is a result of increasing the packing efficiency of the fatty acids at the oil and water interface. Product form is a critical design goal and part of the consumer experience during cosmetic product usage and purchase. Product stability ensures proper intended product form specified in the design goals. Fatty acids and fatty alcohols contribute significantly during emulsion design to ensure product stability. Tactile sensory goals are also a critical part of product design. In many cases, it is the single most influential product attribute that consumers will use when making a purchase choice or evaluating the performance of a product. The influence fatty acids and fatty alcohols play in tactile sensory perception is an important role during and after product usage. Fatty acids impart sensory aspects such as slip, tact, glide, and shine.

The film left on the skin or hair after product application of products containing both fatty acids and alcohols are described as "waxy" during sensory evaluation. The waxy film is mostly perceived after a short period of time after product application. During product application fatty acids and alcohols are thought to contribute to tactile sensory attributes such as "dry time" and "play time" (Condea 2000; Harry and Rieger 2000; Schlossman 2009).

Esters

Fatty acids and alcohols when combined by the chemical reaction referred to esterification are called esters. The reaction equilibrium is controlled by the removal of water during processing and is therefore considered a condensation process. The reverse of the esterification reaction is referred to as hydrolysis and is catalyzed by heat and extreme pH conditions. Process temperatures and pH requirements of formulations are a critical consideration during the design of cosmetics containing esters (Reusch 1977).

Triglycerides sourced from naturally occurring oils are considered natural esters. Triglycerides are a combination of glycerol and three fatty acids. Cosmetic application of natural esters almost always includes esters from plant sources (Lehninger 1982; Noweck and Grafahrend 2006; Anneken et al. 2006). The lower molecular weight and cyclic esters have aromatic application in the cosmetic industry (Gottschalck and Bailey 2008; Harry and Rieger 2000). Many possible combinations of fatty alcohols and acids result in a plentitude of synthetic esters available to the cosmetic product designer. Currently there are hundreds of ester choices ranging from simple linear to complex branched, and low to high molecular weight and polarity (Gottschalck and Bailey 2008). The dynamic range of polarity includes esters that are water dispersible to esters that are insoluble in some hydrocarbons. Melting points range to yield both liquid and solid physical forms at room temperatures. Cosmetic application includes all categories and product forms. Typical usage level is 1.0–20 %. Esters are one of the most popular and major cosmetic ingredients and in some product categories, such as skin lotions, the major ingredient (Harry and Rieger 2000; Schlossman 2009). Monoglycerol esters function as emulsifiers for both oil-in-water and water-in-oil designs. Typical usage levels for this application range from 1.0 to 5.0 %. The moderate polar esters have application as pigment dispersants and organic sunscreen boosting properties (Gottschalck and Bailey 2008; Harry and Rieger 2000; Schlossman 2009). Very-high molecular weight and extensively branched esters which are synthesized from guerbet alcohol are known as guerbet esters (Reusch 1977). INCI: Trioctyldodecyl citrate is one example of a guerbet ester which despite its high molecular weight

(66 carbons) has a low-melting point due to the extensive branching. Guerbet esters also have suggested applications as pigment dispersants and organic sunscreen boosters (Gottschalck and Bailey 2008). The linear and branched esters function as tactile sensory modifiers giving short- and long-term sensory benefits due to their emollient nature. Long-term benefits are consumer perceived as skin softening or emollient. This benefit is believed to be achieved through the plasticization effect esters have on the proteins in the epidermis and hair upon penetration of the surface. The natural esters have an even higher tendency to be absorbed into the skin and hair. This phenomenon is believed to be due to the similarity in polarity, molecular weight, and geometry of the natural esters and sebum (Mackenna et al. 1950; Camera et al. 2010). Natural oils such as coconut and olive have a long history of cosmetic application supported by this belief (Kiritsakis 1998; Nutritional composition of Mediterranean crops 1980; Gode et al. 2012; Keis et al. 2005). Linear and branched esters contribute to the immediate short-term sensory attributes referred to as "cushion" and "slip" during product application. Some high-molecular weight branched esters are perceived to contribute an abundant amount of cushion and slip are suggested as mineral oil replacements. It should be noted that there is a trend in skin care formulations to avoid mineral oil use due to impurities (polyaromatic and polycyclic hydrocarbons), but overall, mineral oil use has been plentiful and safe. It can also be very occlusive preventing other skin care ingredients from penetrating the skin and many consumers are mindful of materials that originate from the petroleum industry as mineral oil does. Molecular weight and/or chain length correlate with "cushion," while increase branching results in more "slip" (Gorcea and Laura 2010). Reduction of skin friction is considered at least one cause of improved "slip" effect (Sivamani et al. 2005). INCI: Neopentylglycol dicaprate/dicaprylate is an example of a complex high-molecular weight branched ester with application as a mineral oil replacement. Short chain lower-molecular weight simple linear esters deliver a "silky" and "dry" tactile sensory profile. Some have suggested application as a silicone replacement. This is due to the very "silky" and silicone like tactile sensory profile they deliver. INCI: Ethyl palmate and ethyl oleate are two examples of low-molecular weight simple linear esters with suggested application as silicone replacements as well. The popular application of esters in cosmetic formulation across all product categories and forms is partially due to the abundance of these desirable and deliverable sensory benefits. The global cosmetic industry consumption of esters is estimated in the thousands of metric tons (Gorcea and Laura 2010; Gottschalck and Bailey 2008).

Glycerin/Glycerol

Glycerol or INCI: Glycerin is referred to as a "humectant" when considering cosmetic application. The three free hydroxyl functional groups can potentially bond to water changing the water dynamics of products during application and usage. Glycerol functions as an osmotic balancing ingredient and contributes directly to skin health. Glycerol, therefore, is considered a biologically functional cosmetic ingredient. Sourced as a byproduct in biodiesel production, the global cosmetic industry consumption of glycerol is estimated in the millions of metric tons (Gorcea and Laura 2010; Gottschalck and Bailey 2008). Cosmetic applications for glycerol also include the entire range of product forms found in personal care. Although glycerin is the major ingredient in deodorant sticks and glycerin soaps, utilization in emulsions with intended use on hair and skin is the primary application (Smolinske 1992). Glycerol can also act as a tactile sensory modifier in this application, both during and after product application. The sensory benefits are not so obvious during the design process and a general trend is difficult to define due to the dynamic nature and potential association with other ingredients in the formulation. A more obvious tactile sensory attribute is more of a concern when contributing to "tack" during and after product application. The source of the tack is usually through hydrogen bonding to the skin during association with other ingredients in the formulation. Additionally glycerin is one of the most commonly used raw materials used in cosmetics and is used as a humectant/moisturizer in various formulation mediums for the relief of dry skin and conditions that require the attraction of water to the stratum corneum environment. Such conditions are aimed to treat or prevent dry, rough, scaly, itchy skin, and minor skin irritations. In addition, glycerin has long been utilized in pharmaceutical applications and medicine for its humectant film forming properties. The osmotic action of glycerin allows it to be used therapeutically in conjunction with salts and minerals that are used to draw water from skin ulcers, lesions, and burned skin. Transdermal therapeutic systems also use glycerin as a method of drug administration, allowing for rate-controlled drug delivery and avoidance of first-pass metabolism in the liver (Jenerowicz et al. 2012). This modality is used along with urea and other ingredients to increase skin permeability in reduced hydration conditions (Bjorklund et al. 2013). Additionally, it is used in the herbal community in the preservation of poultices to prevent from drying out during storage (Sheu et al. 2002). Glycerin's primary mechanism of action is a humectant and it is extremely hygroscopic, meaning that it readily absorbs water from other sources. So, in part, glycerin works because of its ability to attract water from the environment and from the lower layers of skin (dermis) increasing the amount of water in the surface layers of skin (Wells 1957; Appa and Orth 1999). Another aspect of glycerin's benefit is that it is a skin-identical ingredient, meaning it is a substance found naturally in skin. In that respect, it is one of the many substances in skin that help maintain the outer barrier and prevent dryness or scaling. This property of glycerin also makes it useful for hair and scalp treatments as well. Male skin care needs in particular are heavily influenced by the need to remove facial hair on a regular basis. Facial skin issues associated with poor hair removal practices are common and include razor burn and irritation. The importance of maintaining hair softness during the shave and restoring facial hydration post-shave are indications for the use of glycerin. Post-shave moisturizers containing glycerin and emollients can create an environment for improved barrier function which can be further improved by incorporating specific ingredients such as niacinamide.

Glycerin combined with other emollients and/or oils is a fundamental cornerstone of most moisturizers. Many skin care formulations on the market today try to maximize the amount of glycerin in formulation. There is no research showing that higher amounts of glycerin have any increased benefit for skin. A major drawback of glycerin when used in pure form is that they can increase water loss by attracting water from the lower layers of skin (dermis) into the surface layers of skin (epidermis) where the water can easily be lost into the environment. As a result, glycerin at high concentrations by itself is not recommended. For this reason, glycerin and other humectants in general are always combined with other ingredients such as amino acids, fatty acids, and carbohydrates to soften skin. The research shows a combination of ingredients including glycerin, dimethicone, petrolatum, antioxidants, fatty acids, lecithin, among many others, are excellent for helping skin heal, reduce associated dermatitis, and restore normal barrier function if used on an ongoing basis (Lodén 2003; Hara-Chikuma and Verkman 2008; Fluhr et al. 2008; Breternitz et al. 2008).

The sensory benefits are not so obvious during the design process and a general trend is difficult to define due to the dynamic nature and potential association with other ingredients in the formulation. A more obvious tactile sensory attribute is more of a concern when contributing to "tack" during and after product application. The source of the tack is usually through hydrogen bonding to the skin during association with other ingredients in the formulation (Gorcea and Laura 2010). However, humectants such as glycerin can provide undesirable effects if used in excess. Aesthetically, they can become sticky. It may compromise the application of cosmetics (powders and films) over the top of them creating pilling and spottiness.

When properly formulated, glycerin fortifies the skin's natural defense properties by filling in the intercellular matrix surrounding the keratinocytes in the epidermis and attracting the right amount of water to maintain the skin's homeostasis. There is also research indicating that the presence of glycerin in the intercellular layers of the skin helps other skin lipids do their jobs better (Fowler 2000; Fluhr et al. 1999).

The Consumer Experience

Overall, the consumer experience will influence product acceptance and perception of performance. Ultimately this experience will influence the marketing success of the product. Fatty acids, alcohols, esters, and glycerol all contribute to the consumer experience through tactile sensory modification which is one of the most critical and valuable tool in product design. Consideration to how these materials change the appearance of the hair and skin is almost of equal importance when evaluating the consumer experience. The cause of the visual effect of changing the appearance of skin is mostly due to changing the optical properties of the skin. The immediate and longer-term effects caused by optical property modifications of the skin are in part modulated by water. Long term and immediate tactile sensory attributes are also influenced by water. Controlling water in skin and hair is the foundation of product design in the cosmetic industry. The insight of understanding how and when to do this will determine a positive or negative consumer experience. Except for safety, consumer wellness is rarely considered during the cosmetic design process due to the regulatory restrictions put on cosmetic claims. Although cosmetics generally are designed to mostly affect the sensory aspects of hair and skin, it is important to note that a key function of all hair and skin products is to keep outside water out and inside water in. This is following and supporting one of the natural intended functions of skin and hair. The water dynamics of the hydrophobic films and osmotic gradients left on skin and hair by products containing fatty acids, alcohols, synthetic esters, and glycerin further support this natural intention (Gorcea and Laura 2010; Harry and Rieger 2000; Schlossman 2009).

References

Adkins CJ. Equilibrium thermodynamics. London: McGraw-Hill; 1968.

Alberts B, Johnson A, Lewis J, et al. Molecular biology of the cell. 4th edn. New York: Garland Science; 2002. ISBN 0-8153-3218-1.

Anneken DJ, Both S, Christoph R, Fieg G, Steinberner U, Westfechtel A. “Fatty acids”. In: Elvers B, editor. Ullmann’s encyclopedia of industrial chemistry. Weinheim: Wiley-VCH; 2006.

Appa Y, Orth D. Glycerin: a natural ingredient for moisturizing skin, dry skin and moisturizers: chemistry and function. Boca Raton: CRC Press; 1999.

Bjorklund S, et al. Glycerol and urea can be used to increase skin permeability in reduced hydration conditions. Eur J Pharm Sci. 2013;50(5):638–45.

Brenner, J. Applications of essential fatty acids in skin care, cosmetics and cosmeceuticals. Cosmet Toiletries. 2004;119(3):75.

Breternitz M, Kowatzki D, Langenauer M, Elsner P, Fluhr JW. Placebo-controlled, double-blind, randomized, prospective study of a glycerol-based emollient on eczematous skin in atopic dermatitis: biophysical and clinical evaluation. Skin Pharmacol Physiol. 2008;21:39–45.

Budin I, Devaraj NK. Membrane assembly driven by a biomimetic coupling reaction. J Am Oil Chem Soc. 2011;134(2):751–3.

Camera E, Ludovici M, Galante M. Comprehensive analysis of the major lipid classes in sebum by rapid resolution high-performance liquid chromatography and electrospray mass spectrometry. J Lipid Res. 2010;51(11):3377–88. First Published on Aug 17, 2010.

Condea. All about Fatty Alcohols, Online digital library by Eskazzi.

Fluhr JW, Gloor M, Lehmann L, Lazzerini S, Distante F, Berardesca E. Glycerol accelerates recovery of barrier function in vivo. Acta Derm Venereol. 1999;79(6);418–21.

Fluhr JW, Darlenski R, Surber C. Glycerol and the skin: holistic approach to its origin and functions. Br J Dermatol. 2008;159(1):23–34.

Fowler JF Jr. Efficacy of a skin-protective foam in the treatment of chronic hand dermatitis. Am J Contact Dermatitis. 2000;11(3):165–9.

Gode V, Bhalla N, Shirhatti V, Mkaskar S, Kamath YK. Quantitative measurement of the penetration of coconut oil into human hair using radiolabeld cocnut oil. J Cosmet Sci. 2012;63(1):27–31.

Gorcea M, Laura D. International Specialty Products (ISP). Evaluating the physiochemical properties of emollient esters for cosmetic use. Cosmet Toiletries. 2010 Dec issue.

Gottschalck TE, Bailey JE. International cosmetic ingredient dictionary and handbook. 12th edn. Washington: The Cosmetic, Toiletry, and Fragrance Association; 2008.

Gray J, Groeschler S, Le T, Gonzalez Z. “Membrane structure” (SWF). Davidson College; 2002. Retrieved: 11 Jan 2007.

Hara-Chikuma M, Verkman AS. Aquaporin-3 facilitates epidermal cell migration and proliferation during wound healing. J Mol Med. 2008;86(2):221–31.
Harry RG, Rieger MM. Harry's cosmeticology. 8th edn. Weymouth: Micelle Press; 2000.
Jenerowicz D, et al. Skin hypersensitivity reactions to transdermal therapeutic systems-still an important clinical problem. Ginekol Pol. 2012;83(1):46–50.
Keis K, Peraud D, Kamath YK, Rele AS. Investigation of penetration abilities of various oils into human hair fibers. J Cosmet Sci. 2005;56:283–95.
Kiritsakis AK. Olive oil, from the tree to the table. 2nd edn. Trumbull Connecticut: Food and Nutrition Press; 1998.
Lehninger AL. Principles of biochemistry. New York: Worth Publishing; 1982. pp. 583–614.
Lodén M. Role of topical emollients and moisturizers in the treatment of dry skin barrier disorders. Am J Clin Dermatol. 2003;4(11);771–88.
Mackenna RMB, Wheatley VR, Wormall A. The composition of the surface skin fat ('sebum') from the human forearm. J Invest Dermatol. 1950;15(1):33–47.
National Toxicology Program. Toxicology and carcinogenesis studies of coconut oil acid diethanolamine condensate (CAS No. 68603-42-9) in F344/N rats and B6C3F1 mice (dermal studies). Natl Toxicol Program Tech Rep Ser. 2001;479:5–226.
Noweck K, Grafahrend W. Fatty alcohols. In: Elvers B, editor. Ullmann's encyclopedia of industrial chemistry. Weinheim: Wiley-VCH; 2006.
Nutritional composition of Mediterranean crops (per 100 g of edible portion). Source: Goulart (1980); Sawaya et al. (1983); Fernandez Diez (1983); IBPGR (1986); Morton (1987); Cantwell (1994).
Pornpattananangkul D, Olson S, Aryal S, Sartor M, Huang C-M, Vecchio K, Zhang L. Stimuli-responsive liposome fusion mediated by gold nanoparticles. ACS Nano. 2010;4(4):1935–42. 100317124523087. doi:10.1021/nn9018587.
Reusch WH. An introduction to organic chemistry. San Francisco: Holden Day; 1977. pp. 407–36.
Richter HJ, Knaut J. Challenges to a mature industry: marketing and economics of oleochemicals in Western Europe. J Am Oil Chem Soc. 1984;61(2):160–75.
Richter HJ, Knaut J. Trends in industrial uses of palm and lauric oils. J Am Oil Chem Soc. 1985;62(2):317–27.
Schlossman ML. Chemistry and manufacture of cosmetics: volume II-formulating. 4th edn. Carol Stream: Allured Books; 2009.
Sheu MT, Chen L-C, Ho H-O. Simultaneous optimization of percutaneous delivery and adhesion for ketoprofen poultice. Int J Pharm. 2002;233(1–2):257–62.
Short RW, Chan JL, Choi JM, Egbert, BM, Rehmus, WE, Kimball, AB. Effects of moisturization on epidermal homeostasis and differentiation. Clin Exp Dermatol. 2007;32(1):88–90.
Sivamani RK, Wu G, Gitis NV, Maibach HI. Skin friction coefficient values. Cosmet Toiletries. 2005;120(9):48.
Smolinske SC. Handbook of food, drug, and cosmetic excipients. Boca Raton: CRC Press; 1992. pp. 75–6. ISBN 0-8493-3585-X.
Voeste T, Buchold H. Production of fatty alcohols from fatty acids. J Am Oil Chem Soc. 1984;61(2):350–2.
Wells F. Glycerin as a constituent of cosmetics and toilet preparations. Drug and Chemical Exports. 1957 Dec.
Williams JJ, Mayurasakorn K, Vannucci SJ, Mastropietro C, Bazan NG, Ten VS, Deckelbaum RJ. N-3 fatty acid rich triglyceride emulsions are neuroprotective after cerebral hypoxic-ischemic injury in Neonatal mice. PLoS ONE. 2013;8(2):e56233. doi:10.1371/journal.pone.0056233.
Windholz M. Merck index. 10th edn. White House Station: Merck & Co; 1983. p. 282.

Chapter 22
Fats and Oils in Personal Care Products

K. Mahmood

Introduction

Cosmetics have been used throughout history and by all known civilizations of the past. It is well documented that Egyptians used goat milk, metal oxides, and coal for various needs of the body, from body wash to decorating body parts. A number of published reviews cover the history and chemistry of cosmetics (Gardner 1962; Berdick 1972; DeNavarre 1978; Armentano 1979; Brand and Brand 1989; Draelos 2000; Kim and Seyke 2002; Nardello-Rataj and Bonte 2008). The focus of this chapter is to highlight cosmetic use of various oils in modern times, which may also include different fats and waxes and has not been the subject of earlier publications. There have been three main sources of oils and fats, namely animal, botanical, and petroleum based. There is also a new emerging sustainable source of oils and fats made available from renewable/bio-based cellulosic feedstock or from algae. Products from these new technologies have not yet made inroads into the cosmetic industry but are knocking at the door. The main thrust in this direction is coming from sustainability initiatives in all industry sectors, plus regulations in food and pharma that specifically require a supply chain traceable to the origin of materials sold commercially and also the adoption of the Convention on Biological Diversity (CBD) (Morand 2010; Robert 2011) by nations across the globe. These initiatives are affecting the manufacturing and supply of natural-origin cosmetic ingredients.

The main purpose of this chapter is to make the reader aware of the current choices of oils available for use in cosmetics. While the US cosmetic industry is regulated by the Food and Drug Administration under the Food, Drug and Cosmetic Act, and the Fair Packaging and Labeling Act, there is no premarket approval of cosmetic products or ingredients, with the exception of color additives. Finished product manufacturers are solely responsible for assuring the safety of the ingredients they use and the final finished product prior to marketing. Since the

K. Mahmood (✉)
Johnson & Johnson Consumer and Personal Products Worldwide, New Jersey, USA
e-mail: kmahmoo1@its.jnj.com

A. Pappas (ed.), *Lipids and Skin Health*, DOI 10.1007/978-3-319-09943-9_22

ingredient labeling of cosmetic products became law in the late 1970s, the industry's trade association, the Personal Care Products Council (PCPC), has worked on the development of an international nomenclature system (INCI names) for cosmetic raw materials. The compilation of INCI names and their reported usage is maintained in a database and published by the PCPC as the *International Cosmetic ingredient Dictionary and Handbook* (Gottschalck and Breslawec 2012). The PCPC's database is the basis for this chapter to access and analyze cosmetic ingredients.

Lipids in Skin Health

Cosmetic products can be a delivery vehicle for oily and fatty ingredients, which are beneficial to skin health and together may also be referred to as lipids. Skin is composed of both hydrophilic and hydrophobic components such as water, proteins, and lipids. Therefore, maintaining a good health for the skin requires enough water and appropriate amounts and types of lipid components. Skin lipids are composed of simple and complex lipids; for example, fatty acids are simple lipids and ceramides are complex lipids. Simple lipids are more accessible and are the main sources of lipid components used in a cosmetic formula necessary to maintain a good balance of hydration and protection needs of the skin.

Oils and fats have various roles in a cosmetic formula; for example, they act as a vehicle for a set of ingredients in formula as lubricants, emollients, cleansers, deodorizers, moisturizers, or simply protecting the skin from the environment (Rabasco and Gonzalez Rodriguez 2000). When ingested they can also be nourishing to the skin, helping to provide a natural balance of skin lipids either initiating repair and/or stimulating endogenous levels by providing with necessary building blocks as needed (Valla 2010; Boelsma et al. 2001).

Oils and Fats

Most of the natural oils from vegetable or animal sources are composed of esters made from glycerol and free fatty acids. Free fatty acids can be fully saturated or with a degree of unsaturation. Oils can also be volatile or nonvolatile, immiscible, or have small miscibility in water. Oil in cosmetic formula serves different purposes based on its structure, volatility, composition, or ability to act as a carrier for other ingredients.

Oils and fats can be grouped in many different ways based on their use, chemistry, origin, preparation, and feedstock deployed to make a specific product. They can be as neutral as saturated hydrocarbons or can have polar endings, as for example, do phospholipids. In this chapter we will review simple natural oils specifically originating from botanicals and animals. One section is also devoted to renewable

Table 22.1 Chemical classes of oils and fats with use in cosmetics

Count	Chemical class	Number of entries	Sources
1	Complex lipids	74	Botanicals, animals, synthetics, and single-cell organisms
2	Essential oils	383	Botanicals
3	Glyceride esters	453	Botanicals, animals, synthetics, and single-cell organisms
3	Fatty acids	50	Animals, botanicals, synthetics
4	Fatty alcohols	31	Synthetics, animals, botanicals
5	Unsaponifiables	27	Botanicals and synthetics
6	Waxes (neutral and synthetic)	97	Animals, botanicals, synthetics
7	Hydrocarbons	149	Synthetics, botanicals, animals
Total	All	1264	A small number of duplications may occur

feedstock as a source of making sustainable equivalent of either vegetable oils or petroleum-based oils.

Oil and Fats with Use in Cosmetics

According to the *International Cosmetic Ingredient Dictionary & Handbook* there are more than 1100 entries from various sources and for different uses. They are classified based on chemical functional groups into at least seven classes with little overlap, resulting in the listing of an ingredient sometimes in multiple classes. Table 22.1 below summarizes the names of classes used in the cosmetic dictionary. The naming and classification of cosmetic ingredients is well defined by using common names as much as possible or as appropriate. This helps the consumer with easy-to-understand labels but also helps the manufacturers to keep labels as simple as possible.

Complex Lipids

Complex lipids are present in all living tissues where they play important structural and functional roles. An earlier review published on the topic by Hanahan and Thompson (1963) is a good starting point for beginners to learn about the role of complex lipids in various species and tissues. Alibardi (2003) has published on the role of complex lipids along with the role of other chemical entities during the evolutionary process while species were adapting to land. Human skin is a layered structure and its main function is to protect our internal vital organs. The stratum corneum being the outermost layer, exposed to environmental aggressions, is

Table 22.2 Source and functions of select complex lipids

Complex lipid type	Source	Functions[a]
Ceramides	Mostly synthetic with some botanical origin	Hair and skin-conditioning agents
Sphingosine	Synthetic, botanical, and animal	Skin protecting, anti-acne, anti-microbial, hair, and skin-conditioning agents
Glycerylphosphocholine	Synthetic, botanical, and animal	Anti-oxidant, emulsifier, skin protecting, surfactant, and skin conditioning
Phosphatidic acid	Botanical, synthetic, and animal	Humectant, skin protecting, bleaching, emollient, hair conditioning, surfactant, and emulsifier

[a] Listed functions do not form an exhaustive list; rather they are indicators of target function; functions identified are those that were reported by ingredient suppliers

composed of lipids and other structural elements, including dead cells. The stratum corneum also regulates the entry of external ingredients in to the skin, for example, any cosmetic ingredient for nourishing needs or for hydration purposes. Some of the complex lipids already in cosmetic use may include sphingolipids, choline, ceramides, sophora lipids, phosphatidic acid, glycolipids, and phospholipids. Sphingolipids are functionally important for the skin as barrier components. A few more common types of complex lipids available are listed in the following table along with their reported functions (Table 22.2).

Essential Oils

Essential oil or plant oil is composed of water-insoluble compounds containing volatile aroma components. They are used as fragrance, flavor, or as pharmacological actives. Their use in cosmetics is restricted though due to their capacity for skin irritation and side effects associated with the reactive nature of some of the components, which are present in the plant oils. Some of the components are specifically forbidden or limited by regulations within the European Union. A list generally known as the "EU list of 26 Allergens" is available for guidance (http://ec.europa.eu/enterprise/sectors/chemicals/files/legislation/allergenic_subst_en.pdf).

The chemistry of essential oils is diverse and not the same as that of other commonly known oils, for example cooking oils from plants that are composed of triglycerides. Essential oils may contain one or more of terpenes; hydroxyl compounds such as alcohols and phenols; carboxylates such as esters, lactones, and coumarins; ethers; oxides; and carbonyl compounds such as aldehydes and ketones. Many references, reviews, and book chapters have been written on the topic (Baser et al. 2007; Oberdieck 1972; Littlejohn 1940). With such a diverse chemistry it is not surprising that essential oils are used for many purposes from aroma therapy to pharmacological actives and anything in between (Miyazawa 2011). Sometimes they are also prepared and used as a dilute version called waters, for example rose water, white lily flower water, lemongrass oil, citrus peel water, and many others. More than 700 plant oils and waters are available for commercial use in cosmetics. The most common uses for essential oils are undoubtedly fragrance related. However, small uses as antimicrobial, emollient, skin protecting, etc. are also common in cosmetics. For example, lemongrass essential oils are reported to be effective as anti-allergenic and anti-inflammatory (Mizushina et al. 2013).

Glyceride Esters

More than 450 cosmetic agents belong to fats and oil category and are composed of glyceride esters made from a molecule of glycerin and one to three molecules of fatty acids. They are called mono-, di-, or triglycerides depending on the number of fatty acids linked to glycerin. The fatty acids can be of one kind or of different kinds for di- and triglycerides, they can be all saturated or have unsaturated fatty acids in the mix. They can be naturally obtained from plant or animal sources or made synthetically. Animal-sourced fats and oils sometimes come from unusual sources such as cattle, donkey, horse, sheep, whale (illegal in the USA), bear, rabbit, ostrich, and brain lipids among others.

The properties of the fats and oils are depended on various factors such as degree of esterification, degree of un-saturation, its source, purity, length of carbon chain of fatty acids, etc. Some of the uses in cosmetics include humectants, skin-protecting agents, emollients, occlusive, viscosity modifiers, and skin and hair-conditioning agents, etc. They are also commonly used in massage formula preparations. They are also essential for human health in general and brain health in particular; for example, cod liver oil is traditionally consumed and touted as a healthy supplement.

Fatty Acids

Fatty acids are straight chain hydrocarbons having at least seven carbons with a carboxylic function at one end. Fatty acids with only even carbon atoms exist in eukaryotic organisms and are either fully saturated systems or possess some unsaturation. Unsaturated fatty acids are also known as omega fatty acids, which are essential for nutrition and health, for example, linoleic acid, arachidonic acid, and docosahexaenoic acid. Saturated fatty acids are mostly deployed in cosmetics, e.g. stearic acid, palmitic acid, myristic acid, etc. The most common source of fatty acids is hydrolysis of oil from vegetable sources such as grains, seeds, and nuts. They can also be obtained from animal fats, for example, arachidonic acid, beeswax acid, and tallow acid. A few of them are synthetic in the cosmetic industry, for example, pelargonic acid, undecanoic acid, arachidic acid, and hydrogenated fatty acids.

Traditionally, the most common products with fatty acids are soaps, which are cleansing products. Cleansing remains a major function of fatty acids but they have additional important functions in cosmetic products as emulsifying agents, fragrances, antifungals, emollients, skin bleaches, opacifiers, pH adjusters, and occlusive agents.

Fatty Alcohols

Fatty alcohols are produced from fatty acids by reducing the carboxylic end moiety into primary alcohol. They are neutral molecules with a basic skeleton similar to the parent fatty acids. They can also be made completely by synthetic chemistry. Their main functions in cosmetics are emulsification and viscosity modifiers. They are also raw materials for making new cosmetic ingredients with a new set of properties. Some examples of cosmetic ingredients derived from fatty alcohols are alkyl sulfates, alkyl ether sulfates, and alkoxylated alcohols.

Unsaponifiables

Just like their name implies, unsaponifiables are mixtures of natural compounds that are not hydrolyzed during the processing of oils and fats of plant origin. They are part of process waste stream from vegetable oil refineries. The composition varies considerably and it depends not only on the source of materials but also the type and extent of processing. A typical unsaponifiable material may comprise glycerides of all types, free fatty acids, sterols, tocopherols, tocotrienols, waxes, fatty alcohols, etc. These materials are deployed for skin and hair-conditioning purposes with additional antioxidant properties. The exact function of the material is depended on further purification and extent of neutralization making it suitable for a specific cosmetic use. It is worth noting that unsaponifiables are a rich source of vitamins such as vitamin E and its unsaturated cousins, tocotrienols. These vitamins in natural form are isolated from unsaponifiables of soy and palm oil refineries.

Waxes

Waxes are high molecular weight natural or synthetic lipids. Natural waxes can come from plant, animal, or petroleum sources. Notable examples of waxes used in cosmetics are lotus wax, beeswax, citrus wax, carnauba wax, lanolin, paraffin, propolis, and shellac. They act as occlusive, emollient, or skin-protecting agents in a cosmetic formula. They may also be used as viscosity adjusters and binders in a cosmetic formula.

Table 22.3 A brief summary of nontraditional methods of making hydrocarbons and triglycerides

Product	Traditional source	New sources	Comments
Hydrocarbons	Petroleum	Biobased feedstock, e.g. wood, corn syrup, sugar cane, triglycerides, etc.	Fermentation or metal catalysis produced hydrocarbons of various types
Triglycerides	Botanicals	Microorganisms, e.g. microalgae	Triglycerides with a control on carbon length, saturation, and composition

Hydrocarbons

This category comprises mostly synthetic source materials specifically petroleum based. A few of them are also sourced from animals, for example, squalene. Hydrocarbons are generally inert chemicals made entirely from two types of atoms, carbon and hydrogen. They can be of complex structure containing aromatics or simple straight chain hydrocarbons. Lower molecular weight hydrocarbons can be used as fuel, for example, natural gas, diesel, and automobile and jet fuels. Hydrocarbons with medium molecular weights are used in cosmetic formulas, for example, mineral oil and petrolatum. The most common cosmetic uses include skin-protecting agents, emollients, viscosity modifiers, solvents, and skin or hair conditioning.

Renewable Materials

Source of Hydrocarbons and Triglycerides

The traditional source of hydrocarbons is petroleum process streams. New sources are being developed, which are relatively more sustainable with higher biobased contents and does not depend on fossil fuel (Thomas et al. 2002) e.g. biobased fuel production. New technologies have been developed utilizing biotechnology principles combined with green chemistry principles to convert a source of cellulosic material such as wood (Goldstein 1975) to new or known chemical diversity with or without oxygen functionalities. Ethanol produced from corn is one of the well-known examples of such transformations. In another example, triglycerides are produced from unicellular algae in variable compositions depending on the culture conditions (Meltzger et al. 1985). These processes allow bulk production of known and unknown triglycerides. In yet other transformations triglycerides, which are more readily available from renewable sources, can also be converted into hydrocarbons (Kubickova and David 2010). In general fats and oils irrespective of their origin are being used downstream in the production of other raw materials, for example, surfactants and emulsifiers for use in personal care products. Table 22.3 presents a nonspecific short summary of new sources that are under development specifically to produce hydrocarbons and triglycerides.

Acknowledgements I would like to acknowledge helpful feedback and prereview of this chapter by Ms Joanne Nikitakis of the Personal Care Products Council.

References

Alibardi L. Adaptation to the land: the skin of reptiles in comparison to that of amphibians and endotherm amniotes. J Exp Zool Mol Dev Evol. 2003:298B(1):12–41.
Armentano MA. Origin and development of cosmetic science and technology. Rev Bras Quimica. (1979);87(527):143–8.
Baser K, Husnu C, Demirci F Chemistry of essential oils. In: Berger RG, editor Flavours and Fragrances. 2007:43–86.
Berdick M Role of fats and oils in cosmetics. J Oil Chem Soc. (1972);49(7):406–9.
Boelsma E, Hendriks HF, Roza L. Nutritional skin care: health effects of micronutrients and fatty acids. Am J Clin Nutr. 2001;73(5):853–64.
Brand EE, BrandHM. The science of beauty. Natur Techniek (Ultrecht). (1989);57(2):86–99.
DeNavarre MG. Oils and fats, the historical cosmetics. J Oil Chem Soc. 1978;55:435–7.
Draelos ZD. Cosmetics and skin care products. A historical perspective. Dermatol Clin. 2000;18(4):557–9.
Gardner LT. The cosmetic industry. J R Soc Arts. (1962);110:892-901, Disc. 901–3.
Goldstein IS. Potential for converting wood into plastics: chemicals from wood may regain importance as the cost of petroleum continues to rise. Science (New York). 1975;189(4206):847–52.
Gottschalck TE, Breslawec HP. International cosmetic ingredient dictionary and handbook. 14th ed. Washington, DC: The Personal Care Products Council; 2012.
Hanahan DJ, Thompson GA, Jr. Complex lipids. Annu Rev Biochem. 1963;32:215–40.
Kim JO, Seyke H. Understanding of chemistry in cosmetics. (2002);42(7):50–55.
Kubickova I, Kubicka D. Utilization of triglycerides and related feedstocks for production of clean hydrocarbon fuels and petrochemicals. Waste Biomass Valorization. 2010;1(3):293–308.
Littlejohn WR. Terpeneless and sesquiterpeneless essential oils. Their characteristics, advantages and mode of employment. Flavours. 1940;3(4):7–18.
Morand S. Biodiversity: an international perspective. Rev Sci Tech. 2010;29(1):65–72.
Meltzger P, Tabache M, Cassadevall E. Triacylglycerol production by three unicellular green algae. Variation of a lipid extract during different culture phases. Agrochimica. 1985;29(2-3-4):256–64.
Miyazawa M. Oreo Saiensu. Studies on metabolism and biological activities of chemical components in essential oil. 2011;11(12):463–80.
Mizushina Y, Sugimoto K, Mitoshi M, Jippo T. Effects of essential oils from herbal plants on anti-allergic and anti-inflammatory activities and their application for functional cosmetics. Frag J. 2013;41(6):42–50.
Nardello-Rataj V, Bonte F. Chemistry and cosmetics. A long history punctuated by innovations. Actual Chim. 2008;323–4:10–2.
Oberdieck R. Riechstoffe Aromen Koerperpflegemittel. Flavors in spices and spicy herbs. 1972;22(390):393–5; (129):293–4, 296; (11):431–2 (1973;23(1):3–4, 6; (2):29–30, 32; (4):107–8, 110, 113).
Rabasco AM. Gonzalez Rodriguez ML. Lipids in Pharmaceutical and cosmetic preparations. Grasas Aceites. 2000;51(1–2):74–96.
Robert B. A new beginning for biodiversity. C R Biol. 2011;2010;334(5–6):483–8.
Thomas S, DiCosimo R, Nagarajan V. Biocatalysis: applications and potentials for the chemical industry. Trends Biotechnol. 2002;20(6):238–42.
Valla C. Omega-3 fatty acids from fish oil: effects on the skin. Nutrafoods. 2010;9(2):33–9.

Part VIII
Clinical Crossover Lipids and Skin

Chapter 23
Acne and Lipid Pathways

Christos C. Zouboulis, Eric Jourdan and Mauro Picardo

Core Messages

- Hyperseborrhea has been considered as a major etiopathogenetic factor of acne.
- Changes in sebaceous gland activity not only correlate with seborrhea but also with alterations in sebum fatty acid composition.
- Sebum lipid fractions with proinflammatory properties are associated in the process of the development of acne lesions.
- The oxidant/antioxidant ratio of the skin surface lipids and alterations of lipid composition are the main players in the induction of acne inflammation.
- Nutrition may influence the development of seborrhea, the fractions of sebum lipids, and acne.

Abstract Hyperseborrhea has been considered as a major etiopathogenetic factor of acne. However, changes in sebaceous gland activity not only correlate with seborrhea but also with alterations in sebum fatty acid composition. Current findings

An article with similar content has been published in: Zouboulis CC, Jourdan E, Picardo M. Acne is an inflammatory disease and alterations of sebum composition initiate acne lesions. J Eur Acad Dermatol Venereol. 2014;28:527–32.

C. C. Zouboulis (✉)
Departments of Dermatology, Venereology, Allergology and Immunology, Dessau Medical Center, Auenweg 38, 06847 Dessau, Germany
e-mail: christos.zouboulis@klinikum-dessau.de

E. Jourdan
Laboratoire dermatologique BIODERMA, Lyon, France
e-mail: e.jourdan@bioderma.com

M. Picardo
Laboratory of Cutaneous Physiopathology and Integrated Center of Metabolomics Research, San Gallicano Dermatological Institute, IRCCS, Via Elio Chianesi 53, 00144 Rome, Italy
e-mail: picardo@ifo.it

A. Pappas (ed.), *Lipids and Skin Health,* DOI 10.1007/978-3-319-09943-9_23

indicate that sebum lipid fractions with proinflammatory properties and inflammatory tissue cascades are associated in the process of the development of acne lesions. The oxidant/antioxidant ratio of the skin surface lipids and alterations of lipid composition are the main players in the induction of acne inflammation. Nutrition may influence the development of seborrhea, the fractions of sebum lipids, and acne. Acne is an inflammatory disease probably triggered, among others, by proinflammatory sebum lipid fractions.

Abbreviations

5-LOX	5-lipoxygenase
COX-2	Cyclooxygenase-2
CRH	Corticotropin-releasing hormone
FADS	Fatty acid desaturase
FoxO1	Forkhead box transcription factor O1
IL	Interleukin
K	Keratin
P. acnes	*Propionibacterium acnes*
PPAR	Peroxisome proliferator-activated receptor
mTORC1	Mammalian target of rapamycin complex 1
SCD	Stearoyl-CoA desaturase
TLR	Toll-like receptor

Introduction: Inflammation in Acne

Induction of inflammatory signalling in the pilosebaceous unit is a major component in the process of the initiation of acne lesions (Zouboulis 2001, 2004a; Zouboulis et al. 2005). Among the 211 upregulated and the 18 downregulated genes in lesional skin of acne patients—compared to normal human skin—a significant proportion is involved in pathways that regulate inflammation (Trivedi et al. 2006a). Scarce inflammatory infiltrates around the ductus seboglandularis, and later on perifollicular infiltration and enhanced cytokine expression at the mRNA and protein levels are closely associated with comedone formation (Jeremy et al. 2003); comedones do not develop in a later stage leading to "inflammatory comedones," as previously reported (Chiba et al. 2000). NF-κB, a transcription factor critical for upregulation of several proinflammatory cytokine genes, has been shown to be activated in acne lesions (Kang et al. 2005). Interestingly, interleukin (IL)-1α is strongly expressed in comedones (Chiba et al. 2000; Anttila 1992; Ingham et al. 1992). It induces hyperproliferation, assessed by the enhanced expression of the hyperproliferative markers keratin (K) 6 and K16 (Hughes et al. 1996) and disturbs terminal differentiation of infundibular keratinocytes, which is related to increased filaggrin expression (Kurokawa et al. 1988), leading to hyperkeratinization in the follicular infundibulum detected in vivo and ex vivo (Guy et al. 1996). IL-1α activates basal keratinocytes by autocrine production inducing K16 expression in suprabasal cells in the active state.

Sebaceous Glands and Innate Immunity

Follicular keratinocytes and sebocytes, the major components of the pilosebaceous unit, may act as symbiotically or immune responding cells capable of microbia recognition and abnormal lipid presentation (Koreck et al. 2003). Innate immunity molecules, such as toll-like receptor (TLR)2, TLR4, CD1d, and CD14, are expressed in human keratinocytes (Song et al. 2002) and SZ95 sebocytes (Koreck et al. 2003; Oeff et al. 2006). Acting that way, keratinocytes and sebocytes may be activated by *Propionibacterium acnes* (*P. acnes*) and recognize altered lipid content in sebum, followed by the production of proinflammatory cytokines. In addition, antimicrobial peptides, such as defensin-1, defensin-2, and cathelicidin, are expressed and are active in the sebaceous gland (Chronnell et al. 2001; Nagy et al. 2006; Nakatsuji et al. 2010; Chen et al. 2011). Human β-defensin-2 is expressed upon exposure to lipopolysaccharides and *P. acnes* (Nagy et al. 2006) and upregulated by sebum free fatty acids (Nakatsuji et al. 2010).

Sebum and Acne

The most obvious function of the sebaceous gland is to excrete sebum (Zouboulis 2004a). Sebum is a mixture of relatively nonpolar lipids, most of which are synthesized de novo by the sebaceous glands (Nikkari 1974). The composition of sebum is remarkably species- and age-specific (Nikkari 1974; Ramasastry et al. 1970; Picardo et al. 2009; Pappas 2009a). Human sebaceous glands secrete a lipid mixture containing squalene and wax esters, as well as cholesterol esters, triglycerides, and possibly some free cholesterol and fatty acids.

For a long time, hyperseborrhea has been considered as a major etiopathogenetic factor for acne. However, emerging data on alterations of sebum composition in acne patients (Picardo et al. 2009; Makrantonaki et al. 2011; Pappas et al. 2009) indicate that sebum composition may be more important for the development of acne lesions than the secreted amount. Indeed, bacterial hydrolases convert some of the triglycerides to free fatty acids on the skin surface (Nicolaides and Wells 1957). On the other hand, there is also evidence that sebaceous glands can also synthesize considerable amounts of free fatty acids (Zouboulis 2001).

Indeed, the oxidant/antioxidant ratio of the skin surface lipids (Stewart et al. 1986) has been taken into consideration in the etiopathogenesis of acne and other skin diseases. Oxygen and micro-organisms transform “native” sebum, with lysis of triglycerides into fatty acids being their most pronounced activity on the skin (Patel and Noble 1992; Saint-Léger 2003). The quantities of lipid peroxide, IL-1α, and NF-κB were found significantly higher in the content of comedones than those in the stratum corneum (Tochio et al. 2009). Certain components of this complex mixture of molecules present in the sebum are clearly cytotoxic or irritant, provoking reactive follicular hyperkeratosis and comedone formation—the first step to acne. Particular attention has been focused on peroxidation of squalene, a sebaceous gland-specific lipid, e.g., by ultraviolet radiation, which led to comedogenesis on the rabbit ear skin (Chiba et al. 2000). Squalene peroxide has been shown to induce an inflammatory response in HaCaT keratinocytes through lipoxygenase activation and increase in the proinflammatory cytokine IL-6 production (Ottaviani et al. 2006). Inflammation of acne-involved sebaceous glands is also associated with lipoxygenase activation and intracellular IL-6 increase (Alestas et al. 2006). Therefore, lipoxygenase activity products may contribute to an implementation of the inflammatory reaction with a concomitant anti-inflammatory feedback response of noninvolved cells of the pilosebaceous unit, as demonstrated by the concomitant increase of peroxisome proliferator-activated receptor (PPAR)α mRNA and protein levels. The clinical relevance of these findings were corroborated by the antiacne activity of zileuton, a 5-lipoxygenase (5-LOX) inhibitor (Zouboulis 2009; Zouboulis et al. 2010).

Stearoyl-CoA desaturase (SCD) and fatty acid desaturase (FADS)-2 are enzymes responsible for the biosynthesis of monounsaturated fatty acids in human sebocytes. In a feedback mode, their expression is downregulated by their products and upregulated by the unspecific Gram+bacterial antigen and TLR-2 ligand macrophage-activating lipopeptide-2 (MALP-2; Georgel et al. 2005; Zouboulis et al. 2011). Interestingly, while *P. acnes* is unable to induce IL-1α expression in the pilosebaceous unit (Ingham et al. 1998; Seltmann et al. 2000), oleate (C18:1)—through keratinocyte toxicity—causes increased IL-1α mRNA levels. Therefore, alterations of saturated and unsaturated fatty acid composition in sebum have currently been taken into consideration as initiators of follicular inflammation and regulators of innate symbiotic and immunity response (Makrantonaki et al. 2011; Ottaviani et al. 2006). Among the sebum lipids, the ones produced by the sebaceous glands are of great importance for the development of acne. Lower essential fatty acid levels were found in wax esters in twins with acne than in twins without acne (Stewart 1992). Several free fatty acids were detected to express pro-inflammatory and anti-inflammatory properties (Nakatsuji et al. 2010; Alestas et al. 2006; Wróbel et al. 2003; Makrantonaki and Zouboulis 2007). For example, high levels of linoleate (C18:2), an essential ω6-fatty acid (Stewart et al. 1986), may protect from the development of comedonal acne (Nicolaides et al. 1972) and its topical application reduces microcomedones and inhibits steroid 5α-reductase activity (Letawe et al. 1998; Namazi 2004). On the other hand, low linoleate levels have been observed in skin surface lipids of acne patients (Downing et al. 1986). However, neither all ω6-fatty acids are comedogenic, not all ω9-fatty acids inhibit comedogenesis (Fig. 23.1). For example, oleate alters the calcium dynamics in epidermal keratinocytes and induces abnormal follicular keratinization leading to comedogenesis in rabbit skin (Choi et al. 1997; Katsuta et al. 2005) but to minor irritation in human skin (Boelsma et al. 1996). SZ95 sebocytes in vitro were currently shown to produce the same amount of lipids after incubation with linoleate and palmitate (16:0). However, their effects on sebocyte inflammatory signaling were strikingly different (Seltmann et al. 2013).

On the other hand, fatty acids exhibit strong antimicrobial activity. The sebaceous ω9-fatty acids sapienate (C16:1δ6), palmitate (C16:0), and oleate (C18:1) are very effective against *Staphylococcus aureus* (Chen et al. 2011; Georgel et al.

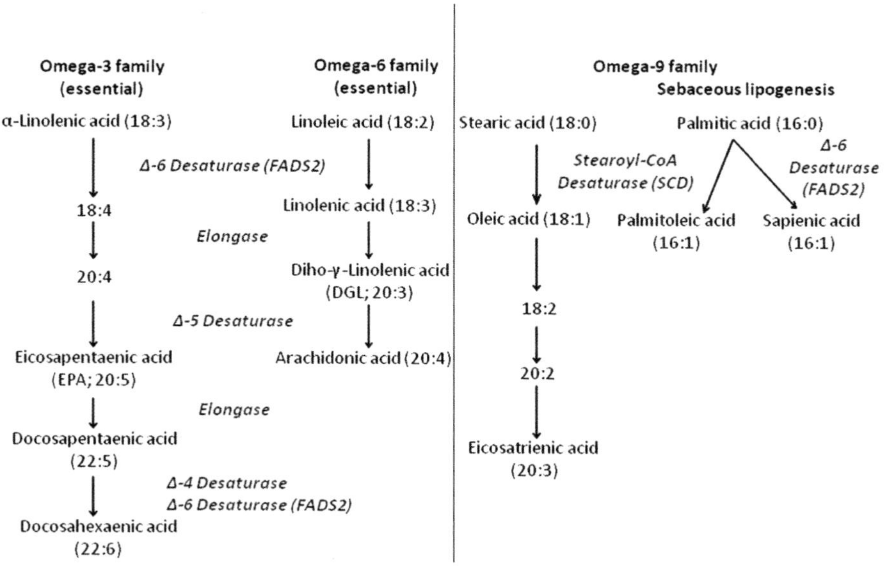

Fig. 23.1 Sebaceous lipogenesis is dependent on the fatty acids available. While unsaturated ω3 and ω6 fatty acids are essential, ω9 can be synthetized by the sebaceous glands. The saturated/unsaturated fatty acid ratio defines inflammatory triggering and can initiate comedogenesis

2005; Wille and Kydonieus 2003; Drake et al. 2008). Moreover, dysfunction of the upstream lipidogenic enzymes SCD and FADS-2 is associated with skin infection and inflammation (Georgel et al. 2005; Seltmann et al. 2000). Lipids at the skin surface, mostly secreted from the sebaceous glands (90 %) and transported through the follicular canal, are part of the symbiotic and innate immunity of the skin and contribute to the antimicrobial skin barrier.

PPAR and Acne

Certain lipid mediators, which are able to interfere with sebocyte differentiation and lipogenesis, have been shown to activate and/or be ligands of PPAR (Makrantonaki et al. 2011; Ottaviani et al. 2006; Alestas et al. 2006; Chen et al. 2003; Zhang et al. 2006). Importantly, lipid peroxidation products are also capable of inducing PPAR activation and production of proinflammatory cytokines. In particular, PPARα seems to be related to β-oxidation of fatty acids and lipid catabolism, whereas PPARγ activation has been linked to lipogenesis (Ferré 2004). Eicosanoid metabolites originated from the arachidonic acid cascade, namely leukotriene B_4 and 15-HETE, have been shown to be ligands of PPARα and PPARγ, respectively (reviewed in Zouboulis et al. 2005; Alestas et al. 2006). Interestingly, the enzymes involved in their formation, including 5-LOX, have been implicated in inflammatory skin diseases characterized by keratinocyte hyperproliferation (Ottaviani et al. 2006) and have been found to be expressed at higher extent in acne-involved skin in comparison to the skin of healthy subjects (Alestas et al. 2006). Activation of 5-LOX results, among other effects, in induced IL-6 and IL-8 expression in human sebocytes, whereas enhanced expression of IL-6 and IL-8 has also been found in acne-affected skin (Alestas et al. 2006). Systemic treatment of acne patients with the 5-LOX inhibitor Zileuton reduces the inflammatory lesion count and the synthesis of sebum lipids, in particular, of those with proinflammatory potential (Zouboulis 2009) through an inflammation-preventive mechanism (Zouboulis et al. 2010). 5-LOX inhibitors may also downregulate the inflammatory activity of lymphocytes and macrophages resulting in cumulative beneficial effects (Jeremy et al. 2003).

Prostaglandins are further proinflammatory mediators thought to be involved in acne lesion development (Zhang et al. 2006). Mice with increased cyclooxygenase-2 (COX-2) expression and prostaglandins E2 levels showed sebaceous gland hyperplasia and enhanced sebum production (Neufang et al. 2001) suggesting an important role for COX-2 signaling pathway in sebocyte biology. Expression and activation of COX-2 has been shown in in vitro models to be PPARγ-mediated. General oxidative stressors, including lipid oxidizing agents, activate PPARγ and induce lipogenesis in sebocytes (Trivedi et al. 2006a, b; Zhang et al. 2006; Ottaviani et al. 2010). All these findings allow the hypothesis that sebocyte proliferation and/or lipogenesis as well as inflammatory reaction may be regulated by PPARγ-mediated pathways.

Neuropeptides

Corticotropin-releasing hormone (CRH), the most proximal element of the hypophysis-pituitary-adrenal axis, acts as a central coordinator for neuroendocrine and behavioral responses to stress. CRH, CRH-binding protein, CRH-receptor 1, and CRH-receptor 2 are expressed in SZ95 sebocytes at mRNA and protein level, whereas CRH-receptor 1 is the predominant type (Zouboulis et al. 2002). In addition, CRH significantly upregulates mRNA levels of 3β-hydroxysteroid dehydrogenase/Δ^{5-4} isomerase and induces sebaceous lipogenesis and IL-6 and IL-8 synthesis (Zouboulis et al. 2002; Krause et al. 2007). In acne-involved skin, the complete CRH system is abundant, especially in the sebaceous glands, possibly activating lipid pathways, which affect immune and inflammatory processes leading to the development and stress-induced exacerbation of acne (Ganceviciene et al. 2009).

Diet

Evidence suggests that diet may influence acne (Rasmussen 1997; Pappas 2009b; Liakou et al. 2013; Smith et al. 2007), whereas it is also an important source of substrate for the synthesis of sebaceous lipids (Rasmussen 1997). This notion is

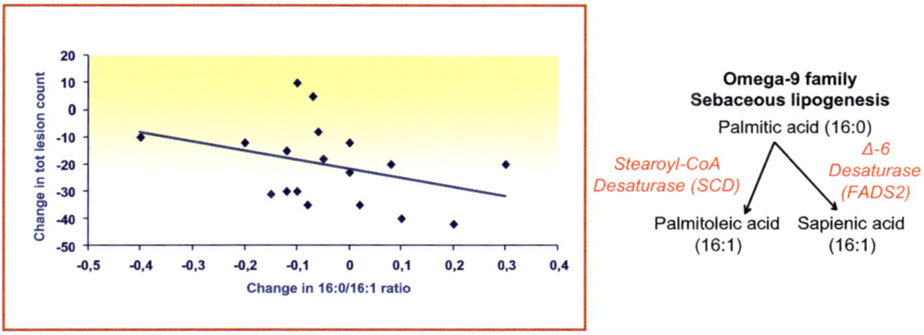

Fig. 23.2 Skin surface lipid composition under a 12-week acne diet. Decrease in the enzymatic desaturation of fatty acids correlates with the clinical improvement in acne

supported also by the observation that sebum contains essential fatty acids, such as linoleate and oleate. On the other hand, extreme caloric restriction dramatically decreases the sebum excretion rate and these changes can be reversed when a normal diet is resumed (Pochi et al. 1970; Downing et al. 1972). Other studies have demonstrated that increased consumption of dietary fat or carbohydrate increases sebum production and modifications to the type of carbohydrate can also alter sebum composition (Macdonald 1964). Typical western diet, comprised of milk and hyperglycaemic foods, may have potentiating effects on serum insulin and insulin-like growth factor-1 levels, thereby promoting the development of acne (Melnik and Schmitz 2009). In contrast, a low-glycemic-load diet for 12 weeks in acne patients reduced parallelly the acne lesion count and increased the C16:0/C16:1 fatty acid ratio (Smith et al. 2007, 2008) suggesting an increased enzymatic desaturation of fatty acids in the sebaceous glands of patients with acne (Fig. 23.2).

The nutritional cell status is primarily sensed by the forkhead box transcription factor O1 (FoxO1) and the serine/threonine kinase mammalian target of rapamycin complex 1 (mTORC1) (Wang et al. 2011). FoxO1 attenuates androgen signaling, interacts with regulatory proteins important for sebaceous lipogenesis, regulates the activity of innate and adaptive immunity, antagonizes oxidative stress, and most importantly functions as a rheostat of mTORC1, the master regulator of cell growth, proliferation, and metabolic homoeostasis. Thus, FoxO1 links nutrient availability to mTORC1-driven processes in the skin: increased protein and lipid synthesis, cell proliferation, cell differentiation including hyperproliferation of acroinfundibular keratinocytes, sebaceous gland hyperplasia, and increased sebaceous lipogenesis (Melnik and Zouboulis 2013). Deeper insights into the molecular interplay of FoxO1/mTORC1-mediated nutrient signaling are thus of critical importance to understand the impact of western diet on the promotion of epidemic acne.

Table 23.1 Sebaceous gland functions, which are possibly involved in the development of acne

Production of sebum (Zouboulis et al. 2003)
Regulation of cutaneous steroidogenesis (Thiboutot et al. 2003; Zouboulis 2004b; Chen et al. 2010; Samson et al. 2010; Slominski et al. 2013)
Regulation of local androgen synthesis (Fritsch et al. 2001)
Interaction with neuropeptides (Zouboulis et al. 2002)
Synthesis of specific lipids with antimicrobial activity (Wille and Kydonieus 2003)
Exhibition of pro- and anti-inflammatory properties (Zouboulis 2001; Zouboulis 2004a; Böhm et al. 2002)

Conclusions

Increased sebum excretion, alteration of lipid composition and the oxidant/antioxidant ratio of the skin surface lipids are major concurrent events associated with the development of acne (Zouboulis 2004a; Table 23.1). Current evidence indicates that sebum composition (lipid quality), and not quantity, plays a central role in the development of acne. This concept is supported by the mode of action of new antiacne compounds, such as the 5-LOX inhibitor Zileuton, which reduces acne lesions by inhibiting proinflammatory lipids (Zouboulis 2009; Zouboulis et al. 2010), and the current evidence of the effect of diet on acne (Melnik and Schmitz 2009). Moreover, old data on in vivo and in vitro modulation of sebaceous lipid composition by isotretinoin, the most potent antiacne drug, can be approached from this new perspective (Stewart et al. 1984; Strauss et al. 1987; Melnik et al. 1988; Zouboulis et al. 1991).

References

Alestas T, Ganceviciene R, Fimmel S, Müller-Decker K, Zouboulis CC. Enzymes involved in the biosynthesis of leukotriene B4 and prostaglandin E2 are active in sebaceous glands. J Mol Med. 2006;84:75–87.

Anttila HS, Reitamo S, Saurat J-H. Interleukin 1 immunoreactivity in sebaceous glands. Br J Dermatol. 1992;127:585–8.

Boelsma E, Tanojo H, Boddé HE, Ponec M. Assessment of the potential irritancy of oleic acid on human skin: Evaluation in vitro and in vivo. Toxicol In Vitro. 1996;10:729–42.

Böhm M, Schiller M, Ständer S, et al. Evidence for expression of melanocortin-1 receptor in human sebocytes in vitro and in situ. J Invest Dermatol. 2002;118:533–9.

Chen W, Yang C-C, Sheu E-M, Seltmann H, Zouboulis CC. Expression of peroxisome proliferator-activated receptor and CCAAT/enhancer binding protein transcription factors in cultured human sebocytes. J Invest Dermatol. 2003;121:441–7.

Chen W, Tsai S-J, Sheu H-M, Tsai J-C, Zouboulis CC. Testosterone synthesized in cultured human SZ95 sebocytes mainly derives from dehydroepiandrosterone. Exp Dermatol. 2010;19:470–2.

Chen CH, Wang Y, Nakatsuji T, et al. An innate bactericidal oleic acid effective against skin infection of methicillin-resistant staphylococcus aureus: a thera-py concordant with evolutionary medicine. J Microbiol Biotechnol. 2011;21:391–9.

Chiba K, Yoshizawa K, Makino I, Kawakami K, Onoue M. Comedogenicity of squalene monohydroperoxide in the skin after topical application. J Toxicol Sci. 2000;25:77–83.

Choi EH, Ahn SK, Lee SH. The changes of stratum corneum interstices and calcium distribution of follicular epithelium of experimentally induced comedones (EIC) by oleic acid. Exp Dermatol. 1997;6:29–35.

Chronnell CM, Ghali LR, Ali RS, et al. Human beta defensin-1 and -2 expression in human pilosebaceous units: upregulation in acne vulgaris lesions. J Invest Dermatol. 2001;117:1120–5.

Downing DT, Strauss JS, Pochi PE. Changes in skin surface lipid composition induced by severe caloric restriction in man. Am J Clin Nutr. 1972;25:365–7.

Downing DT, Stewart ME, Wertz PW, Strauss JS. Essential fatty acids and acne. J Am Acad Dermatol. 1986;14:221–5.

Drake DR, Brogden KA, Dawson DV, Wertz PW. Antimicrobial lipids at the skin surface. J Lipid Res. 2008;49:4–11.

Ferré P. The biology of peroxisome proliferators-activated receptors: relationship with lipid metabolism and insulin sensitivity. Diabetes. 2004;53:43–50.

Fritsch M, Orfanos CE, Zouboulis CC. Sebocytes are the key regulators of androgen homeostasis in human skin. J Invest Dermatol. 2001;116:793–800.

Ganceviciene R, Graziene V, Fimmel S, Zouboulis CC. Involvement of the corticotropin-releasing hormone system in the pathogenesis of acne vulgaris. Br J Dermatol. 2009;160:345–52.

Georgel P, Crozat K, Lauth X et al. A toll-like receptor 2-responsive lipid effector pathway protects mammals against skin infections with Gram-positive bacteria. Infect Immun. 2005;73:4512–21.

Guy R, Green MR, Kealey T. Modeling acne in vitro. J Invest Dermatol. 1996;106:176–82.

Hughes BR, Morris C, Cunliffe WJ, Leigh IM. Keratin expression in pilosebaceous epithelia in truncal skin of acne patients. Br J Dermatol. 1996;134:247–56 .

Ingham E, Eady EA, Goodwin CE, Cove JH, Cunliffe WJ. Pro-inflammatory levels of interleukin-1 alpha-like bioactivity are present in the majority of open comedones in acne vulgaris. J Invest Dermatol. 1992;98:895–901.

Ingham E, Walters CE, Eady EA, Cove JH, Kearney JN, Cunliffe WJ. Inflammation in acne vulgaris: failure of skin micro-organisms to modulate keratinocyte inter-leukin 1 alpha production in vitro. Dermatology. 1998;196:86–8.

Jeremy AH, Holland DB, Roberts SG, Thomson KF, Cunliffe WJ. Inflammatory events are involved in acne lesion initiation. J Invest Dermatol. 2003;121:20–7.

Kang S, Cho S, Chung JH, Hammerberg C, Fisher, GJ, Voorhees JJ. Inflammation and extracellular matrix degradation mediated by activated transcription fac-tors nuclear factor-kappaB and activator protein-1 in inflammatory acne lesions in vivo. Am J Pathol. 2005;166:1691–9.

Katsuta Y, Iida T, Inomata S, Denda M. Unsaturated fatty acids induce calcium influx into keratinocytes and cause abnormal differentiation of epidermis. J Invest Dermatol. 2005;124:1008–13.

Koreck A, Pivarcsi A, Dobozy A, Kémeny L. The role of innate immunity in the pathogenesis of acne. Dermatology. 2003;206:96–105.

Krause K, Schnitger A, Fimmel S, Glass E, Zouboulis CC. Corticotropin-releasing hormone skin signaling is receptor-mediated and is predominant in the sebaceous glands. Horm Metab Res. 2007;39:166–70.

Kurokawa I, Mayer-da-Silva A, Gollnick H, Orfanos CE. Monoclonal antibody labeling for cytokeratins and filaggrin in the human pilosebaceous unit of normal, seborrhoeic and acne skin. J Invest Dermatol. 1988;91:566–71.

Letawe C, Boone M, Piérard GE. Digital image analysis of the effect of topically applied linoleic acid on acne microcomedones. Clin Exp Dermatol. 1998;23:56–8.

Liakou AI, Theodorakis MJ, Melnik BC, Pappas A, Zouboulis CC. Nutritional clinical studies in dermatology. J Drugs Dermatol. 2013;12:1104–9.

Macdonald I. Changes in the fatty acid composition of sebum associated with high carbohydrate diets. Nature. 1964;203:1067–8.

Makrantonaki E, Zouboulis CC. Testosterone metabolism to 5α-dihydrotestosterone and synthesis of sebaceous lipids is regulated by the peroxisome proliferators-activated receptor ligand linoleic acid in human sebocytes. Br J Dermatol. 2007;156:428–32.

Makrantonaki E, Ganceviciene R, Zouboulis CC. An update on the role of the sebaceous gland in the pathogenesis of acne. Dermatoendocrinology. 2011;3:41–9.
Melnik BC, Schmitz G. Role of insulin, insulin-like growth factor-1, hyperglycaemic food and milk consumption in the pathogenesis of acne vulgaris. Exp Dermatol. 2009;18:833–41.
Melnik BC, Zouboulis CC. Potential role of FoxO1 and mTORC1 in the pathogenesis of Western diet-induced acne. Exp Dermatol. 2013;22:311–5.
Melnik B, Kinner T, Plewig G. Influence of oral isotretinoin treatment on the composition of comedonal lipids. Implications for comedogenesis in acne vulgaris. Arch Dermatol Res. 1988;280:97–102.
Nagy I, Pivarcsi A, Kis K, et al. Propionibacterium acnes and lipopolysaccharide induce the expression of antimicrobial peptides and proinflammatory cytokines/chemokines in human sebocytes. Microbes Infect. 2006;8:2195–205.
Nakatsuji T, Kao MC, Zhang L, Zouboulis CC, Gallo RL, Huang C-M. Sebum free fatty acids enhance the innate immune defense of human sebocytes by upregulating β-defensin-2 expression. J Invest Dermatol. 2010;130:985–94.
Namazi MR. Further insight into the pathomechanism of acne by considering the 5-alpha-reductase inhibitory effect of linoleic acid. Int J Dermatol. 2004;43:701.
Neufang G, Fürstenberger G, Heidt M, Marks F, Müller-Decker K. Abnormal differentiation of epidermis in transgenic mice constitutively expressing cyclooxy-genase-2 in skin. Proc Natl Acad Sci U S A. 2001;98:7629–34.
Nicolaides N, Wells GC. On the biogenesis of the free fatty acids in human skin surface fat. J Invest Dermatol. 1957;29:423–33.
Nicolaides N, Fu HC, Ansari MNA, Rice GR. The fatty acids of esters and sterol esters from vernix caseosa and from human surface lipid. Lipids. 1972;7:506–17.
Nikkari T. Comparative chemistry of sebum. J Invest Dermatol. 1974;62:257–67.
Oeff MK, Seltmann H, Hiroi N, et al. Differential regulation of toll-like receptor and CD14 pathways by retinoids and corticosteroids in human sebocytes. Dermatology. 2006;213:266.
Ottaviani M, Alestas T, Flori E, Mastrofrancesco A, Zouboulis CC, Picardo M. Peroxidated squalene induces the production of inflammatory mediators in HaCaT keratinocytes: a possible role in acne vulgaris. J Invest Dermatol. 2006;126:2430–7.
Ottaviani M, Camera E, Picardo M. Lipid mediators in acne. Mediators Inflamm. 2010;2010:pii:858176.
Pappas A. The relationship of diet and acne: a review. Dermatoendocrinol. 2009a;1:262–7.
Pappas A. Epidermal surface lipids. Dermatoendocrinology. 2009b;1:72–6.
Pappas A, Johnsen S, Liu JC, Eisinger M. Sebum analysis of individuals with and without acne. Dermatoendocrinology. 2009;1:157–61.
Patel SD, Noble WC. Changes in skin surface lipid composition during therapy for severe acne vulgaris and relation to colonisation with propionibacteria. Microb Ecol Health Dis. 1992;5:291–7.
Pochi PE, Downing DT, Strauss JS. Sebaceous gland response in man to prolonged total caloric deprivation. J Invest Dermatol. 1970;55:303–9.
Picardo M, Ottaviani M, Camera E, Mastrofrancesco A. Sebaceous gland lipids. Dermatoendocrinology. 2009;1:68–71.
Ramasastry P, Downing DT, Pochi PE, Strauss JS. Chemical composition of human skin surface lipids from birth to puberty. J Invest Dermatol. 1970;54:139–44.
Rasmussen JE. Diet and acne. Int J Dermatol. 1997;16:488–92.
Saint-Léger D. Fonction sébacée normale et pathologique. Des recherches au milieu du gué? Pathol Biol (Paris). 2003;51:275–8.
Samson M, Labrie F, Zouboulis CC, Luu-The V. Biosynthesis of dihydrotestosterone by a pathway that does not require testosterone as intermediate in the SZ95 sebaceous gland cell line. J Invest Dermatol. 2010;130:602–4.
Seltmann H, Rudawski IM, Holland KT, Orfanos CE, Zouboulis CC. Propionibacterium acnes does not influence the interleukin-1/interleukin-8 cascade in immortalized human sebocytes in vitro. J Invest Dermatol. 2000;114:816.

Seltmann H, Nikolakis G, Zouboulis CC. Novel pattern of sebaceous differentiation and lipogenesis induced by the ω-9 fatty acid palmitic acid. Exp Dermatol. 2013;22:e18.
Slominski A, Zbytek B, Nikolakis G, et al. Steroidogenesis in the skin: implications for local immune functions. J Steroid Biochem Mol Biol. 2013;137:107–123.
Smith RN, Mann NJ, Braue A, Mäkeläinen H, Varigos GA. A low-glycemic-load diet improves symptoms in acne vulgaris patients: a randomized controlled trial. Am J Clin Nutr. 2007;86:107–15.
Smith RN, Braue A, Varigos GA, Mann NJ. The effect of a low glycemic load diet on acne vulgaris and the fatty acid composition of skin surface triglycerides. J Dermatol Sci. 2008;50:41–52.
Stewart ME. Sebaceous gland lipids. Semin Dermatol. 1992;11:100–5.
Stewart ME, Benoit AM, Downing DT, Strauss JS. Suppression of sebum secretion with 13-cis-retinoic acid: effect on individual skin surface lipids and implications for their anatomic origin. J Invest Dermatol. 1984;82:74–8.
Stewart ME, Grahek MO, Cambier LS, Wertz PW, Downing DT. Dilutional effect of increased sebaceous gland activity on the proportion of linoleic acid in sebaceous wax esters and in epidermal acylceramides. J Invest Dermatol. 1986;87:733–6.
Strauss JS, Stewart ME, Downing DT. The effect of 13-cis-retinoic acid on sebaceous glands. Arch Dermatol. 1987;123:1538a–41.
Song PI, Park YM, Abraham T, et al. Human keratinocytes express functional CD14 and toll-like receptor 4. J Invest Dermatol. 2002;119:424–32.
Thiboutot D, Jabara S, McAllister JM, et al. Human skin is a steroidogenic tissue: steroidogenic enzymes and cofactors are expressed in epidermis, normal sebocytes, and an immortalized sebocyte cell line (SEB-1). J Invest Dermatol. 2003;120:905–14.
Tochio T, Tanaka H, Nakata S, Ikeno H. Accumulation of lipid peroxide in the content of comedones may be involved in the progression of comedogenesis and inflammatory changes in comedones. J Cosmet Dermatol. 2009;8:152–8.
Trivedi NR, Gilliland KL, Zhao W, Liu W, Thiboutot DM. Gene array expression profiling in acne lesions reveals marked upregulation of genes involved in inflammation and matrix remodeling. J Invest Dermatol. 2006a;126:1071–9.
Trivedi NR, Cong Z, Nelson AM, et al. Peroxisome proliferator-activated receptors increase human sebum production. J Invest Dermatol. 2006b;126:2002–9.
Wang H, Brown J, Gu Z, et al. Convergence of the mammalian target of rapamycin complex 1- and glycogen synthase kinase 3-β-signaling pathways regulates the innate inflammatory response. J Immunol. 2011;186:5217–26.
Wille JJ, Kydonieus A. Palmitoleic acid isomer (C16:1δ6) is the active antimicrobial fatty acid in human skin sebum. Skin Pharmacol Appl Skin Physiol. 2003;16:176–87.
Wróbel A, Seltmann H, Fimmel S, et al. Differentiation and apoptosis in human immortalized sebocytes. J Invest Dermatol. 2003;120:175–81.
Zhang Q, Seltmann H, Zouboulis CC, Konger RL. Involvement of PPAR-gamma in oxidative stress-mediated prostaglandin E2 production in SZ95 human sebaceous gland cells. J Invest Dermatol. 2006;126:42–8.
Zouboulis CC. Is acne vulgaris a genuine inflammatory disease? Dermatology. 2001;203:77–9.
Zouboulis CC. Acne and sebaceous gland function. Clin Dermatol. 2004a;22:360–6.
Zouboulis CC. The human skin as a hormone target and an endocrine gland. Hormones. 2004b;3:9–26.
Zouboulis CC. Zileuton, a new efficient and safe systemic anti-acne drug. Dermatoendocrinology. 2009;1:188–92.
Zouboulis CC, Korge B, Akamatsu H, et al. Effects of 13-cis-retinoic acid, all-trans-retinoic acid and acitretin on the proliferation, lipid synthesis and keratin expression of cultured human sebocytes in vitro. J Invest Dermatol. 1991;96:792–7.
Zouboulis CC, Seltmann H, Hiroi N, et al. Corticotropin releasing hormone: an autocrine hormone that promotes lipogenesis in human sebocytes. Proc Natl Acad Sci U S A. 2002;99:7148–53.

Zouboulis CC, Fimmel S, Ortmann J, Turnbull JR, Boschnakow A. Sebaceous glands. In Hoath SB, Maibach HI, editors. Neonatal skin: structure and function. 2nd ed. New York: Marcel Dekker; 2003. pp. 59–88.
Zouboulis CC, Eady A, Philpott M, et al. What is the pathogenesis of acne? Exp Dermatol. 2005;14:143–52.
Zouboulis CC, Seltmann H, Alestas T. Zileuton prevents the activation of the leukotriene pathway and reduces sebaceous lipogenesis. Exp Dermatol. 2010;19:148–50.
Zouboulis CC, Angres S, Seltmann H. Regulation of stearoyl-CoA desaturase and fatty acid desaturase 2 expression by linoleic acid and arachidonic acid in human sebocytes leads to enhancement of proinflammatory activity but does not affect lipogenesis. Br J Dermatol. 2011;165:269–76.

Chapter 24
Atopic Dermatitis (AD) and Lipids

Yoshinori Masukawa

Core Messages

- The stratum corneum (SC) of atopic dermatitis (AD) skin contains unusual lipids, particularly with respect to ceramides (CER).
- Abnormalities in the lipid bilayer at intercellular spaces of the SC cause an impaired barrier function as seen in higher values of transepidermal water loss.
- The mechanism underlying these lipid abnormalities could be altered enzymatic activities relevant to the *de novo* synthesis of CER in the epidermis of AD skin.
- Whether the abnormalities are primary or secondary to AD has not been fully elucidated although the outside–inside view has become more probable.

Introduction

Atopic dermatitis (AD) is a chronic relapsing inflammatory skin disease characterized by pruritic and eczematous skin lesions. AD is thought to be caused by multiple pathogenic factors, such as genetic susceptibility, environmental triggers, cutaneous barrier dysfunction, bacterial infection, and/or immune dysregulation. About 20% of Caucasian children as well as 2–10% of adults are affected by AD (Alanne et al. 2011; Slattery et al. 2011). One of the biggest discoveries in recent studies of AD was an apparent loss-of-function that genetic variants in the gene encoding filaggrin demonstrated; those are a strong predisposing factor for the development of AD with very-high significance (Palmer et al. 2006). Up to 60% of European AD patients have loss-of-function mutations in the filaggrin gene (Elias and Wakefield 2011).

Y. Masukawa (✉)
Global R&D, Kao Germany GmbH, Pfungstadter Strasse 98-100, 64297 Darmstadt, Germany
e-mail: yoshinori.masukawa@kao.com

A. Pappas (ed.), *Lipids and Skin Health*, DOI 10.1007/978-3-319-09943-9_24

The relationship between AD and skin lipids was first reported by Melnik et al. 1988; Imokawa et al. 1991 then published an epoch-making article reporting lipid abnormalities with special reference to ceramides (CER) in the stratum corneum (SC) of AD skin. Since then, great attention has been paid to CER in the SC of patients with AD as well as their abnormal immune system. Until the beginning of the twenty-first century, thin-layer chromatography (TLC) was the only tool available to analyze CER in the SC, but the appearance of a new powerful technique, liquid chromatography-mass spectrometry (LC-MS), shifted the paradigm of dermatological studies regarding CER (Vietzke et al. 2001; Farwanah et al. 2005a). Progress worthy of special mention was that LC-MS analysis of CER in the SC of human skin revealed as many as 350 species that were structurally characterized (Masukawa et al. 2008), and that each of those diverse species could be quantified precisely and comprehensively using a newly developed LC-MS method (Masukawa et al. 2009). This method allowed the detailed features of the CER composition of the SC to be delineated.

The aim of this chapter is to clarify what has been known and unknown about the relationship between AD and skin lipids by answering the following four questions: (1) Are SC lipids in AD skin different from the lipids found in normal skin? (2) Do the lipid abnormalities affect the structures and/or properties of AD skin? (3) Is the mechanism underlying the lipid abnormalities known? (4) Are the lipid abnormalities primary or secondary to the development of AD? This chapter focuses on the relationships with skin lipids and not on relationships with skin proteins, such as filaggrin and cornified envelopes, except for those relevant to skin lipids. Readers who are interested in relationships between AD and skin proteins should consult other reviews (Proksch et al. 2008; Kypriotou et al. 2012; Nishifuji and Yoon 2013).

Are SC Lipids in AD Skin Different from the Lipids Found in Normal Skin?

Skin barrier function strongly relies on the SC (outermost layers of the skin), which consists of stacked layers of corneocytes (enriched proteins) "bricks" embedded in an intercellular lipid mixture "mortar" (Michaels et al. 1975). CER, cholesterol and free fatty acids (FFA) are the three abundant lipid classes in the free intercellular lipids of the SC of human skin and CER accounts for 40–50 % of the total lipid mass (Wertz 1992). There are 12 CER subclasses (Robson et al. 1994; Ponec et al. 2003; Masukawa et al. 2008; Van Smeden et al. 2011), which can be expressed based on previous terminology (Motta et al. 1993; Robson et al. 1994) as depicted in Fig. 24.1; CER[ADS] consisting of α-hydroxy fatty acids and dihydrosphingosines (corresponding to CER 5 in the TLC analysis); CER[AH] consisting of α-hydroxy fatty acids and 6-hydroxysphingosines (CER 6II); CER[AP] consisting of α-hydroxy fatty acids and phytosphingosines (CER 6I); CER[AS] consisting of α-hydroxy fatty acids and sphingosines (CER 5); CER[EODS] consisting of ester-linked ω-hydroxy fatty acids and dihydrosphingosines (CER 1); CER[EOH] consisting of ester-linked ω-hydroxy fatty acids and 6-hydroxysphingosines (CER 4);

Fatty acid / Sphingoid	Non-hydroxy fatty acid [N] 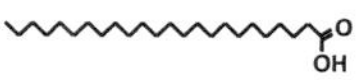	α-hydroxy fatty acid [A] 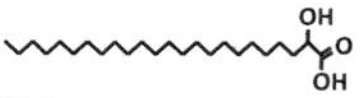	Esterified ω-hydroxy fatty acid [EO] 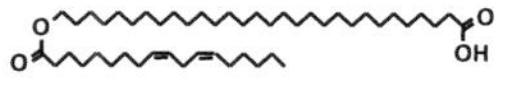
Dihydrosphingosine [DS] 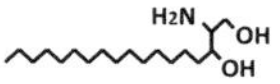	CER[NDS]	CER[ADS]	CER[EODS]
Sphingosine [S] 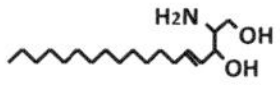	CER[NS]	CER[AS]	CER[EOS]
Phytosphingosine [P] 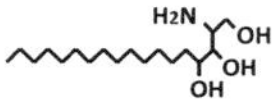	CER[NP]	CER[AP]	CER[EOP]
6-hydroxy sphingosine [H] 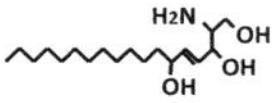	CER[NH]	CER[AH]	CER[EOH]

Fig. 24.1 Structures and nomenclature of ceramides (CER) in human stratum corneum (SC). (Note: this research was originally published in J. Lipid Res. (Masukawa et al. 2008). © the American Society for Biochemistry and Molecular Biology)

CER[EOP] consisting of ester-linked ω-hydroxy fatty acids and phytosphingosines (CER 2); CER[EOS] consisting of ester-linked ω-hydroxy fatty acids and sphingosines (CER 1); CER[NDS] consisting of nonhydroxy fatty acids and dihydrosphingosines (CER 2); CER[NH] consisting of nonhydroxy fatty acids and 6-hydroxysphingosines (CER 6I); CER[NP] consisting of nonhydroxy fatty acids and phytosphingosines (CER 3); and CER[NS] consisting of nonhydroxy fatty acids and sphingosines (CER 2). In addition to the 12 free CER subclasses, two subclasses of protein-bound CER are found in the SC, CER[OH] consisting of ω-hydroxy fatty acids and 6-hydroxysphingosines and CER[OS] consisting of ω-hydroxy fatty acids and sphingosines (Robson et al. 1994).

Table 24.1 shows a summary of intercellular lipids reported in the SC of AD lesional, AD nonlesional and controlled healthy nonlesional skin. Although numerous studies have emphasized diverse results due to the different subjects tested and the different methods used, there are common features for AD lesional skin as follows: (1) the level and/or wt.% of total CER is lower; (2) the CER composition is altered; and (3) the chain length of CER species is shortened. The first feature was confirmed by analyses done by Imokawa et al. 1991; Matsumoto et al. 1999; and Ishikawa et al. 2010. The second feature, i.e., that the balance of CER[EOS], other EO-containing CER subclasses and CER[NP] is commonly altered, was reported by Imokawa et al. (1991); Di Nordo et al. (1998); Matsumoto et al. (1999); Ishikawa

Table 24.1 Skin lipids for AD lesional (AL), AD nonlesional (ANL), and controlled healthy nonlesional (HNL)

Authors	Materials	Methods	Results
Melnik et al. 1988	SC from 10 ANL and 10 HNL	TLC	Lower wt.% of total CER in ANL
Imokawa et al. 1991	Cyanoacrylate-stripped SC from 35 AL, 35 ANL, and 65 HNL	TCL	Lower level of total CER in AL and ANL
			Lower wt.% of CER 1 (CER[EOS]) in AL and ANL
Yamamoto et al. 1991	Extracted SC lipids from 6 ANL and 6 HNL	TLC	Lower wt.% of CER 1 (CER[EOS]) in ANL
Di Nardo et al. 1998	Cyanoacrylate-stripped SC from 28 AL, 19 ANL, and 20 HNL	TLC	Lower levels of CER 1 (CER[EOS]) and CER 3 (CER[NP]) in AL
			Higher wt.% of CH in AL and ANL
			Intermediated levels in ANL between AL and HNL
Matsumoto et al. 1999	Extracted SC lipids from 14 AL,30 ANL, and 25 HNL	TLC	Lower levels of total CER and CER 1 (CER[EOS]) in AL
			Not different between ANL and HNL
Bleck et al. 1999	Cyanoacrylate-stripped SC from 10 ANL and 10 HNL	TLC MALDI-MS	Lower wt.% of CER[EOS] and CER[NP], higher wt% of CER[EOP], and shorter chain in CER[AS] in ANL
Macheleidt et al. 2002	Biopsied epidermis from 10 AL, 8 ANL, and 5 HNL	TLC	Lower wt.% of ω-hydroxy CER in AL and ANL
		GC	Lower wt.% of very-long-chain FFA in AL and ANL
Arikawa et al. 2002	Tape-stripped SC from 73 AL, 83 ANL, and 69 HNL	TLC	Lower level of sphingosine in AL and ANL
Okamoto et al. 2003	Tape-stripped SC from 44 AL, 47 ANL, and 40 HNL	TLC	Lower level of sphingosylphosphorylcholine in AL and ANL
Ishibashi et al. 2003	Tape-stripped SC from 92 AL, 105 ANL, and 81 HNL	TLC	Lower level of glucosphingosine in AL and ANL
Farwanah et al. 2005b	Extracted SC lipids from 7 ANL and 7 HNL	TLC LC-MS	Not different in CER between ANL and HNL
Ishikawa et al. 2010	Tape-stipped SC from 7 AL, 7 ANL, and 7 HNL	LC-MS	Lower levels of total CER, CER[NH], CER[NP], CER[EOS], CER[EOH], and CER[EOP], higher level of CER[AS], lower levels of longer chain in CER[NS], CER[NDS], CER[NH], CER[AS], and CER[AH], and higher levels of shorter chain in CER[NS] (especially with C34), CER[NDS] and CER[AS] in AL Intermediated level in ANL between AL and HNL

Table 24.1 (continued)

Authors	Materials	Methods	Results
Jungersted el al. 2010	Cyanoacrylate-stripped SC from 12 ANL-FLGm, 19 ANL-FLGw, 6 HNL-FLGm, and 12 HNL-FLGw	TLC	Lower wt.% of CER[EOP] and higher wt.% of CER[AP] in ANL-FLGm than HNL-FLGm and HNL-FLGw
			Lower wt.% of CER[EOS] and CER[AP] in ANL-FLGw than HNL-FLGm and HNL-FLGw
Angelova-Fischer et al. 2011	Cyanoacrylate-stripped SC from 14 AL/ANL-FLGm, 23 AL/ANL-FLGw, and 20 HNL	TLC	Lower levels of CER[EOH] and fatty acids in AL-FLGm than AL-FLGw
			Higher level of CH in ANL-FLGm
Janssens et al. 2011	Tape-stripped SC from 6 ANL and 6 HNL	LC-MS	Lower wt.% of CER[NP] and (CER[EODS]+CER[EOS]+CER[EOP]+CER[EOH]) in ANL
Janssens et al. 2012	Tape-stripped SC from 14 ANL-FLGm, 14 ANL-FLGw, and 15 HNL	LC-MS	Higher wt.% of shorter C34-CER[NS], C34-CER[NH], C34-CER[AS], and C34-CER[AH] and lower wt.% of (CER[EODS]+CER[EOS]+CER[EOP]+CER[EOH]) in ANL with no differences between FLGm and FLGw

TLC thin layer chromatography, *MALDI-MS* matrix-associated laser desorption-mass spectrometry, *GC* gas chromatography, *LC-MS* liquid chromatography-mass spectrometry, *FLGm* filaggrin gene mutation, *FLGw* no filaggrin mutation et al. (2010); and Angelova-Fischer et al. (2011). The third feature, most recently unveiled, comes from the significantly higher levels of CER[NS], CER[NDS], and CER[AS] with shorter chain lengths, as represented in C34-CER[NS] (Ishikawa et al. 2010). The validity of the third feature is corroborated by the fact that there were significantly lower levels of CER[NS], CER[NDS], CER[NH], CER[AS], and CER[AH] with longer chain lengths in the AD lesions (Ishikawa et al. 2010), the fact that a CER[AS] species with a shorter chain length was detected in AD nonlesional skin but not in healthy skin (Bleck et al. 1999), and the fact that significantly higher wt.% of C34-CER[NS], C-34CER[NH], C34-CER[AS], and C-34CER[AH] were found in AD nonlesional skin (Janssens et al. 2012). Macheleidt et al. (2002) found a lower wt.% of very-long-chain FFA in the SC of AD lesional skin although this is not for CER.

Compared with the lipid abnormalities in AD lesional skin, AD nonlesional skin looks somewhat indefinite in terms of the levels and composition of CER. As shown in Table 24.1, the intermediate features of AD nonlesional skin between AD lesional skin and healthy skin were described in some articles (Di Nardo et al. 1998; Ishikawa et al. 2010) and characteristics in the lipid abnormalities similar to AD lesions were shown in AD nonlesional skin by others (Bleck et al. 1999; Janssens et al. 2011, 2012). On the other hand, Matsumoto et al. (1999) and Farwanah et al. (2005b) reported no differences between AD nonlesional skin and healthy skin. Those inconsistencies in results obtained for nonlesional SC of AD skin would be due to the varieties of subjects tested (severity, progress, and degree of nonlesions), sampling sites/procedures and analytical methods used. Filaggrin gene mutations do not appear to directly influence the lipid abnormalities for the nonlesional SC of AD skin. No significant differences at the nonlesional sites were found between individuals carrying and not carrying the mutations (Jungersted et al. 2010). In another study undertaken by Janssens et al. (2012), the nonlesional SC of AD subjects carrying filaggrin mutations did not have any differences in lipids with those not carrying the mutations. However, the lower level of CER[EOH] in the lesional SC of AD patients carrying the mutations than those not carrying them was pointed out (Angelova-Fischer et al. 2011). To define characteristics of the lipids in AD nonlesional skin and the impact of filaggrin gene mutations on the lipids in the SC of AD skin, much larger-scaled studies are needed.

Collectively, the answer for the question “are SC lipids in AD skin different from the lipids found in normal skin?” is likely “yes” for the SC of AD lesional skin, as indicated by the lower level of total CER, the altered CER composition and the CER species with shorter chain lengths. For AD nonlesional skin, the abnormalities may be present with slight but similar characteristics to AD lesional skin, but further studies are required in a way that the subjects tested are standardized in terms of severity, progress, and degree of nonlesional skin. The filaggrin gene mutations do not seem to directly affect the lipid abnormalities, at least for the nonlesional SC of AD skin but this also remains to be defined.

Do the Lipid Abnormalities Affect the Structures and/or Properties of Skin?

No studies have been reported that characterized structures of the lipid bilayer at intercellular spaces in the SC of AD lesional skin, and only structures in nonlesional SC have been investigated. The long-periodicity phase in the lipid bilayer in the nonlesional SC was found to be slightly but significantly reduced in the repeat distance or repeat quantity compared to healthy SC (Janssens et al. 2012). Regarding the lateral lipid packing, it was found that the nonlesional SC of AD patients has an increased percentage of hexagonal lattice, gel phase, compared to healthy skin which is characterized by a larger presence of orthorhombic packing, crystalline phase (Pilgram et al. 2001; Janssens et al. 2012, 2013). These differences could be interpreted as originating from the lipid abnormalities, such as a lower level of total CER, an altered CER composition, and/or CER species with shorter chain lengths.

The diminished level of total CER in the SC of AD skin had a negative correlation with transepidermal water loss (TEWL), which is an index of impaired barrier function (Ishikawa et al. 2010). Also, there was a significantly negative correlation of the TEWL value versus the level of each CER subclass other than CER[AS] and CER[NS]. The subclass of CER[AS] had a significantly positive correlation with TEWL (Ishikawa et al. 2010). Only the subclass CER[AS] seems to have a different nature in terms of the involvement with the barrier function in AD skin. An effect of chain lengths of CER species on the TEWL has also been revealed. Thus, the more abundant the CER species with shorter chain lengths are, the higher the TEWL values (Ishikawa et al. 2010; Joo et al. 2010; Janssens et al. 2012). Since the level of C34 CER species sounds strongly correlated with TEWL (Ishikawa et al. 2010; Janssens et al. 2012), it may be a characteristic marker for the diagnosis of AD. Janssens et al. (2013) showed that the change in tendency in the lateral packing is correlated with the TEWL value. That correlation could be interpreted by the physicochemical nature that the hexagonal lattice is a less-packed structure in the lateral direction, where water can be less disturbed through the lipid bilayer.

The structure of the lipid bilayer in the SC of AD skin is likely to be changed into a bilayer with the reduced repeat distance or repeat quantity in the long-periodicity phase and with an increase in the hexagonal lattice, which may be due to the lipid abnormalities. This change in structure would cause a higher TEWL value corresponding to the impaired barrier function of AD skin.

Is the Mechanism Underlying the Lipid Abnormalities Known?

An ultrastructural study of AD skin versus healthy skin indicated the immature formation of lipid lamellae at the border between the stratum granulosum and the SC of AD skin (Fartasch et al. 1992). Thus, in AD skin, lamellar body-discs remained

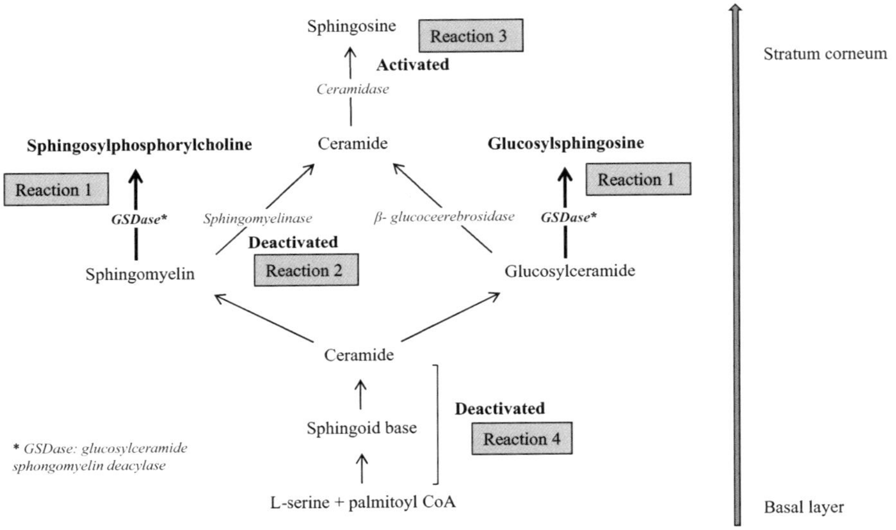

Fig. 24.2 Metabolism of ceramides (CER) in human skin

undelivered and were found even within the horny cells, in contrast to healthy skin where the body-discs completely disappeared. This suggested an abnormal keratinization coming from the unusual lipid metabolism in AD skin. The deficiency of CER in the SC of patients with AD can be explained by the extraordinary upregulation of glucosylceramide sphingomyelin deacylase (GSDase), which hydrolyzes glucosylceramide (GlcCER) or sphingomyelin (SM) at an acyl site to yield sphingosylphosphorylcholine (SPC) or glucosylsphingosine (GSP), respectively, instead of CER (Imokawa 2009), as illustrated in Reaction 1 of Fig. 24.2. The substantiality of the enzyme is supposed to be the β-subunit of acid ceramidase (CDase) based on a study using rat skin (Nogami-Itoh et al. 2010). At first, it was found that in the skin of patients with AD, the activities of three sphingolipid hydrolysis enzymes, β-glucocerebrosidase, sphingomyelinase (SMase), and CDase were not changed (Jin et al. 1994; Murata et al. 1996) whereas SM hydrolysis was increased with the occurrence of SPC as a reaction product and this hitherto undiscovered enzyme was tentatively termed SM deacylase (Murata et al. 1996; Hara et al. 2000). In a subsequent study, this enzyme was then termed GSDase because it hydrolyzes not only SM but also GlcCER in AD skin (Higuchi et al. 2000). The fact that the levels of SPC and GSP were both significantly higher in the epidermis of AD patients (Okamoto et al. 2003; Ishibashi et al. 2003), as listed in Table 24.1, corroborates the mechanism that the upregulation of GSDase generates the CER deficiency.

Other possible mechanisms underlying the diminished level of CER were proposed regarding reduced SMase activity (Reaction 2 of Fig. 24.2) and the involvement of bacterial CDase (Reaction 3 of Fig. 24.2). Acid SMase as well as neutral SMase, which produce CER from SM in the epidermis, were decreased both in the lesional and nonlesional skin from AD patients compared to control healthy skin

(Jensen et al 2014). The involvement of bacteria secreting CDase by which CER would be decomposed in the SC of AD skin was proposed (Ohnishi et al. 1999). Those mechanisms might be responsible in part for the diminished level of total CER. However, the altered CER composition cannot be explained only by the SMase activity or bacterial CDase because CER[NS] and CER[AS] are derived in part from the corresponding SM precursors while other subclasses such as EO-containing CER are derived only from GlcCER (Uchida et al. 2000; Hamanaka et al. 2002). Those mechanisms are not enough to explain the altered CER composition in the selective changes in the balances of CER[EOS], other EO-containing CER subclasses, and CER[NP].

Macheleidt et al. (2002) compared the *de novo* synthesis of GlcCER and CER in lesional AD skin with healthy skin using a metabolic labeling technique, which revealed remarkable decreases of newly biosynthesized ClcCER and CER in lesional AD skin. An experimental system using a reconstructed human epidermal keratinization model suggested that the Th2 type of inflammation evoked in AD skin may be one factor involved in the downregulated biosynthesis of CER, which results in the reduced levels of CER in the SC (Sawada et al. 2012). Therefore, the deficiency of CER in the SC of AD skin is likely to be caused not only by the abnormal pathway from GlcCER and SM to CER, such as the upregulation of GSDase (Reaction 1 of Fig. 24.2), but also reduced the *de novo* synthesis of CER skeletons themselves (Reaction 4 of Fig. 24.2). As for the chain length, elongases in the epidermis seem to be involved. Although the results were obtained in an experimental system using mice but not humans, some elongases that synthesize very-long-chain FFA were downregulated in a hapten-induced AD model (Park et al. 2012). It could be assumed that the downregulated elongases resulted in the decreased levels of CER species with longer -chain lengths (Bleck et al. 1999; Ishikawa et al. 2010; Janssens et al. 2012)very-long-chain FFA (Macheleidt et al. 2002).

Based on evidence accumulated to date, the mechanism underlying the lipid abnormalities for the SC in AD skin can most probably be explained by a combination of events, as follows: (1) the lower level of total CER would be caused by both the upregulation of GSDase and the reduced *de novo* synthesis of CER in the epidermis of AD skin, (2) the altered CER composition might be ascribed to changes in activities of enzymes relevant to the production of CER in the SC although this remains to be clarified, and (3) CER species with shorter chain lengths might originate from downregulated elongases although that also remains to be elucidated.

Are the Lipid Abnormalities Primary or Secondary to the Development of AD?

Traditionally, it was thought that the primary cause of AD was an immunological abnormality that led to the secondary barrier dysfunction (inside–outside view of AD pathogenesis). Many reports on the pathogenesis of AD focused on the primary role of abnormalities in the immune system, as reviewed by Leung (2006) and Ong and Leund (2006). In fact, therapy for AD was largely directed toward ameliorating Th2-mediated inflammation and/or pruritus using steroids or immunomodulators in spite of concerns about their side effects. A new paradigm of the outside–inside view (or outside–inside–outside), however has proposed that the primary inherited and acquired barrier abnormalities are followed by immune system activation, which further exacerbates the barrier function with a vicious cycle (Elias 2008; Cork et al. 2009; Elias and Schmuth 2009; Elias and Wakefield 2011).

Elias and Schmuth (2009) insist on the probability of the outside–-inside view because specific replacement therapy, which targets the prominent lipid abnormalities that account for the barrier abnormality in AD, corrects not only the barrier impairment but also comprises an effective anti-inflammatory therapy for AD. Topical application of a CER or pseudoceramide-dominant physiological lipid-base barrier repair emulsion has demonstrated clinical efficacies to improve the impaired barrier function in AD skin as well as to ameliorate AD symptoms (Mao-Qiang et al. 1996; Berardesca et al. 2001; Chamlin et al. 2002; Jensen and Elias 2006; Madaan 2008; Bikowski 2009; Park et al. 2010; Kircik et al. 2011). The availabilities of skin care products containing lipids/oils for AD skin have also been evidenced in other cases. Topical application of emollients, moisturizers, or creams containing lipids/oils, such as CER (Hon et al. 2013), pseudoceramide (Hon et al. 2011), petrolatum (Matsumoto et al. 2007), and paraffin oil and vegetable oil (Patzelt et al. 2012), efficiently improved the impaired skin conditions. Those clinical efficacies could be considered in part to be caused by the lipids/oils which provide an exogenous barrier to water loss from the inside and to the penetration of foreign material from the outside. This corroborates the validity of the outside–inside view as well as the important role of CER. However, the outside–inside view and the role of CER are still speculative. Before getting a true answer for the question, the further detailed studies are required.

The current conclusion to answer the question "are the lipid abnormalities primary or secondary to the development of AD?" is that it has not been fully elucidated whether the impaired barrier function due to the lipid abnormalities is primary or secondary to AD. Recent studies showing the availabilities of lipids/oils therapy to treat AD skin certainly suggest that the outside–inside view might be more likely and that CER might be a key. However, no technologies seem to have reached a scientifically convincing proof from the viewpoint of the mechanism underlying the clinical efficacies to AD skin, although there have been some reports of technologies that can upregulate levels of endogenous CER in *in vivo* and *in vitro* studies (Rawlings et al. 1996; Tanno et al. 2000; Ishikawa et al. 2012). We need to wait some time to get a true answer which would be obtained from our future studies.

Summary

It is well known that lipid abnormalities occur in the SC of lesional AD skin, as seen in the lower level of total CER, the altered CER composition and the CER species with shorter chain lengths compared to control healthy skin. It is also known that the lipid abnormalities cause an altered structure of the lipid bilayer, which further causes the impaired barrier function of AD skin, as seen in the higher values of TEWL. The reason for the lower level of total CER is likely to involve a mechanism where GSDase is upregulated and the *de novo* synthesis of CER is deactivated in the epidermis of AD skin. Based on accumulated evidence showing the clinical efficacies of lipids/oils therapy for patients with AD, the outside–inside view seems more likely than the traditional inside–outside view. On the other hand, there are still several unknown factors as follows; the lipid abnormalities for the nonlesional SC of AD, the reasons why the composition of CER subclasses is altered and why the chain lengths of CER species become shorter in AD lesions, the mechanism underlying the development of AD, and its relationship to the lipid abnormalities, that is, whether the lipid abnormalities are primary or secondary for AD. Those unknown factors will be clarified by our future work, which should help improve the quality of life for patients who suffer from AD symptoms.

References

Alanne S, Nermes M, Soderlund R, Laitinen K. Quality of life in infants with atopic dermatitis and healthy infants: a follow-up from birth to 24 months. Acta Paediatr. 2011;100:e65–70.

Angelova-Fischer I, Mannheimer AC, Hinder A, Ruether A, Franke A, Neubert RH, Fischer TW, Zillikens D. Distinct barrier integrity phenotypes in filaggrin-related atopic eczema following sequential tape stripping and lipid profiling. Exp Dermatol. 2011;20:351–6

Arikawa J, Ishibashi M, Kawashima M, Takagi Y, Ichikawa Y, Imokawa G. Decreased levels of sphingosine, a natural antimicrobial agent, may be associated with vulnerability of the stratum corneum from patients with atopic dermatitis to colonization by *Staphylococcus aureus*. J Invest Dermatol. 2002;119:433–9.

Berardesca E, Barbareschi M, Veraldi S, Pimpinelli N. Evaluation of efficacy of a skin lipid mixture in patients with irritant contact dermatitis, allergic contact dermatitis or atopic dermatitis: a multicenter study. Contact Dermatitis. 2001;45:280–5.

Bikowski J. Case studies assessing a new skin barrier repair cream for the treatment of atopic dermatitis. J Drugs Dermatol. 2009;8:1037–41.

Bleck O, Abeck D, Ring J, Hoppe U, Vietzke JP, Wolber R, Brandt O, Schreiner V. Two ceramide subfractions detectable in Cer(AS) position by HPTLC in skin surface lipids of non-lesional skin of atopic eczema. J Invest Dermatol. 1999;113:894–900.

Chamlin SL, Kao J, Frieden IJ, Sheu MY, Fowler AJ, Fluhr JW, Williams ML, Elias PM. Ceramide-dominant barrier repair lipids alleviate childhood atopic dermatitis: changes in barrier function provide a sensitive indicator of disease activity. J Am Acad Dermatol. 2002;47:198–208.

Cork MJ, Danby SG, Vasilopoulos Y, Hadgraft J, Lane ME, Moustafa M, Guy RH, Macgowan AL, Tazi-Ahnini R, Ward SJ. Epidermal barrier dysfunction in atopic dermatitis. J Invest Dermatol. 2009;129:1892–908.

Di Nardo A, Wertz P, Giannetti A, Seidenari S. Ceramide and cholesterol composition of the skin of patients with atopic dermatitis. Acta Derm Venereol. 1998;78:27–30.

Elias PM. Barrier repair trumps immunology in the pathogenesis and therapy of atopic dermatitis. Drug Discov Today Dis Mech. 2008;5:e33–8.

Elias PM, Schmuth M. Abnormal skin barrier in the etiopathogenesis of atopic dermatitis. Curr Opin Allergy Clin Immunol. 2009;9:437–46.

Elias PM, Wakefield JS. Therapeutic implications of a barrier-based pathogenesis of atopic dermatitis. Clin Rev Allergy Immunol. 2011;41:282–95.

Fartasch M, Bassukas ID, Diepgen TL. Disturbed extruding mechanism of lamellar bodies in dry non-eczematous skin of atopics. Br J Dermatol. 1992;127:221–7.

Farwanah H, Wohlrab J, Neubert RH, Raith K. Profiling of human stratum corneum ceramides by means of normal phase LC/APCI-MS. Anal Bioanal Chem. 2005a;383:632–7.

Farwanah H, Raith K, Neubert RH, Wohlrab J. Ceramide profiles of the uninvolved skin in atopic dermatitis and psoriasis are comparable to those of healthy skin. Arch Dermatol Res. 2005b;296:514–21.

Hamanaka S, Hara M, Nishio H, Otsuka F, Suzuki A, Uchida Y. Human epidermal glucosylceramides are major precursors of stratum corneum ceramides. J Invest Dermatol. 2002;119:416–23.

Hara J, Higuchi K, Okamoto R, Kawashima M, Imokawa G. High-expression of sphingomyelin deacylase is an important determinant of ceramide deficiency leading to barrier disruption in atopic dermatitis. J Invest Dermatol. 2000;115:406–13.

Higuchi K, Hara J, Okamoto R, Kawashima M, Imokawa G. The skin of atopic dermatitis patients contains a novel enzyme, glucosylceramide sphingomyelin deacylase, which cleaves the N-acyl linkage of sphingomyelin and glucosylceramide. Biochem J. 2000;350:747–56.

Hon KL, Wang SS, Lau Z, Lee HC, Lee KK, Leung TF, Luk NM. Pseudoceramide for childhood eczema: does it work? Hong Kong Med J. 2011;17:132–6.

Hon KL, Pong NH, Wang SS, Lee VW, Luk NM, Leung TF. Acceptability and efficacy of an emollient containing ceramide-precursor lipids and moisturizing factors for atopic dermatitis in pediatric patients. Drugs R D. 2013;13:37–42.

Imokawa G. A possible mechanism underlying the ceramide deficiency in atopic dermatitis: expression of a deacylase enzyme that cleaves the N-acyl linkage of sphingomyelin and glucosylceramide. J Dermatol Sci. 2009;55:1–9.

Imokawa G, Abe A, Jin K, Higaki Y, Kawashima M, Hidano A. Decreased level of ceramides in stratum corneum of atopic dermatitis: an etiologic factor in atopic dry skin? J Invest Dermatol. 1991;96:523–6.

Ishibashi M, Arikawa J, Okamoto R, Kawashima M, Takagi Y, Ohguchi K, Imokawa G. Abnormal expression of the novel epidermal enzyme, glucosylceramide deacylase, and the accumulation of its enzymatic reaction product, glucosylsphingosine, in the skin of patients with atopic dermatitis. Lab Invest. 2003;83:397–408.

Ishikawa J, Narita H, Kondo N, Hotta M, Takagi Y, Masukawa Y, Kitahara T, Takema Y, Koyano S, Yamazaki S, Hatamochi A. Changes in the ceramide profile of atopic dermatitis patients. J Invest Dermatol. 2010;130:2511–4.

Ishikawa J, Shimotoyodome Y, Chen S, Ohkubo K, Takagi Y, Fujimura T, Kitahara T, Takema Y. Eucalyptus increases ceramide levels in keratinocytes and improves stratum corneum function. Int J Cosmet Sci. 2012;34:17–22.

Janssens M, van Smeden J, Gooris GS, Bras W, Portale G, Caspers PJ, Vreeken RJ, Kezic S, Lavrijsen AP, Bouwstra JA. Lamellar lipid organization and ceramide composition in the stratum corneum of patients with atopic eczema. J Invest Dermatol. 2011;131:2136–8.

Janssens M, van Smeden J, Gooris GS, Bras W, Portale G, Caspers PJ, Vreeken RJ, Hankemeier T, Kezic S, Wolterbeek R, Lavrijsen AP, Bouwstra JA. Increase in short-chain ceramides correlates with an altered lipid organization and decreased barrier function in atopic eczema patients. J Lipid Res. 2012;53:2755–66.

Janssens M, Mulder AA, van Smeden J, Pilgram GS, Wolterbeek R, Lavrijsen AP, Koning RI, Koster AJ, Bouwstra JA. Electron diffraction study of lipids in non-lesional stratum corneum of atopic eczema patients. Biochim Biophys Acta. 2013;1828:1814–21.

Jensen JM, Elias PM. The stratum corneum of the epidermis in atopic dermatitis. In: Elias PM, Feingold KR, editors. Skin barrier. New York: Taylor and Francis; 2006. pp. 569–89.

Jensen JM, Fölster-Holst R, Baranowsky A, Schunck M, Winoto-Morbach S, Neumann C, Schütze S, Proksch E. Impaired sphingomyelinase activity and epidermal differentiation in atopic dermatitis. J Invest Dermatol. 2014;122:1423–31.

Jin K, Higaki Y, Takagi Y, Higuchi K, Yada Y, Kawashima M, Imokawa G. Analysis of b-glucocerebrosidase and ceramidase actibities in atopic and aged dry skin. Acta Derm Venereol (Stockh). 1994;74:337–40.

Joo KM, Jeong HJ, Lee SY. Relationship between cutaneous barrier function and ceramide species in human stratum corneum. J Dermatol Sci. 2010;60:47–50.
Jungersted JM, Scheer H, Mempel M, Baurecht H, Cifuentes L, Høgh JK, Hellgren LI, Jemec GB, Agner T, Weidinger S. Stratum corneum lipids, skin barrier function and filaggrin mutations in patients with atopic eczema. Allergy. 2010;65:911–8.
Kircik LH, Del Rosso JQ, Aversa D. Evaluating clinical use of a ceramide-dominant, physiologic lipid-based topical emulsion for atopic dermatitis. J Clin Aesthet Dermatol. 2011;4:34–40.
Kypriotou M, Huber M, Hohl D. The human epidermal differentiation complex: cornified envelope precursors, S100 proteins and the 'fused genes' family. Exp Dermatol. 2012;21:643–9.
Leung DY. New insights into the complex gene-environment interactions evolving into atopic dermatitis. J Allergy Clin Immunol. 2006;118:37–45.
Macheleidt O, Kaiser HW, Sandhoff K. Deficiency of epidermal protein-bound omega-hydroxyceramides in atopic dermatitis. J Invest Dermatol. 2002;119:166–73.
Madaan A. Epiceram for the treatment of atopic dermatitis. Drugs Today (Barc). 2008;44:751–5.
Mao-Qiang M, Feingold KR, Thornfeldt CR, Elias PM. Optimization of physiological lipid mixtures for barrier repair. J Invest Dermatol. 1996;106:1096–101.
Masukawa Y, Narita H, Shimizu E, Kondo N, Sugai Y, Oba T, Homma R, Ishikawa J, Takagi T, Kitahara T, Takema Y, Kita K. Characterization of overall ceramide species in human stratum corneum. J Lipid Res. 2008;49:1466–76.
Masukawa Y, Narita H, Sato H, Naoe A, Kondo N, Sugai Y, Oba T, Homma R, Ishikawa J, Takagi Y, Kitahara T. Comprehensive quantification of ceramide species in human stratum corneum. J Lipid Res. 2009;50:1708–19.
Matsumoto M, Umemoto N, Sugiura H, Uehara M. Difference in ceramide composition between "dry" and "normal" skin in patients with atopic dermatitis. Acta Derm Venereol. 1999;79:246–7.
Matsumoto T, Yuasa H, Kai R, Ueda H, Ogura S, Honda Y. Skin capacitance in normal and atopic infants, and effects of moisturizers on atopic skin. J Dermatol. 2007;34:447–50.
Melnik B, Hollmann J, Plewig G. Decreased stratum corneum ceramides in atopic individuals—a pathobiochemical factor in xerosis? Br J Dermatol. 1988;119:547–9.
Michaels AS, Chandrasekaran SK, Shaw JE. Drug penetration through human skin. Theory and in vitro experimental measurements. AIChE J. 1975;21:985–96.
Motta S, Monti M, Sesana S, Caputo R, Carelli S, Ghidoni R. Ceramide composition of the psoriatic scale. Biochim Biophys Acta. 1993;1182:147–51.
Murata Y, Ogata J, Higaki Y, Kawashima M, Yada Y, Higuchi K, Tsuchiya T, Kawainami S, Imokawa G. Abnormal expression of sphingomyelin acylase in atopic dermatitis: an etiologic factor for ceramide deficiency? J Invest Dermatol. 1996;106:1242–9.
Nishifuji K, Yoon JS. The stratum corneum: the rampart of the mammalian body. Vet Dermatol. 2013;24:60–72.
Nogami-Itoh M, Teranishi Y, Kuwahara H, Kusumoto M, Nakamura K, Matsumoto M, Sakai J, Kimura T, Kawashima M. Purification and identification of sphingomyelin deacylase from rat skin. J Invest Dermatol. 2010;130(suppl 2):s24.
Ohnishi Y, Okino N, Ito M, Imayama S. Ceramidase activity in bacterial skin flora as a possible cause of ceramide deficiency in atopic dermatitis. Clin Diagn Lab Immunol. 1999;6:101–4.
Okamoto R, Arikawa J, Ishibashi M, Kawashima M, Takagi Y, Imokawa G. Sphingosylphosphorylcholine is upregulated in the stratum corneum of patients with atopic dermatitis. J Lipid Res. 2003;44:93–102.
Ong PY, Leung DY. Immuno desregulation in atopic dermatitis. Curr Allergy Asthema Rep. 2006;6:384–9.
Palmer CN, Irvine AD, Terron-Kwiatkowski A, Zhao Y, Liao H, Lee SP, Goudie DR, Sandilands A, Campbell LE, Smith FJ, O'Regan GM, Watson RM, Cecil JE, Bale SJ, Compton JG, DiGiovanna JJ, Fleckman P, Lewis-Jones S, Arseculeratne G, Sergeant A, Munro CS, Houate BE, McElreavey K, Halkjaer LB, Bisgaard H, Mukhopadhyay S, McLean WI. Common loss-of-function variants of the epidermal barrier protein filaggrin are a major predisposing factor for atopic dermatitis. Nat Genet. 2006;38:441–6.

Park KY, Kim DH, Jeong MS, Li K, Seo SJ. Changes of antimicrobial peptides and transepidermal water loss after topical application of tacrolimus and ceramide-dominant emollient in patients with atopic dermatitis. Prevention of AD with emollient. J Korean Med Sci. 2010;25:766–71.

Park YH, Jang WH, Seo JA, Park M, Lee TR, Park YH, Kim DK, Lim KM. Decrease of ceramides with very long-chain fatty acids and downregulation of elongases in a murine atopic dermatitis model. J Invest Dermatol. 2012;132:476–9.

Patzelt A, Lademann J, Richter H, Darvin ME, Schanzer S, Thiede G, Sterry W, Vergou T, Hauser M. *In vivo* investigations on the penetration of various oils and their influence on the skin barrier. Skin Res Technol. 2012;18:364–9.

Pilgram GS, Vissers DC, van der Meulen H, Pavel S, Lavrijsen SP, Bouwstra JA, Koerten HK. Aberrant lipid organization in stratum corneum of patients with atopic dermatitis and lamellar ichthyosis. J Invest Dermatol. 2001;117:710–7.

Ponec M, Weerheim A, Lankhorst P, Wertz PW. New acylceramide in native and reconstructed epidermis. J Invest Dermatol. 2003;120:581–8.

Proksch E, Brandner JM, Jensen JM. The skin: an indispensable barrier. Exp Dermatol. 2008;17:1063–72.

Rawlings AV, Davies A, Carlomusto M, Pillai S, Zhang K, Kosturko R, Verdejo P, Feinberg C, Nguyen L, Chander P. Effect of lactic acid isomers on keratinocyte ceramide synthesis, stratum corneum lipid levels and stratum corneum barrier function. Arch Dermatol Res. 1996;288:383–90.

Robson KJ, Stewart ME, Michelsen S, Lazo ND, Downing DT. 6-hydroxy-4-sphingenine in human epidermal ceramides. J Lipid Res. 1994;35:2060–8.

Sawada E, Yoshida N, Sugiura A, Imokawa G. Th1 cytokines accentuate but Th2 cytokines attenuate ceramide production in the stratum corneum of human epidermal equivalents: an implication for the disrupted barrier mechanism in atopic dermatitis. J Dermatol Sci. 2012;68:25–35.

Slattery MJ, Essex MJ, Paletz EM, Vanness ER, Infante M, Rogers GM, Gern JE. Depression, anxiety, and dermatologic quality of life in adolescents with atopic dermatitis. J Allergy Clin Immunol. 2011;128:668–71.

Tanno O, Ota Y, Kitamura N, Katsube T, Inoue S. Nicotinamide increases biosynthesis of ceramides as well as other stratum corneum lipids to improve the epidermal permeability barrier. Br J Dermatol. 2000;143:524–31.

Uchida Y, Hara M, Nishio H, Sidransky E, Inoue S, Otsuka F, Suzuki A, Elias PM, Holleran WM, Hamanaka S. Epidermal sphingomyelins are precursors for selected stratum corneum ceramides. J Lipid Res. 2000;41:2071–82.

Van Smeden J, Hoppel L, van der Heijden R, Hankemeier T, Vreeken RJ, Bouwstra JA. LC/MS analysis of stratum corneum lipids: ceramide profiling and discovery. J Lipid Res. 2011;52:1211–21.

Vietzke JP, Brandt O, Abeck D, Rapp C, Straßner M, Schreiner V, Hintze U. Comparative investigation of human stratum corneum ceramides. Lipids. 2001;36:299–304.

Wertz PW. Epidermal lipids. Semin Dermatol. 1992;11:106–13.

Yamamoto A, Serizawa S, Ito M, Sato Y. Stratum corneum lipid abnormalities in atopic dermatitis. Arch Dermatol Res. 1991;283:219–23.

Index

A. Pappas (ed.), *Lipids and Skin Health*, DOI 10.1007/978-3-319-09943-9

34959463R00219

Printed in Poland
by Amazon Fulfillment
Poland Sp. z o.o., Wrocław